Advances in Photoassisted and Photocatalytic Processes for Water Remediation

Advances in Photoassisted and Photocatalytic Processes for Water Remediation

Guest Editors

Jaime Carbajo
Patricia García-Muñoz

MDPI

Basel • Beijing • Wuhan • Barcelona • Belgrade • Novi Sad • Cluj • Manchester

Guest Editors
Jaime Carbajo
Chemical and Materials
Engineering Department
Universidad Autónoma de
Madrid
Madrid
Spain

Patricia García-Muñoz
Department of Industrial
Chemical and Environmental
Engineering
Universidad Politécnica de
Madrid (UPM)
Madrid
Spain

Editorial Office
MDPI AG
Grosspeteranlage 5
4052 Basel, Switzerland

This is a reprint of the Special Issue, published open access by the journal *Catalysts* (ISSN 2073-4344), freely accessible at: https://www.mdpi.com/journal/catalysts/special_issues/photoassisted_Photocatalytic_Water_Remediation.

For citation purposes, cite each article independently as indicated on the article page online and as indicated below:

Lastname, A.A.; Lastname, B.B. Article Title. *Journal Name* **Year**, *Volume Number*, Page Range.

ISBN 978-3-7258-8481-0 (Hbk)
ISBN 978-3-7258-8482-7 (PDF)
https://doi.org/10.3390/books978-3-7258-8482-7

Contents

About the Editors

Jaime Carbajo

Jaime Carbajo is a Professor in the Department of Chemical Engineering and Materials at the Faculty of Chemistry, Universidad Complutense de Madrid, Spain. His research interests focus on environmental engineering, particularly advanced wastewater treatment technologies, including photocatalysis and advanced oxidation processes (AOPs). He has participated in numerous national and European research projects related to the removal of emerging contaminants and the development of catalytic materials, as well as in collaborations with industrial partners for process development and scale-up. He obtained his PhD in Chemical Engineering from Universidad Autónoma de Madrid, where he carried out research at the Instituto de Catálisis y Petroleoquímica (CSIC), earning the highest distinction and the Special Doctorate Award and the European Doctorate Mention in 2013. He has co-authored more than 30 scientific publications in high-impact journals and several books, most of them ranked in the first quartile, with over 1,500 citations, and he actively contributes to teaching, supervision, and scientific evaluation activities in the field of chemical and environmental engineering.

Patricia García-Muñoz

Patricia García-Muñoz is Associate Professor of the Chemical and Environmental Engineering Department at ETSI Industriales of Technical University of Madrid where she leads a new line of research focused on wastewater treatment for the removal of micropollutants, nanoplastics, and antibiotic resistance genes and on the development of new macroscopic catalysts for process scale-up. She has led projects at the international, national, and regional levels of different relevant calls. She has international research stays in prestigious centers like University of California Santa Barbara (USA) or University of Strasbourg (France) where she worked as postdoctoral researcher during more than two years. To date, she has published 63 JCR-indexed articles in leading journals in the fields of chemical and environmental technology, 3 editorials, 2 patents, 14 book chapters, and over 50 conference presentations with an h-value of 28 (over 2400 citations).

catalysts

MDPI

Editorial

Advances in Photoassisted and Photocatalytic Processes for Water Remediation

Jaime Carbajo [1,*] and Patricia García-Muñoz [2,*]

1 Chemical and Materials Engineering Department, Universidad Complutense de Madrid, Pl. de las Ciencias, 2, 28040 Madrid, Spain
2 Department of Chemical & Environmental Engineering, Escuela Técnica Superior de Ingenieros Industriales, Universidad Politécnica de Madrid, C/José Gutiérrez Abascal 2, 28006 Madrid, Spain
* Correspondence: jaime.carbajo@ucm.es (J.C.); patricia.gmunoz@upm.es (P.G.-M.)

Since heterogeneous photocatalysis emerged as a promising approach able to harness light energy—ideally solar radiation—to drive oxidation and reduction as effective technologies for water remediation under mild conditions [1], research has increasingly focused on bridging the gap between laboratory-scale studies and real-world applications. This includes addressing the challenges posed by complex water matrices, developing visible-light-active and stable photocatalysts, and integrating photocatalytic processes with complementary technologies. In this context, this Special Issue entitled *"Advances in Photoassisted and Photocatalytic Processes for Water Remediation"* presents a collection of contributions that address key aspects such as emerging pollutant removal, real wastewater treatment, catalyst design, and the scale-up of photoassisted AOPs.

The presence of pharmaceuticals and other contaminants in real water matrices, such as greywater, represents a major challenge for photocatalytic processes, as background constituents can significantly affect treatment efficiency. Inorganic ions, natural organic matter, and surfactants may act as radical scavengers or interfere with light penetration, thereby limiting degradation performance. Understanding these matrix effects is therefore essential for advancing photocatalysis toward realistic and scalable applications [2,3].

In this context, (*Contribution 1*) evaluate the photocatalytic degradation of pharmaceuticals—naproxen, metformin, and sulfamethoxazole—in membrane bioreactor-treated greywater, demonstrating that efficient removal can be achieved even in complex matrices. The study highlights the influence of background constituents on degradation kinetics and supports the feasibility of photocatalysis as a post-treatment step for water reuse. Complementarily, (*Contribution 2*) investigate photocatalytic treatment of greywater under realistic conditions, emphasizing the role of operational parameters and matrix composition on process performance. Together, these contributions provide valuable insight into the challenges and opportunities associated with real wastewater applications.

The study of (*Contribution 3*) demonstrates an effective and sustainable approach for naphthalene (NAP) removal from water using persulfate activated by Fe(III)–goethite and LED irradiation at natural pH. Optimal conditions were identified, achieving over 92% NAP removal while reducing toxicity. The use of a solid iron source, mild operating conditions, and energy-efficient LED light highlights the environmental sustainability of this treatment for water remediation as previously described by García-Muñoz et al. [4].

The scale-up of heterogeneous photocatalysis also remains a critical challenge, particularly due to issues related to catalyst recovery and reactor design. Immobilized photocatalytic systems have emerged as a promising alternative to slurry reactors, as they enable continuous operation and facilitate integration into treatment systems. However, their practical implementation depends on the development of scalable fabrication methods that ensure both activity and durability [5,6].

(*Contribution 4*) report the fabrication of titania–siloxane photocatalytic coatings using low-temperature UV and plasma curing methods, enabling deposition on temperature-sensitive substrates. The resulting films exhibit effective photocatalytic activity while maintaining structural integrity, and their compatibility with roll-to-roll manufacturing provides a clear pathway toward large-scale production. This work directly addresses key engineering challenges associated with the practical deployment of photocatalytic technologies.

One of the main limitations of conventional photocatalysts such as TiO_2 is their restricted activity under UV irradiation. To overcome this limitation, band-gap engineering strategies, particularly anionic doping, have been widely explored to extend light absorption into the visible region, with nitrogen doping being the most studied anion to effectively modify the electronic structure and enhance photocatalytic activity [7,8].

Another key and extended strategy involves modifying titania to increase its activity and to avoid the recombination of photogenerated charges. In this sense, (*Contribution 5*) hydrothermally synthesized Sr-doped TiO_2 for the electrocatalytic application of water splitting; 1% Sr-doped TiO_2 nanoparticles showed photocatalytic activity, and 1% and 5% Sr-doped TiO_2 depicted maximum current density for both HER and OER.

The contribution of (*Contribution 6*) showed the process to obtain a $ZnIn_2S_4$ photocatalyst with the aim of Cr(VI) photoreduction in water, another typical use of photocatalysis. The employed solvents in the synthesis protocol and the methods were critical for the removal yields of Cr(VI), not just because of the band-gap-associated changes but also because of the resulting surface morphology.

In another work, a zirconia-based photocatalyst was synthesized by (*Contribution 7*) via a simple thermal decomposition of UiO-66, with Ti(IV) and Hf(IV) cations used to tailor its photocatalytic performance. Among all samples, tetragonal ZrO_2 showed the highest activity, achieving 89% methyl orange degradation under UV irradiation in 300 min, outperforming monoclinic and mixed oxides and comparable to commercial anatase TiO_2.

Other strategies to overcome intrinsic limitations such as wide band gaps and rapid charge recombination include composite formation and the use of alternative semiconductors, which have been widely explored to enhance photocatalytic performance under visible light [3,9].

Another contribution to this Special Issue is a study carried out by (*Contribution 8*), who obtained $Bi_2O_3/CoFe_2O_4$ nanocomposites. The heterojunction was applied for the photocatalytic degradation of methyl orange in the water phase. The $Bi_2O_3/CoFe_2O_4$ exhibited narrowed band gaps (1.58–1.62 eV) and enhanced activity compared to the individual oxides. An equimolar nanocomposite showed the best performance due to a synergistic type-II heterojunction effect.

In this context, (*Contribution 9*) investigate N-doped photocatalysts, demonstrating that nitrogen incorporation enhances light absorption and improves photocatalytic performance compared to undoped materials. The study highlights the relationship between electronic structure modification and degradation efficiency, reinforcing the importance of rational catalyst design for the development of visible-light-driven systems in water remediation. For their part, (*Contribution 10*) present a comprehensive review of photocatalytic systems for CEC removal, focusing on catalyst modification strategies and process intensification. The authors discuss alternative materials such as g-C_3N_4 and WO_3, as well as the benefits of combining photocatalysis with ozonation to enhance degradation efficiency. This contribution provides a critical overview of current strategies to improve

https://doi.org/10.3390/catal16050474

photocatalytic performance and supports the development of more efficient and scalable treatment systems.

Photocatalytic performance is fundamentally governed by material properties such as band-gap energy, band-edge positions, and charge carrier dynamics. These factors determine light absorption, redox capability, and recombination rates, and are therefore critical for both water splitting and pollutant degradation processes. Transition metal oxides remain among the most studied materials, although challenges related to efficiency and stability persist [3,10]. In this regard, (*Contribution 11*) provide a detailed analysis of photocatalytic water splitting based on transition metal oxides, focusing on the relationship between physicochemical properties and performance. The work identifies key limitations such as recombination losses and highlights design criteria for improving efficiency. These insights are directly relevant to the development of advanced photocatalysts for water remediation.

On the other hand, an increasing trend consists of modeling the newly developed processes to validate the technology. In this sense, (*Contribution 12*) developed a model for simulating the degradation of pollutants via a combined bio-chemical treatment with good fittings ($R^2 = 0.998$).

Finally, the increasing complexity of wastewater treatment has driven the development of hybrid systems combining photocatalysis with complementary technologies. Such integrated approaches can overcome the limitations of individual processes and enhance overall efficiency, particularly in real water matrices [2,6]. In this framework, (*Contribution 13*) present a comprehensive review of photoassisted AOPs for micropollutant removal, with particular emphasis on their integration with other treatment technologies. The authors critically analyze combined systems, discussing their mechanisms, advantages, and limitations, and identify key factors for optimization and scale-up. This contribution provides a valuable overview of current trends and future directions in hybrid photocatalytic systems.

The contributions collected in this Special Issue highlight the rapid progress and increasing maturity of photoassisted and photocatalytic processes for water remediation. From fundamental material design to process integration and real wastewater applications, these works collectively address key challenges associated with efficiency, scalability, and practical implementation. In particular, the importance of understanding real matrix effects, developing visible-light-active materials, and integrating photocatalysis with complementary technologies emerges as a unifying theme. These advances contribute to paving the way toward sustainable and efficient water treatment solutions.

Funding: MAINSTREAM PROJECT (PCI2025-167062-2) funded by MCIN/AEI/10.13039/501100011033/ and co-funded by the European Union.

Conflicts of Interest: The authors declare no conflicts of interest.

List of Contributions:

1. Ojobe, B.; Zouzelka, R.; Satkova, B.; Vagnerova, M.; Nemeskalova, A.; Kuchar, M.; Bartacek, J.; Rathousky, J. Photocatalytic Removal of Pharmaceuticals from Greywater. *Catalysts* **2021**, *11*, 1125. https://doi.org/10.3390/catal11091125.
2. Aslam, M.; Fazal, D.B.; Ahmad, F.; Fazal, A.B.; Abdullah, A.Z.; Ahmed, M.; Qamar, M.; Rafatullah, M. Photocatalytic Degradation of Recalcitrant Pollutants of Greywater. *Catalysts* **2022**, *12*, 557. https://doi.org/10.3390/catal12050557.
3. Sánchez-Yepes, A.; Lorenzo, D.; Sáez, P.; Romero, A.; Santos, A. Abatement of Naphthalene by Persulfate Activated by Goethite and Visible LED Light at Neutral pH: Effect of Common Ions and Organic Matter. *Catalysts* **2022**, *12*, 732. https://doi.org/10.3390/catal12070732.

4. Svoboda, T.; Veselý, M.; Bartoš, R.; Homola, T.; Dzik, P. Low-Temperature Mineralisation of Titania-Siloxane Composite Layers. *Catalysts* **2021**, *11*, 50. https://doi.org/10.3390/catal11010050.
5. Fazil, M.; Ahmad, T. Pristine TiO_2 and Sr-Doped TiO_2 Nanostructures for Enhanced Photocatalytic and Electrocatalytic Water Splitting Applications. *Catalysts* **2023**, *13*, 93. https://doi.org/10.3390/catal13010093.
6. Song, Y.; Zhang, H.; Li, Z.; Huang, L.; Xie, J.; Han, L.; Zheng, R.; Zhang, Y.; Zhang, H. Study on Optimum Preparation Conditions of $ZnIn_2S_4$ to Effectively Reduce Cr(VI) under Visible Light Radiation. *Catalysts* **2022**, *12*, 1429. https://doi.org/10.3390/catal12111429.
7. Jannah, I.N.A.; Sekarsari, H.F.; Mulijani, S.; Wijaya, K.; Wibowo, A.C.; Patah, A. Facile Synthesis of Various ZrO_2 Phases and ZrO_2-MO_2 (M = Ti, Hf) by Thermal Decomposition of a Single UiO-66 Precursor for Photodegradation of Methyl Orange. *Catalysts* **2022**, *12*, 609. https://doi.org/10.3390/catal12060609.
8. Naveed, A.B.; Riaz, F.; Mahmood, A.; Shahid, A.; Aqeel, S. A Facile Synthesis of $Bi_2O_3/CoFe_2O_4$ Nanocomposite with Improved Synergistic Photocatalytic Potential for Dye Degradation. *Catalysts* **2021**, *11*, 1180. https://doi.org/10.3390/catal11101180.
9. Jiang, Z.; Wickramasinghe, S.; Tsai, Y.H.; Samia, A.C.S.; Gurarie, D.; Yu, X. Modeling and Experimental Studies on Adsorption and Photocatalytic Performance of Nitrogen-Doped TiO_2 Prepared via the Sol–Gel Method. *Catalysts* **2020**, *10*, 1449. https://doi.org/10.3390/catal10121449.
10. Fernandes, E.; Gomes, J.; Martins, R.C. Semiconductors Application Forms and Doping Benefits to Wastewater Treatment: A Comparison of TiO_2, WO_3, and g-C_3N_4. *Catalysts* **2022**, *12*, 1218. https://doi.org/10.3390/catal12101218.
11. Mikaeili, F.; Gilmore, T.; Gouma, P.-I. Photochemical Water Splitting via Transition Metal Oxides. *Catalysts* **2022**, *12*, 1303. https://doi.org/10.3390/catal12111303.
12. Makgato, S.; Nkhalambayausi-Chirwa, E. An Evaluation of the Kinetic Properties Controlling the Combined Chemical and Biological Treatment of Toxic Recalcitrant Organic Compounds from Aqueous Solution. *Catalysts* **2022**, *12*, 965. https://doi.org/10.3390/catal12090965.
13. Rodríguez-Chueca, J.; Carbajo, J.; García-Muñoz, P. Intensification of Photo-Assisted Advanced Oxidation Processes for Water Treatment: A Critical Review. *Catalysts* **2023**, *13*, 401. https://doi.org/10.3390/catal13020401.

References

1. Fujishima, A.; Honda, K. Electrochemical photolysis of water at a semiconductor electrode. *Nature* **1972**, *238*, 37–38. [CrossRef] [PubMed]
2. Ahmed, S.; Rasul, M.G.; Brown, R.; Hashib, M.A. Influence of parameters on the heterogeneous photocatalytic degradation of pollutants in wastewater: A short review. *J. Environ. Chem. Eng.* **2021**, *9*, 104807.
3. Wang, W.; Yu, H.; Quan, X.; Chen, S.; Zhao, H. Recent advances in photocatalytic removal of emerging contaminants in water: A review. *Chem. Eng. J.* **2022**, *428*, 131145.
4. García-Muñoz, P.; Pliego, G.; Zazo, J.; Bahamonde, A.; Casas, J.A. Ilmenite ($FeTiO_3$) as low cost catalyst for advanced oxidation processes. *J. Environ. Chem. Eng.* **2016**, *4*, 542–548. [CrossRef]
5. Chen, X.; Mao, S.S. Titanium dioxide nanomaterials: Synthesis, properties, modifications, and applications. *Chem. Rev.* **2010**, *110*, 6503–6570. [CrossRef]
6. Kumar, A.; Khan, M.; He, J.; Lo, I.M.C. Recent developments and challenges in photocatalytic water treatment using advanced oxidation processes. *J. Water Process Eng.* **2023**, *51*, 103368.
7. Asahi, R.; Morikawa, T.; Ohwaki, T.; Aoki, K.; Taga, Y. Visible-light photocatalysis in nitrogen-doped titanium oxides. *Science* **2001**, *293*, 269–271. [CrossRef] [PubMed]
8. Di Paola, A.; García-López, E.; Marcì, G.; Palmisano, L. A survey of photocatalytic materials for environmental remediation under visible light. *Catalysts* **2022**, *12*, 117.
9. Ong, W.-J.; Tan, L.-L.; Ng, Y.H.; Yong, S.-T.; Chai, S.-P. Graphitic carbon nitride (g-C_3N_4)-based photocatalysts for artificial photosynthesis and environmental remediation. *Chem. Rev.* **2016**, *116*, 7159–7329. [CrossRef] [PubMed]
10. Kudo, A.; Miseki, Y. Heterogeneous photocatalyst materials for water splitting. *Chem. Soc. Rev.* **2009**, *38*, 253–278. [CrossRef] [PubMed]

https://doi.org/10.3390/catal16050474

catalysts

MDPI

Article

Modeling and Experimental Studies on Adsorption and Photocatalytic Performance of Nitrogen-Doped TiO_2 Prepared via the Sol–Gel Method

Zhuoying Jiang [1], Sameera Wickramasinghe [2], Yu Hsin Tsai [2], Anna Cristina S. Samia [2], David Gurarie [3] and Xiong Yu [1,*]

1 Department of Civil and Environmental Engineering, Case Western Reserve University, Cleveland, OH 44106, USA; zxj45@case.edu

2 Department of Chemistry, Case Western Reserve University, Cleveland, OH 44106, USA; smw151@case.edu (S.W.); yxt368@case.edu (Y.H.T.); axs232@case.edu (A.C.S.S.)

3 Department of Mathematics, Applied Mathematics and Statistics, Case Western Reserve University, Cleveland, OH 44106, USA; dxg5@case.edu

* Correspondence: xxy21@case.edu; Tel.: +1-216-368-6247

Received: 28 October 2020; Accepted: 9 December 2020; Published: 11 December 2020

Abstract: Nitrogen-doped TiO_2 has a great potential as a photocatalyst under visible light irradiation with applications in the removal of air and water pollutants, and the treatment of bacterial contaminations. In this study, nitrogen-doped TiO_2 nanoparticles were synthesized via the sol–gel method and a post-annealing heat treatment approach. The effects of annealing treatment on the photocatalyst crystalline size and degree of crystallinity were analyzed. Methylene blue dye was used as the model water contaminant for the evaluation of the photoactivity of the synthesized nitrogen-doped TiO_2 nanoparticles. The degradation of methylene blue was attributed to three mechanisms, i.e., adsorption, photocatalysis, and direct light photolysis. A kinetic model was developed to distinguish the impact of these three different mechanisms on the removal of contaminants. Adsorption and photocatalysis are heterogeneous processes for removing water organic contaminants. The characterization analysis demonstrates that they are relevant to the microstructures and surface chemical compositions of nitrogen-doped TiO_2 photocatalysts. The processing–structure–performance relationship helped to determine the optimal processing parameters for nitrogen-doped TiO_2 photocatalyst to achieve the best performance. While we used methylene blue as the model contaminant, the generalized quantitative model framework developed in this study can be extended to other types of contaminants after proper calibration.

Keywords: nitrogen-doped TiO_2 nanoparticles; photocatalysis; adsorption; sol–gel method; dynamic modeling

1. Introduction

TiO_2 as a photocatalyst has attracted extensive research interest as a promising material for eliminating air and water contaminants due to its many advantages, i.e., nontoxicity, good chemical stability, and relatively low cost [1–7]. However, the large bandgap of TiO_2 limits its absorption spectrum to the ultraviolet range. This limitation prevents the use of TiO_2 in the indoor lighting environment, since commonly used lamps, such as LEDs, fluorescent lamps, incandescent lamps, etc., emit low amount of ultraviolet light [8,9]. To extend the absorption spectrum into the visible light range, TiO_2 has been modified by doping with impurity elements, e.g., nitrogen, carbon, iron [10–14]. The doped TiO_2 has a narrower bandgap and can be responsive to the visible light excitation. Among those doped TiO_2, nitrogen-doped TiO_2 (N-TiO_2) has been intensively studied and widely used in the environmental protection applications, including air purification and water treatment [14–19]. Figure 1 illustrates the possible mechanisms of degradation of organic contaminants in the presence of N-TiO_2 photocatalysts. The N-TiO_2 photocatalyst facilitates the degradation of the organic contaminants via heterogeneous processes [20–22]. The organic molecules are first adsorbed on the photocatalyst surfaces denoted as path 1 in Figure 1, and then degradation reaction denoted as path 2 occurs. With light irradiance, electron and hole pairs are generated in the photocatalyst, and they are further reacted with the oxygen and water in the surroundings, respectively. Superoxide anion (O_2^-) and hydroxyl radical ($^{\cdot}OH$) are consequently produced. The generated reactive oxygen species (ROS) have a strong oxidizing ability to degrade the adsorbed organic contaminants. In addition to adsorption and photocatalysis, some organic contaminants can undergo direct photolysis under light irradiation without photocatalyst, denoted as path 3 in Figure 1. For example, tricholocarbon (TCC), a disinfectant [23], and β-blocker as a pharmaceutical ingredient [24], can be efficiently degraded via direct photolysis under sunlight. In practice, adsorption, photocatalysis, and direct photolysis occur simultaneously. The kinetics of these chemical processes are of practical importance for industrial applications. Industries are keen to speed up the reactions and promote economic growth. The knowledge of these reaction mechanisms can provide the guidelines to the materials development, reactor designs, and optimization of reaction conditions. Therefore, it is critical to elucidate the effects of these three mechanisms on the removal of organic contaminants.

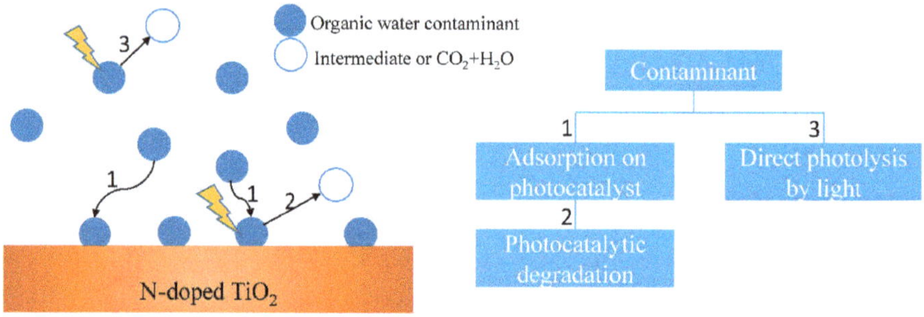

Figure 1. Schematic of the different mechanisms involved in the degradation of contaminants by N-TiO_2 photocatalysts.

The Langmuir–Hinshelwood equation is a commonly used model to explain the heterogeneous catalytic process, combining adsorption and reaction processes [25–27]. However, this model has limitations as it requires an equilibrium state between the adsorbate and adsorbent at all times during the reaction. In reality, adsorption and reaction are both dynamic processes. Additionally, direct photolysis by light irradiation was neglected in most reported experiments, where light sources with high intensity were used. In such cases, there could be a noticeable degradation of contaminants through the direct photolysis mechanism, but this effect is commonly incorporated in the interpretation of the photocatalysis mechanism and should be decoupled. Here, we proposed a kinetic model that combines adsorption, photocatalysis, and direct photolysis. We also carried out related experiments with respect to the model. By fitting experimental data to the proposed model, the effects of these three mechanisms on degradation of contaminants are quantitatively evaluated, and one can access their dominance in the photocatalytic experiments.

This work represents a comprehensive quantitative model for the nitrogen-doped titania (N-TiO_2) nanoparticle-assisted photocatalytic degradation of organic dyes in aqueous environments. The presented model holistically considers the kinetics of critical processes involved in the photocatalytic degradation of a representative water-born organic dye contaminant, including the processes involved in dye adsorption, photolysis, and photocatalytic reactions. The generalized quantitative model framework can be extended to other types of contaminants after proper calibration. Moreover, this study experimentally investigated the effects of synthesis and processing conditions on the structural features of titania nanoparticle-based photocatalyst as well as their performance. This sheds light on the processing–structure–performance relationship in visible-light-driven N-doped TiO_2 photocatalysts and will help to determine the optimal processing parameters that achieve the best performance.

2. Result and Discussion

2.1. Characterization of N-TiO_2 Photocatalysts

The pXRD patterns of the N-TiO_2 photocatalyst annealed at different temperatures and time are shown in Figure 2. The crystallite size and degree of crystallinity were further studied. The crystallite size was estimated from the Debye–Scherrer equation:

$$L = \frac{K\lambda}{d \cdot cos\theta} \quad (1)$$

where L is the crystallite size, K is the crystallite shape factor, and K is usually taken as about 0.9 [28], λ is the wavelength of the X-ray, θ is the Braggs' angle, and d is the full width at half maximum (FWHM) intensity of the pXRD peak in the unit of rad. Here, the first peak, which is the most reliable peak, was utilized in the calculation. The degree of crystallinity was estimated using Rietveld analysis that is integrated in the X-ray powder diffraction (PDXL) software:

$$\text{Degree of crystallinity} = \frac{\text{Area of crystal peaks}}{\text{Area of crystal peaks} + \text{Area of amorphous halo}} \quad (2)$$

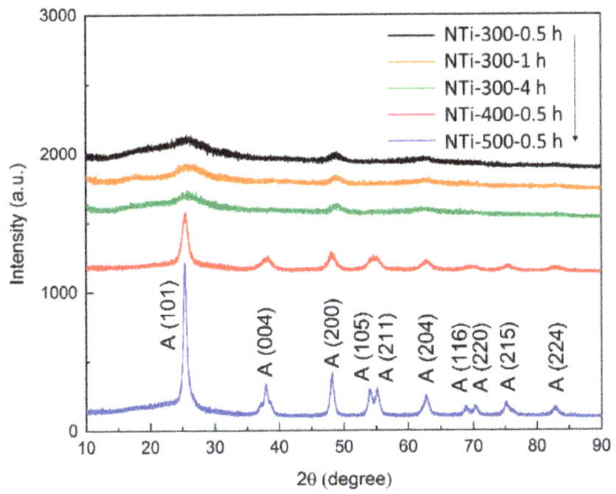

Figure 2. The pXRD spectra of the N-TiO_2 photocatalysts. Black: NTi-300-0.5 h, Orange: NTi-300-1 h, Green: NTi-300-4 h, Red: NTi-400-0.5 h, Blue: NTi-500-0.5 h.

The calculated crystallite size and degree of crystallinity are listed in Table 1. The pXRD results illustrate that as the annealing temperature increases, the crystallinity of N-TiO_2 increases as well, and they show a single anatase phase. In the meantime, the crystallite size increases significantly with increased annealing temperature. This indicates that a higher temperature not only makes the grain crystallized, but also facilitates the crystallite to grow larger. However, when holding the annealing temperature at 300 °C and increasing the annealing time, the crystallinity slightly increases, but the crystallite does not grow much.

Table 1. The crystallite size, degree of crystallinity, Brunauer–Emmett–Teller (BET) surface area, and Barrett–Joyner–Halenda (BJH) pore size of N-TiO_2.

Photocatalysts	Crystallite Size (nm)	Degree of Crystallinity (%)	BET Surface Area (m^2 g^{-1})	Average Pore Size (nm)	Pore Volume (cm^3 g^{-1})
NTi-300-0.5 h	1.82	37.0	132.0	2.6	0.154
NTi-300-1 h	2.37	46.4	53.9	2.6	0.054
NTi-300-4 h	1.96	50.9	46.1	2.7	0.048
NTi-400-0.5 h	8.44	68.3	143.0	3.8	0.177
NTi-500-0.5 h	13.74	71.8	57.4	6.0	0.095

XPS was performed to analyze the chemical compositions and chemical states of the N-TiO_2 powders. Figure 3a shows a comparison of the full spectra of N-TiO_2 under various different annealing temperatures and time. The spectra indicate the presence of titanium, oxygen, carbon, and nitrogen elements in the three samples. The concentration of nitrogen decreases with elevated annealing temperature. This may be due to the nitrogen loss from TiO_2 with a high temperature treatment.

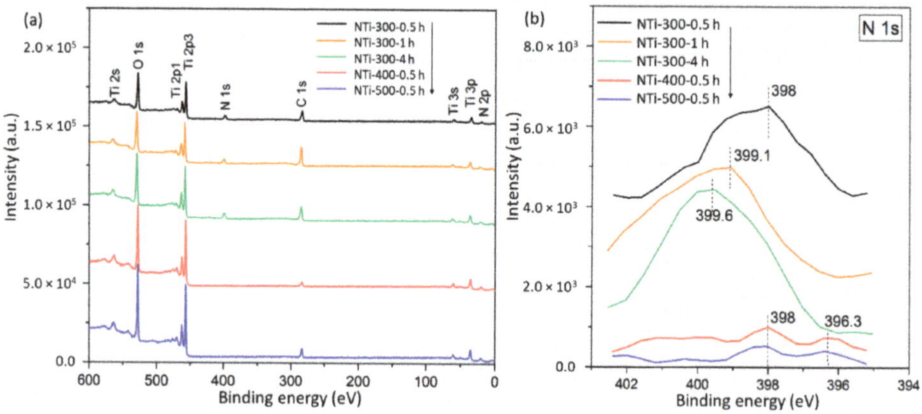

Figure 3. The XPS results of N-TiO_2. (**a**) Full spectra and (**b**) N 1 s spectra. Black: NTi-300-0.5 h, Orange: NTi-300-1 h, Green: NTi-300-4 h, Red: NTi-400-0.5 h, Blue: NTi-500-0.5 h.

Figure 3b reveals the N 1s spectra of five N-TiO_2 photocatalysts prepared with increased annealing temperatures. For N-TiO_2 annealed at 300 °C, all the three samples have a peak at a binding energy of 398 to 399.6 eV, but for NTi-400-0.5 h and NTi-500-0.5 h, they have another weaker peak at a binding energy of 396.3 eV. According to previous work [20,29], the peak between 398 and 400 eV can be attributed to the interstitial N bonding with lattice O in N-TiO_2. For the peak at 396.3 eV, it comes from Ti-N bonding, which can be interpreted as substitutional N at the O lattice [20,30]. For NTi-300-0.5 h, there is a single peak at 398 eV, indicating that N is present in the N-TiO_2 as interstitials. When annealed at higher temperatures, the major peak between 398 and 400 eV reduces and a second peak at 396.3 eV appears. This explains that some interstitial N is lost, and some is transformed to substitutional N during the annealing process. However, the total N content in N-TiO_2, including interstitial and substitutional N, decreases with increased annealing temperature as shown in Figure 3b.

The Brunauer–Emmett–Teller (BET) surface area and Barrett–Joyner–Halenda (BJH) pore size distributions were also measured. Figure 4 shows the N_2 gas adsorption–desorption isotherms of the prepared N-TiO_2 photocatalysts and their corresponding pore size distributions. The BET surface area, BJH pore size, and pore volume are listed in Table 1. Figure 5a–c shows the SEM images of N-doped

TiO_2 annealed for 1 h under 300 °C, 400 °C, and 500 °C. With increasing annealing temperature, more crystalline particles appear in the surface of the catalyst materials.

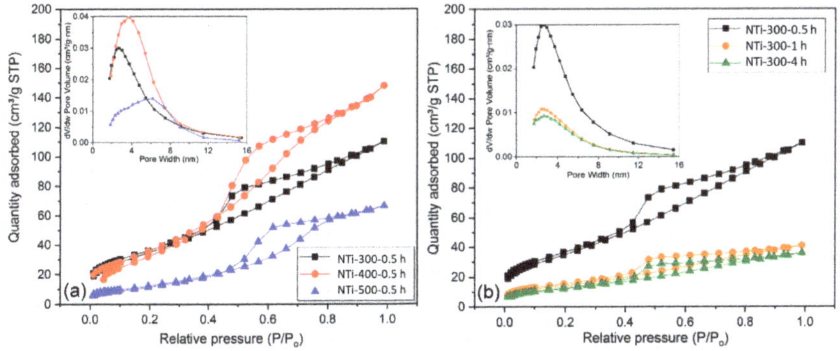

Figure 4. The N_2 adsorption–desorption isotherms: (**a**) N-TiO_2 with different annealing temperatures, and (**b**) N-TiO_2 with different annealing time. Inserted in (**a**,**b**) are pore size distributions. Black: NTi-300-0.5 h, Orange: NTi-300-1 h, Green: NTi-300-4 h, Red: NTi-400-0.5 h, Blue: NTi-500-0.5 h.

(**a**)

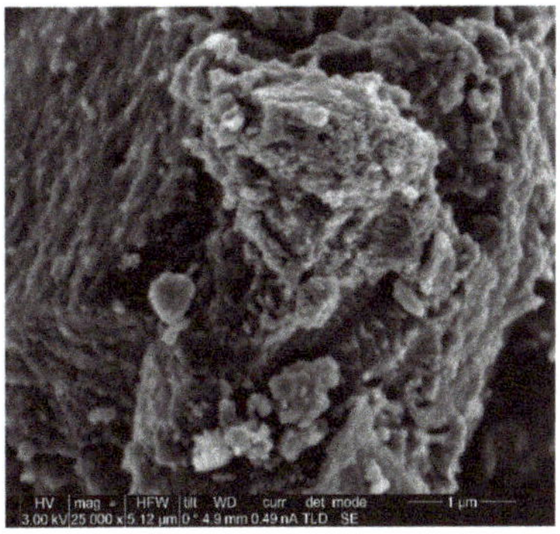

(**b**)

(**c**)

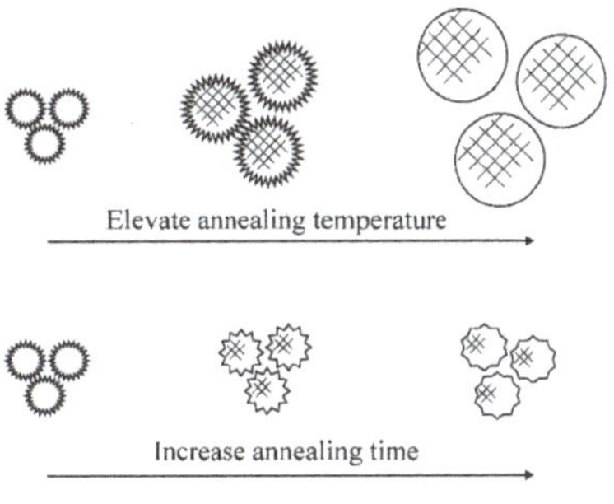

(**d**)

Figure 5. (**a**–**c**) SEM image of N-doped TiO_2 annealed for 1 h under 300 °C, 400 °C, and 500 °C respectively. (**d**) Schematic illustration of the effect of annealing temperature and time on the size and morphology of N-TiO_2.

The following observations were obtained from the experimental data via the pXRD, BET, and SEM analysis as summarized in Table 1. The effects of annealing temperature: with increasing annealing temperature, the crystallite size increases, and pore size also increases (Table 1). This implies that higher annealing temperature helps the crystalline grain size to grow. The increase of crystallite size also leads to larger intergranular pores. It was observed that the BET surface area remains similar for samples annealed at 300 °C and 400 °C. This might be due to the fact that the grain surface remains porous and unordered even though the grain grows larger and more crystalline. When the annealing temperature is increased to 500 °C, the BET surface area drops dramatically because the grains grow larger, and the grain surface becomes smooth and nonporous with high temperature treatment. The effects of annealing time: when holding the annealing temperature at 300 °C and increasing the annealing time, the BET surface area decreases, but the pore size remains similar. This implies that under the same temperature, increasing annealing time increases the degree of crystallinity but does not increase the crystallite size. As a result, more crystallized particles have a smaller surface area. Figure 5d presents the conceptual illustration of the effects of annealing temperature and annealing time on the size, crystalline morphology transformation based on the pXRD, BET, and SEM analysis. That is, increasing annealing temperature increases both the crystal size and degree of crystallinity, while increasing annealing time increases the degree of crystallinity but not the crystal size.

2.2. *Measurement of Photocatalytic Activity and Kinetic Modeling Results*

The photocatalytic activity of the N-TiO_2 photocatalysts was measured with the methylene blue decomposition method, and the results are shown in Figure 6. Three experiments were carried out to measure the adsorption, photocatalysis, and direct photolysis effects. First, the photolysis rate was determined by measuring the degradation rate of methylene blue under light irradiation without N-TiO_2, as shown in Figure 6a. Second, the adsorption rates of each N-TiO_2 photocatalyst were measured in the dark environment, shown as black symbols in Figure 6b–f. Third, the degradation rates of methylene blue were measured with N-TiO_2 photocatalysts under light irradiation, shown as red symbols in Figure 6b–f. These sets of experimental data are called "photocatalysis" but actually contain photolysis, adsorption, and photocatalytic effects.

To fit the experimental data into the kinetic model, different parameters were used in the experiments. Let us recall the meaning of rate coefficients: k_1 is the adsorption rate on the photocatalyst surface, k_2 is the degradation rate of the adsorbate by photocatalysis, and k_3 is the photolysis rate by light irradiation. For pure photolysis rate, there are no adsorption and photocatalytic degradations, so k_1 and k_2 are equal to zero, and k_3 can be fitted directly to experimental data. The adsorption rate k_1 and total available adsorption sites C_{P0} of the five photocatalysts (see Figure 6) can be obtained by setting k_2 and k_3 as zero and fitting the adsorption experimental data to the kinetic model. After these two operations, k_1, k_3, and C_{P0} can be obtained. Finally, the photocatalytic rate k_2 can be obtained from the known k_1, k_3, and C_{P0} and fitting the photocatalytic experimental data. The best-fitted parameters k_1, k_2, k_3, and C_{P0} of the different N-TiO_2 photocatalysts are shown in Table 2.

Table 2. The fitted parameters of the kinetics model k_1, k_2, k_3, and C_{P0} of N-TiO_2.

Photocatalysts	k_1 (m^3 mol^{-1} h^{-1})	C_{P0} (mol m^{-3})	$k_1 \cdot C_{P0}$ (h^{-1})	k_2 (h^{-1})	k_3 (h^{-1})
NTi-300-0.5 h	30.32	0.0660	2.0011	≈0	
NTi-300-1 h	30.12	0.0447	2.3464	≈0	
NTi-300-4 h	14.09	0.0383	0.5396	0.1669	0.0753
NTi-400-0.5 h	79.55	0.0141	1.1217	0.3861	
NTi-500-0.5 h	25.10	0.0068	0.1707	0.0996	

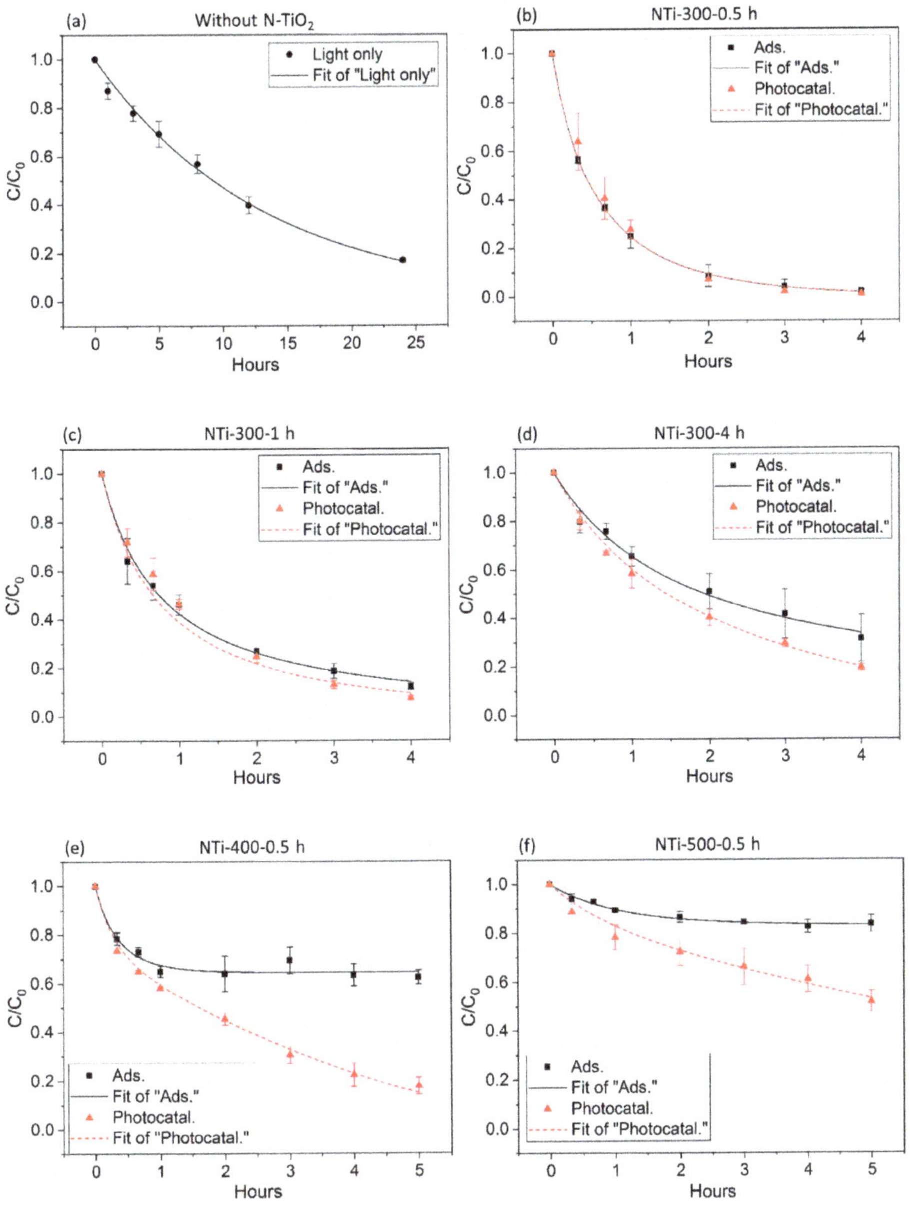

Figure 6. The adsorption and photocatalytic measurement of N-TiO_2 photocatalysts. (**a**) Methylene blue solution without N-TiO_2 irradiation by light, (**b**) NTi-300-0.5 h, (**c**) NTi-300-1 h, (**d**) NTi-300-4 h, (**e**) NTi-400-0.5 h, and (**f**) NTi-500-0.5 h.

With increased annealing time but holding the same annealing temperature at 300 °C, the adsorption capacity, indicated by C_{P0}, and adsorption rate, indicated by k_1, both decrease as shown in Figure 6b–d and Table 2, which reveal that the adsorption effect mitigates. This can be due to increased crystallinities of N-TiO_2. Holding the annealing time from 0.5 h to 4 h results in incremental crystallinity from 37% to 50%, but crystallite sizes remain stable with increased annealing time. As a result, the photocatalysts become more ordered and nonporous, which is further confirmed by the BET results. BET shows that the surface area of N-TiO_2 decreases with increased annealing time. On the other hand, N-TiO_2 annealed at 300 °C for a longer annealing time shows a better photocatalytic activity, indicated by k_2. Actually, the adsorption and photocatalysis results as shown in Figure 6 coincide for NTi-300-0.5 h and NTi-300-1 h, which implies that the photocatalytic activities of those two photocatalysts are very limited, and the removal of the contaminants mainly comes from adsorption. With a longer annealing time, N-TiO_2 photocatalysts become more crystalline and ordered, which helps the separation of electrons and holes generated by the photocatalysis reaction. Therefore, the photocatalytic activity increases with increased annealing time.

Upon holding the annealing time for 0.5 h and increasing the annealing temperatures from 300 °C to 500 °C, the adsorption effects of N-TiO_2 dramatically mitigate as shown in Figure 6b,e,f and Table 2. The adsorption capacity, C_{P0}, decreases down to a tenth, which implies fewer adsorption sites on the photocatalyst surface. This is due to higher crystallinity and larger crystallites confirmed by pXRD analysis. However, the photocatalytic activities first increase but later decrease with increased annealing temperatures, because a higher annealing temperature not only assists the crystallization of N-TiO_2 but also leads to a loss of the doped nitrogen content, which weakens the visible-light-driven photocatalytic activity. Therefore, annealing at intermediate temperature, e.g., 400 °C, results in a better photocatalytic activity.

Overall, the adsorption and photocatalytic performance are relevant to the microstructure and doped nitrogen content in the N-TiO_2 photocatalysts. Crystallinity and doped nitrogen amounts are two dominant factors that affect the photocatalytic activity of N-TiO_2. The characteristic properties of N-TiO_2 can be tuned by altering the annealing temperature and time. The annealing parameters should be appropriately selected to reach a compromise between nitrogen content and microstructure to obtain better photocatalytic activity.

3. Experimental Approach

3.1. Photocatalyst Preparation

N-TiO_2 photocatalysts were synthesized via the sol–gel method as described in a previous publication [31,32]. A colorless precursor solution containing 4 mL of titanium isopropoxide (97%, Sigma-Aldrich, St. Louis, MO, USA), 20 mL ethylenediamine (99%, Strem Chemicals, Inc., Newburyport, MA, USA), and 150 mL of 1-hexanol (anhydrous, Sigma-Aldrich) was refluxed for 18 h under Ar gas environment. The yellow-colored solution was cooled to the room temperature. To adjust the OH^- concentration before hydrolysis, acetic acid (Fisher Scientific, Waltham, MA, USA) was added into the solution. The precursor solution needs approximately 30 mL of acetic acid to neutralize the excess ethylenediamine used during reaction, which was confirmed by a pH paper test. The resulting solutions were hydrolyzed by drop-wise adding 58 mL of distilled water. The synthesized slurries were centrifuged at 7000 rpm with 60 mL distilled water and 40 mL ethanol (100% concentration, Decon Labs, Inc., King of Prussia, PA, USA) for 10 min, respectively. The slurries were left to air dry overnight inside the fume hood and then dried in an oven set at 80 °C. After that, the samples were annealed in the air at 300 °C, 400 °C, and 500 °C for 0.5 h, 1 h, and 4 h, respectively. The annealing process was achieved in a tube furnace (Thermo Scientific Lindberg, Columbia, MD, USA). The heating rate was set as 1 °C min^{-1}, and the cooling rate was set as 2 °C min^{-1}. The N-TiO_2 photocatalysts were obtained with different annealing temperatures and time, which are designated NTi-x-y, where x denotes annealing temperature in °C, and y represents annealing timed in h. Finally, five photocatalysts were prepared and studied: NTi-300-0.5 h, NTi-300-1 h, NTi-300-4 h, NTi-400-0.5 h, and NTi-500-0.5 h. All the reagents used in this study were used without further purification.

3.2. Photocatalyst Characterization

The concentration and chemical state of the nitrogen dopant were analyzed with a PHI Versaprobe 5000 Scanning X-ray Photoelectron Spectrometer (XPS) using a monochromatic Al Kα source operated at 45 W. The powder X-ray diffraction (pXRD) analysis was carried out with a Rigaku MiniFlex powder X-ray diffractometer using Cu-Kα radiation (wavelength is 0.154 nm). The Brunauer–Emmett–Teller (BET) specific surface area and Barrett–Joyner–Halenda (BJH) pore size distribution were measured based on N_2 gas adsorption and desorption by using Micromeritics TriStar *II 3020*.

3.3. Measurement of Photocatalytic Activities

The photocatalytic activities of the N-TiO_2 were measured by the degradation of the methylene blue aqueous solution (Sigma-Aldrich). The visible light source was a 52 W LED (Tadd, LLC, Cary, IL, USA) with the wavelength between 400 and 700 nm. The photocatalytic performance was analyzed by adding 10 mg photocatalyst powder into 25 mL of 15 mg L^{-1} methylene blue aqueous solution (0.4 g/L). The reaction suspension was placed 5 cm below the LED light with magnetic stirring. At each given time interval, small amounts of the samples were collected after filtering suspension with 0.22 μm syringe filter, and absorbance spectra of the solutions were measured using a Cary 60 UV-Vis spectrophotometer. The dynamic process of the photocatalyst adsorbing methylene blue was measured in the dark environment as the same step described above. To evaluate the photodegradation of methylene blue by the light only, the experiment was conducted without adding the photocatalyst.

3.4. Model of Adsorption and Photocatalysis

Degradation of water contaminant (*C*) under light irradiance in the presence of photocatalyst proceeds in two steps. The first step involves the adsorption of the contaminant molecules on the photocatalyst surface. The vacant adsorption sites (*P*) are occupied and become occupied sites ($C-P$):

$$C + P \rightarrow C - P \tag{3}$$

In the second step, the adsorbed contaminant is decomposed under the light irradiation, and thus, the occupied adsorption sites turn back to vacancy:

$$C - P \overset{light}{\rightarrow} P \tag{4}$$

In addition to the photocatalysis process, the contaminant can also be decomposed by direct light irradiation. Overall, we define three reaction rate constants in this model: (1) k_1—adsorption rate on the photocatalyst surface; (2) k_2—decomposition rate of adsorbed contaminant by the photocatalysis; and (3) k_3—direct photolysis by light in the solution. The governing equations for describing the reaction rate of this systems yield:

$$\frac{dC_C}{dt} = -k_1 \cdot C_C \cdot C_P - k_3 \cdot C_C \tag{5}$$

$$\frac{dC_P}{dt} = -k_1 \cdot C_C \cdot C_P + k_2 \cdot C_{C-P} \tag{6}$$

where C_C (mol m^{-3}) is the concentration of the methylene blue, C_P (mol m^{-3}) is the concentration of unoccupied adsorption sites in the photocatalyst, and C_{C-P} (mol m^{-3}) is the concentration of occupied adsorption sites by methylene blue in the photocatalyst. The total adsorption sites including occupied and unoccupied sites are fixed when a certain amount of photocatalyst is added:

$$C_{C-P} + C_P = C_{P0} \tag{7}$$

By substituting C in Equation (7) into Equation (6), Equation (6) can be written as:

$$\frac{dC_P}{dt} = -k_1 \cdot C_C \cdot C_P + k_2 \cdot (C_{P0} - C_P) \tag{8}$$

For better interpretation of the variables, nondimensionalization is derived by substituting C_C and C_P with $C_C^* = \frac{C_C}{C_{C0}}$ and $C_P^* = \frac{C_P}{C_{P0}}$. Therefore, Equations (3) and (6) are converted to a coupled system:

$$\frac{dC_C^*}{dt} = -k_1 C_{P0}\, C_C^*\, C_P^* - k_3 C_C^* \tag{9}$$

$$\frac{dC_P^*}{dt} = -k_1 C_{C0}\, C_C^*\, C_P^* + k_2\left(1 - C_P^*\right) \tag{10}$$

where $C_C{}^*$ and $C_P{}^*$ are the dimensionless concentrations of contaminant and unoccupied adsorption sites in the photocatalyst, respectively. C_{C0} and C_{P0} (mol m^{-3}) are corresponding the initial concentrations of contaminant and adsorption sites. We used MATHEMATICA to solve system of Equations (7) and (8), and rate coefficients k_1, k_2, and k_3 by fitting experimental data to the model solutions were estimated.

4. Conclusions

Nitrogen-doped TiO_2 photocatalysts were synthesized via the sol–gel method and a post-annealing heat treatment approach. The effects of annealing temperature and annealing time are characterized for their influence on the microstructure and surface chemical composition of the fabricated photocatalysts. These characteristics are relevant to the adsorption and photocatalysis effects of the nitrogen-doped TiO_2 photocatalysts. The kinetics of adsorption, photocatalysis, and direct photolysis mechanisms were measured with the methylene blue decomposition method. A kinetic model was built to distinguish the impacts of these three mechanisms on the removal of contaminants. The modeling and experimental results show that a higher annealing temperature leads not only to less adsorption, but also loss of interstitial nitrogen. Therefore, annealing temperature should be appropriately selected to reach a tradeoff between nitrogen content and microstructure facilitating catalytic performance.

Author Contributions: Conceptualization, X.Y.; methodology, Z.J., A.C.S.S., D.G., and X.Y.; validation, Z.J.; formal analysis, S.W., Y.H.T., and Z.J.; writing—original draft preparation, Z.J.; writing—review and editing, X.Y.; supervision, X.Y.; funding acquisition, X.Y. and A.C.S.S. All authors have read and agreed to the published version of the manuscript.

Funding: This work is supported by US National Science Foundation with an award number 1563238.

Acknowledgments: We would like to thank Burcu Gurkan and Yun-Yang Lee in the Department of Chemical and Biomolecular Engineering at Case Western Reserve University for helping the BET measurement.

Conflicts of Interest: The authors declare no conflict of interest.

References

1. Ji, J.; Xu, Y.; Huang, H.; He, M.; Liu, S.; Liu, G.; Xie, R.; Feng, Q.; Shu, Y.; Zhan, Y.; et al. Mesoporous TiO_2 under VUV irradiation: Enhanced photocatalytic oxidation for VOCs degradation at room temperature. *Chem. Eng. J.* **2017**, *327*, 490–499. [CrossRef]
2. Ao, C.H.; Lee, S.C.; Yu, J.Z.; Xu, J.H. Photodegradation of formaldehyde by photocatalyst TiO_2: Effects on the presences of NO, SO_2 and VOCs. *Appl. Catal. B Environ.* **2004**, *54*, 41–50. [CrossRef]
3. Jiménez-Tototzintle, M.; Ferreira, I.J.; Duque, S.D.S.; Barrocas, P.R.G.; Saggioro, E.M. Removal of contaminants of emerging concern (CECs) and antibiotic resistant bacteria in urban wastewater using UVA/TiO_2/H_2O_2 photocatalysis. *Chemosphere* **2018**, *210*, 449–457. [CrossRef] [PubMed]
4. Corena, J.R.A.; Bergendahl, J.A.; Hart, F.L. Advanced oxidation of five contaminants in water by UV/TiO_2: Reaction kinetics and byproducts identification. *J. Environ. Manag.* **2016**, *181*, 544–551. [CrossRef] [PubMed]

5. Dong, H.; Zeng, G.; Tang, L.; Fan, C.; Zhang, C.; He, X.; He, Y. An overview on limitations of TiO_2-based particles for photocatalytic degradation of organic pollutants and the corresponding countermeasures. *Water Res.* **2015**, *79*, 128–146. [CrossRef] [PubMed]
6. Ohama, Y.; Van Gemert, D. *Application of Titanium Dioxide Photocatalysis to Construction Materials: State-of-the-Art Report of the RILEM Technical Committee 194-TDP*; Springer Science and Business Media: Berlin/Heidelberg, Germany, 2011.
7. Karácsonyi, É.; Baia, L.; Dombi, A.; Danciu, V.; Mogyorósi, K.; Pop, L.C.; Pap, Z. The photocatalytic activity of TiO_2/WO_3/noble metal (Au or Pt) nanoarchitectures obtained by selective photodeposition. *Catal. Today* **2013**, *208*, 19–27. [CrossRef]
8. Quill, J.; Fedor, G.; Brennan, P.; Everett, E. Quantifying the Indoor Light Environment. Q-Lab Corporation. 2007. Available online: https://www.q-lab.com/documents/public/1e85c327-1aa2-4e87-808e-df04653e4d5e.pdf (accessed on 10 December 2020).
9. Pimputkar, S.; Speck, J.S.; DenBaars, S.P.; Nakamura, S. Prospects for LED lighting. *Nat. Photon.* **2009**, *3*, 180–182. [CrossRef]
10. Gascon, J.; SantaClara, J.; Oar-Arteta, L.; Van Koppen, L.; Osadchii, D.; Gascon, J.; Kapteijn, F. Photocatalytic properties of TiO_2 and Fe-doped TiO_2 prepared by metal organic framework-mediated synthesis. *Chem. Eng. J.* **2019**, *360*, 75–88. [CrossRef]
11. Karafas, E.S.; Romanias, M.N.; Stefanopoulos, V.; Binas, V.; Zachopoulos, A.; Kiriakidis, G.; Papagiannakopoulos, P. Effect of metal doped and co-doped TiO_2 photocatalysts oriented to degrade indoor/outdoor pollutants for air quality improvement. A kinetic and product study using acetaldehyde as probe molecule. *J. Photochem. Photobiol. A Chem.* **2019**, *371*, 255–263. [CrossRef]
12. Wang, N.; Li, X.; Yang, Y.; Guo, T.; Zhuang, X.; Ji, S.; Zhang, T.; Shang, Y.; Zhou, Z. Enhanced photocatalytic degradation of sulfamethazine by Bi-doped TiO_2 nano-composites supported by powdered activated carbon under visible light irradiation. *Sep. Purif. Technol.* **2019**, *211*, 673–683. [CrossRef]
13. Islam, S.Z.; Nagpure, S.; Kim, D.-Y.; Rankin, S.E. Synthesis and Catalytic Applications of Non-Metal Doped Mesoporous Titania. *Inorganics* **2017**, *5*, 15. [CrossRef]
14. Peng, X.; Wang, M.; Hu, F.; Qiu, F.; Dai, H.; Cao, Z. Facile fabrication of hollow biochar carbon-doped TiO_2/CuO composites for the photocatalytic degradation of ammonia nitrogen from aqueous solution. *J. Alloys Compd.* **2019**, *770*, 1055–1063. [CrossRef]
15. Jo, W.-K.; Kim, J.-T. Application of visible-light photocatalysis with nitrogen-doped or unmodified titanium dioxide for control of indoor-level volatile organic compounds. *J. Hazard. Mater.* **2009**, *164*, 360–366. [CrossRef]
16. Chen, C.; Bai, H.; Chang, S.-M.; Chang, C.; Den, W. Preparation of N-doped TiO_2 photocatalyst by atmospheric pressure plasma process for VOCs decomposition under UV and visible light sources. *J. Nanoparticle Res.* **2007**, *9*, 365–375. [CrossRef]
17. Fisher, M.B.; Keane, D.A.; Fernández-Ibáñez, P.; Colreavy, J.; Hinder, S.J.; McGuigan, K.G.; Pillai, S.C. Nitrogen and copper doped solar light active TiO_2 photocatalysts for water decontamination. *Appl. Catal. B Environ.* **2013**, *130*, 8–13. [CrossRef]
18. Choi, H.; Antoniou, M.G.; Pelaez, M.; De La Cruz, A.A.; Shoemaker, J.A.; Dionysiou, D.D. Mesoporous Nitrogen-Doped TiO_2 for the Photocatalytic Destruction of the Cyanobacterial Toxin Microcystin-LR under Visible Light Irradiation. *Environ. Sci. Technol.* **2007**, *41*, 7530–7535. [CrossRef] [PubMed]
19. Reddy, P.A.K.; Reddy, P.V.L.; Sharma, V.M.; Srinivas, B.; Kumari, V.D.; Subrahmanyam, M. Photocatalytic Degradation of Isoproturon Pesticide on C, N and S Doped TiO_2. *J. Water Resour. Prot.* **2010**, *2*, 235–244. [CrossRef]
20. Asahi, R.; Morikawa, T.; Irie, H.; Ohwaki, T. Nitrogen-Doped Titanium Dioxide as Visible-Light-Sensitive Photocatalyst: Designs, Developments, and Prospects. *Chem. Rev.* **2014**, *114*, 9824–9852. [CrossRef]
21. Schneider, J.; Matsuoka, M.; Takeuchi, M.; Zhang, J.; Horiuchi, Y.; Anpo, M.; Bahnemann, D.W. Understanding TiO_2 Photocatalysis: Mechanisms and Materials. *Chem. Rev.* **2014**, *114*, 9919–9986. [CrossRef]
22. Hashimoto, K.; Irie, H.; Fujishima, A. TiO_2 Photocatalysis: A Historical Overview and Future Prospects. *Jpn. J. Appl. Phys.* **2005**, *44*, 8269–8285. [CrossRef]

23. Trouts, T.D.; Chin, Y.-P. Direct and indirect photolysis of triclocarban in the presence of dissolved organic matter. *Elem. Sci. Anthr.* **2015**, *3*, 000050. [CrossRef]
24. Liu, Q.-T.; Williams, H.E. Kinetics and Degradation Products for Direct Photolysis of β-Blockers in Water. *Environ. Sci. Technol.* **2007**, *41*, 803–810. [CrossRef] [PubMed]
25. Kumar, K.V.; Porkodi, K.; Rocha, F. Langmuir–Hinshelwood kinetics—A theoretical study. *Catal. Commun.* **2008**, *9*, 82–84. [CrossRef]
26. Coleman, H.M.; Eggins, B.R.; Byrne, J.A.; Palmer, F.L.; King, E. Photocatalytic degradation of 17-β-oestradiol on immobilised TiO_2. *Appl. Catal. B Environ.* **2000**, *24*, L1–L5. [CrossRef]
27. Yu, T.; Tan, X.; Zhao, L.; Yin, Y.; Chen, P.; Wei, J. Characterization, activity and kinetics of a visible light driven photocatalyst: Cerium and nitrogen co-doped TiO_2 nanoparticles. *Chem. Eng. J.* **2010**, *157*, 86–92. [CrossRef]
28. Holzwarth, U.; Gibson, N.W. The Scherrer equation versus the 'Debye-Scherrer equation'. *Nat. Nanotechnol.* **2011**, *6*, 534. [CrossRef]
29. Zhou, X.; Peng, F.; Wang, H.; Yu, H.; Yang, J. Preparation of nitrogen doped TiO_2 photocatalyst by oxidation of titanium nitride with H_2O_2. *Mater. Res. Bull.* **2011**, *46*, 840–844. [CrossRef]
30. Cheng, X.; Yu, X.; Xing, Z.; Yang, L. Synthesis and characterization of N-doped TiO_2 and its enhanced visible-light photocatalytic activity. *Arab. J. Chem.* **2016**, *9*, S1706–S1711. [CrossRef]
31. Zhao, Y.; Qiu, X.; Burda, C. The Effects of Sintering on the Photocatalytic Activity of N-Doped TiO_2 Nanoparticles. *Chem. Mater.* **2008**, *20*, 2629–2636. [CrossRef]
32. Qiu, X.; Burda, C. Chemically synthesized nitrogen-doped metal oxide nanoparticles. *Chem. Phys.* **2007**, *339*, 1–10. [CrossRef]

catalysts

MDPI

Article

Low-Temperature Mineralisation of Titania-Siloxane Composite Layers

Tomáš Svoboda [1,*], Michal Veselý [1], Radim Bartoš [1], Tomáš Homola [2] and Petr Dzik [1]

[1] Faculty of Chemistry, Brno University of Technology, Purkyňova 118, 612 00 Brno, Czech Republic; vesely-m@fch.vut.cz (M.V.); xcbartosr@fch.vutbr.cz (R.B.); dzik@fch.vut.cz (P.D.)

[2] Department of Physical Electronics, Faculty of Science, Masaryk University, Kotlářská 267/2, 611 37 Brno, Czech Republic; tomas.homola@mail.muni.cz

* Correspondence: xcsvobodato@fch.vut.cz

Abstract: This paper deals with low-temperature mineralisation of coatings made with titania-siloxane compositions (TSC). Methyltriethoxysilane has been adopted as the precursor for the siloxane, and during its synthesis, an oligomeric siloxane condensate with methyl moieties acting as TiO_2 binder has been produced. These methyl moieties, contained in TSC, provide solubility and prevent gelling, but reduce the hydrophilicity of the system, reduce the transfer of electrons and holes generated in the TiO_2. In order to avoid these unfavourable effects, TSC mineralisation can be achieved by nonthermal treatment, for example, by using UV-radiation or plasma treatment. Characterisation of the siloxane was performed by gel permeation chromatography (GPC), which showed the size of the siloxane chain. Thermogravimetric analysis revealed a temperature at which the siloxane mineralises to SiO_2. Printed layers of two types of TSC with different siloxane contents were studied by a scanning electron microscope (SEM), where a difference in the porosity of the samples was observed. TSC on fluorine-doped tin oxide (FTO) coated glass and microscopic glass were treated with non-thermal UV and plasma methods. TSC on FTO glass were tested by voltammetric measurements, which showed that the non-thermally treated layers have better properties and the amount of siloxane in the TSC has a great influence on their efficiency. Samples on microscopic glass were subjected to a photocatalytic decomposition test of the model pollutant Acid orange 7 (AO7). Non-thermally treated samples show higher photocatalytic activity than the raw sample.

Keywords: titanium oxide; methyltriethoxysilane; siloxane; plasma treatment; UV treatment; AO7

1. Introduction

The photocatalytic layers are nowadays of great interest for water [1] and air [2] purification. Titanium dioxide (TiO_2) layers are particularly popular because of their relatively high efficiency, non-toxicity, and affordability. TiO_2 have been widely applied for contaminant remediation and microorganism destruction [3,4].

TiO_2 is often used as a coating on hard and durable materials such as glass. Recently, there has been an interest in depositing it on flexible substrates such as polyurethanes, polyesters, polyvinyl chlorides, and others, for photovoltaic, textile, and paper industries, and others. The main problem is the stabilisation of TiO_2 on the substrate in order not to release it from the photocatalytic layer into the environment [4–6]. Generally, thermal methods of hundreds of degrees Celsius [7] are used to increase the adhesion and photocatalytic activity. Another problem lies in the TiO_2 photocatalytic activity that disturbs the organic materials onto which TiO_2 is deposited. The main directions of the research are to protect the substrates from UV and photocatalytic degradation, to ensure sufficient penetration of the pollutants to the photocatalyst, to provide flexibility, and to develop the process of preparing such a photocatalytic system that could be used on an industrial scale [6,8].

One way to protect the substrate from degradation by thermal treatments and photocatalytic processes is to use a suitable binder as a matrix. The matrix serves as a mechanical support for TiO_2 and as a protective layer for the substrate. However, the bonding of the binder itself may not be a sufficient condition for a functional photocatalytic layer. Other treatments are needed that may also be thermal treatments, but it would be possible to use the photocatalytic processes themselves in the layer using UV-irradiation or short-term exposure to plasma for low-temperature treatment.

Many areas extensively use mesoporous oxide thin films as functional and structural materials. They include protective and low-dielectric constant layers, selective gas permeation membranes, wettability layers, and gas sensors [9–11].

In this work we developed and investigated polysiloxane as the TiO_2 anchor matrix and we obtain the titania-siloxane composition (TSC), which can be used for deposition on flexible substrates without the need for thermal treatments which damage the substrate. Siloxane (oligomer or polymer) can be obtained from organosilicon precursors, which can be further doped withTiO_2 particles. Siloxane serves as a matrix for anchoring TiO_2, and, at the same time, it can retain porosity to successfully adsorb the pollutants to the surface of the photocatalyst grains deposited deeper in the layer [6,12]. The siloxane used in this study contains a certain proportion of methyl moieties that provide solubility and prevent gelling. On the other hand, the methyl moieties reduce the transfer of electrons generated by the TiO_2 and significantly deteriorates the photocatalytic activity of TiO_2. There is also a certain proportion of hydroxyl groups in siloxanes. The hydroxyl group, as a residual group, induces moisture adsorption through hydrogen bonding when exposed to moisture [13]. Moisture (H_2O) is an important part of the photocatalytic processes taking place in titanium dioxide. Moisterure-binding hydroxyl groups improve the wettability of the surface and the photocatalytic activity itself and is a better supply of pollutants in the aqueous medium [14]. In order to remove the organic matter from the siloxane and improve the transfer of the electrons, siloxane must be completely mineralised towards amorphous silica surface. This step is traditionally performed by thermal annealing at hundreds of °C. On the other hand, this approach is not compatible if flexible and thermally-sensitive substrates such as polyethylene terephthalate (PET) and polyethylene naphthalate (PEN) are used. Low-cost substrates are favorable for future generation manufacture of emerging technologies, including flexible and printed electronics. Therefore, it is important to investigate novel low-temperature methods compatible with rapid and low-temperature post-treatment of siloxanes as replacements for traditional thermal annealing that is not compatible with flexible electronics.

In this work we also investigated two non-thermal methods for post-processing of titania-siloxane composition (TSC) layers: UV-irradiation and open-air plasma. UV-irradiation was already successfully tested for the fabrication of amorphous TiO_2 thin films. UV-irradiation at room temperature leads to a higher conduction band minimum level of the film and a smaller amount of hydroxyl group at the film surface, compared to the thermal-assisted (100–250 °C) UV-annealing or the thermal-only annealing (500 °C). [15]. Plasma treatment can be used for calcination and removal of organic residues from sol–gel and generation of mesoporous films [16–20]. The plasma technique is more attractive because it has many advantages, such as low processing temperature, short processing time and inexpensive equipment [9].

We studied the properties of TSC in synergy with UV-irradiation and plasma treatment as the techniques for non-thermal curing of photocatalytic layers to improve photocatalytic activity. The main parameter for the mineralisation of siloxane was the study of the decrease of methyl groups by the Fourier-transform infrared spectroscopy (FT-IR) method. The photocatalytic activity was monitored by voltammetric measurements and photocatalytic degradation of AO7.

2. Results and Discussion

2.1. Viscosity of Siloxane Solutions

Figure 1 shows viscosity of siloxane solutions tested with various concentration of methanol ranging from 10% to 50%. All solutions tested show linear dependence of dynamic viscosity with shear rate. The sample with 50% of ethanol showed a slight deviation from Newtonian behaviour (at the lowest shear rate there is a sign of viscosity increase). The viscosity of the solutions increases considerably as the concentration increases, which is important, especially for the future printing of compositions containing siloxane.

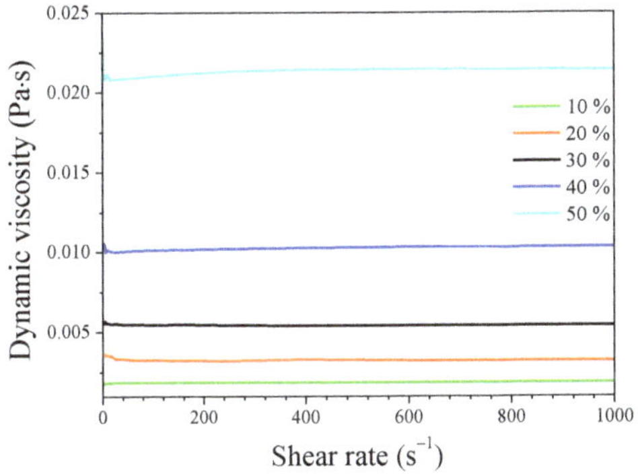

Figure 1. Viscosity of siloxane solutions in absolute ethanol (concentrations 10, 20, 30, 40, 50%).

2.2. Surface Tension of Siloxane Solutions

We also investigated thc effect of applying concentrations of ethanol on the surface tension of siloxane solutions. The pure ethanol has a surface tension of 21.90 $mN{\cdot}m^{-1}$ at 25 °C [21]. The results in Table 1 show that there was no significant change in the surface tension of solutions with various concentration of ethanol. However, with the decreasing concentration of pure ethanol, a slight increase in surface tension occurs when the concentration of siloxane in volume, and hence on the surface of the liquid, increases. We see a certain similarity with ethanol, considering the possible structure of siloxane (Equations (1) and (2)) where R–Si–$(OH)_3$ is present in the solution, and because in our case R means methyl, the hydrocarbon residue will not have the weight to significantly affect the surface tension. It can therefore be assumed that the surface tension of the resulting siloxane-blended compositions will not be significantly affected by the amount of siloxane.

Table 1. Surface tension of siloxane solutions (concentrations 10, 20, 30, 40, 50%).

Concentration (%)	10	20	30	40	50
Surface tension ($mN{\cdot}m^{-1}$)	22.58	22.96	23.52	24.08	24.75

2.3. Gel Permeation Chromatography

Gel permeation chromatography (GPC) results are shown in Table 2. The polydispersity is close to 1, which means that the particles in solution have similar dimensions, expressed by the radius of gyration and the weight average molar mass.

Table 2. The results of GPC. The polydispersity is close to 1, which means that the particles in solution have similar sizes.

Dilution of 10% Solution	Injection Volume (μL)	M_W (kDa)	Polydispersity (M_W/M_N)	Radius of Gyration (nm)
1:1	100	1.497	1.019	12.8
1:1	100	1.534	1.036	11.9
1:1	100	1.475	1.033	12.0
1:1	100	1.506	1.046	12.5
	Diameter M_W (kDa)	1.52	1.032	13
	Selective standard deviation	0.04	0.009	2

2.4. Specific Surface Area (SSA) and SEM of Siloxane/TiO_2

Two siloxane/TiO_2 compositions of ratios 1:1 and 1:3 were created and compared. SSA was measured by by nitrogen adsorption using the BET isotherm (Figure 2). The two different values of siloxane/TiO_2 ratios have a direct impact of the layer textural properties as is evident from Figure 3. The more titania rich formulation exhibited a fluffy texture with many accessible voids while the more binder rich formulation is apparently denser and more compact. SSA for siloxane/TiO_2 compositions of ratios 1:1 and 1:3 was calculated as 12.3 m^2/g and 34.9 m^2/g, respectively. The question of phase composition of both coatings was addressed separately in our previous communication [22] and we found out that binder mineralisation process has no impact on the crystallinity of titania.

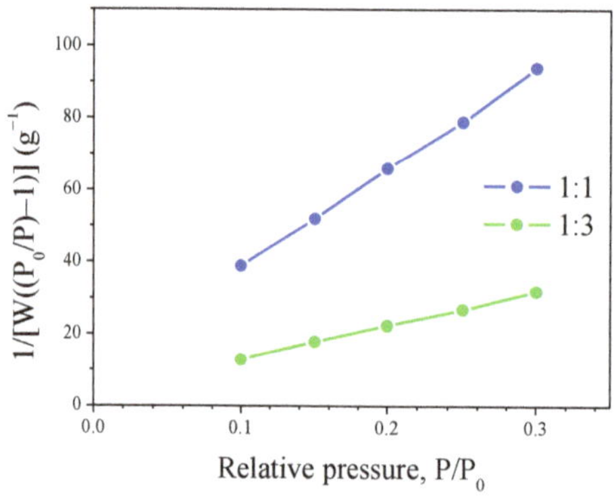

Figure 2. Multi-point BET plot for siloxane/TiO_2 with various ratio 1:1 and 1:3.

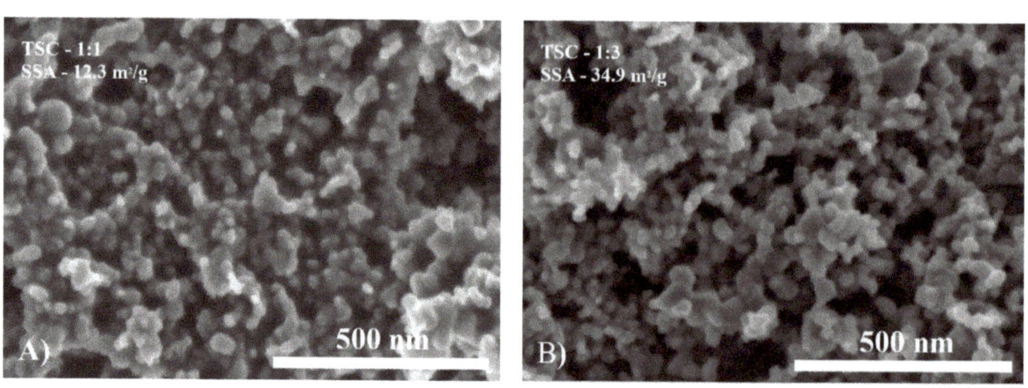

Figure 3. SEM images. (**A**) Shows a less porous layer with a siloxane/TiO_2 ratio of 1:1, and (**B**) shows a much more porous layer of a 1:3 ratio.

2.5. Thermal Treatment of TSC and TGA, DTG

The coatings were further sintered in furnace oven at temperature 450 °C for 30 min to mineralise the binder, i.e., remove the organic methyl moieties and support the transport of electrons from TiO_2. FT-IR results showed in Figure 4 revealed that siloxane was not mineralised completely. Peaks located at 2995–2950 cm^{-1} and 2895–2840 cm^{-1} corresponding to asymmetrical and symmetrical stretching of $-CH_3$ and indicating the presence of methyl groups in polysiloxane binder.

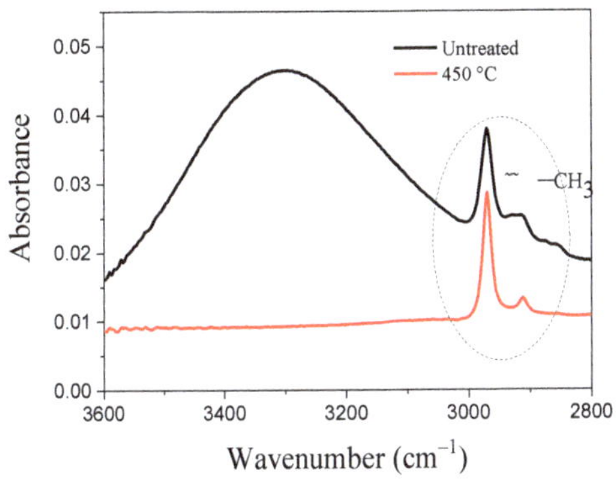

Figure 4. FT-IR measurement of TSC on soda-lime glass. The processing temperature of 450 °C was not sufficient to mineralise the siloxane because the methyl groups are still present after thermal treatment.

The coatings were further analysed by DSC-TGA and heated to 1300 °C. During the thermal process, there were several changes as the temperature rose (Figure 5). The first range was from 20 °C to 477 °C. There was probably a loss of sorbed air moisture or residual solvent. Weight decreases steeply. The second major change occurs at 478 °C, when there was a further rapid drop in weight. Siloxane should be calcined at this temperature and thus degrade the primarily methyl groups bonded to silicon oxide. Another weight change was significantly slowed down from a temperature of 600 °C until it stabilised and no changes occurred. It is clear from the measurements that, for complete calcination, it is necessary to achieve temperatures higher than 478 °C, and therefore, the thermal method of mineralisation is only suitable for resistant materials.

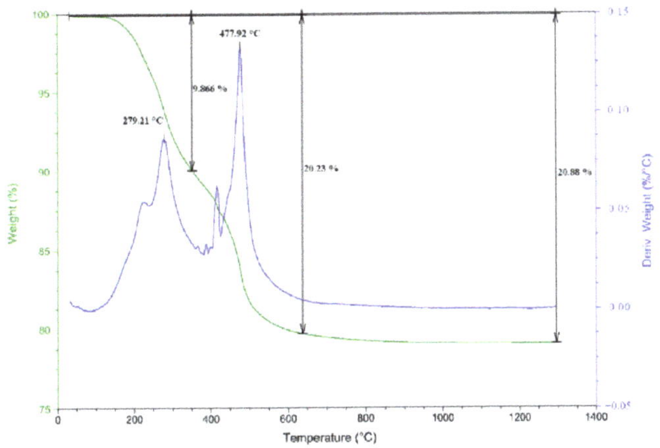

Figure 5. The results of the TGA, DTG. The mineralisation and removal of the methyl groups from the siloxane should occur at 478 °C.

2.6. FT-IR of Nonthermal Curing TSC

UV-irradiation and non-thermal atmospheric-pressure plasma treatment were further investigated in order to replace time-consuming thermal sintering, which is problematic if polysiloxane/TiO_2 is deposited on thermally-sensitive materials. Both siloxane/TiO_2 compositions of ratios 1:1 and 1:3 were investigated by FT-IR before and after UV-irradiation and plasma treatment.

Figure 6A,B shows FT-IR spectra of siloxane/TiO_2 layers prepared by both compositions of ratios 1:1 and 1:3, cured by UV-irradiation for 0–210 min. The decrease of peaks related to asymmetrical and symmetrical stretching of $-CH_3$ (2995–2950 cm^{-1} and 2895–2840 cm^{-1}) indicate the removal of $-CH_3$ groups from siloxane surface and apparently the transformation of siloxane towards amorphous SiO_2. For both coatings (1:1 and 1:3 ratio), UV-irradiation for 150 min was enough to remove all detectable $-CH_3$ groups. This decrease can be further explained by photocatalytical reaction between anatase TiO_2 with bandgap 3.2 eV [23] that can provide additional catalytical reaction on polysiloxane surface and enhance the degradation of methyl groups.

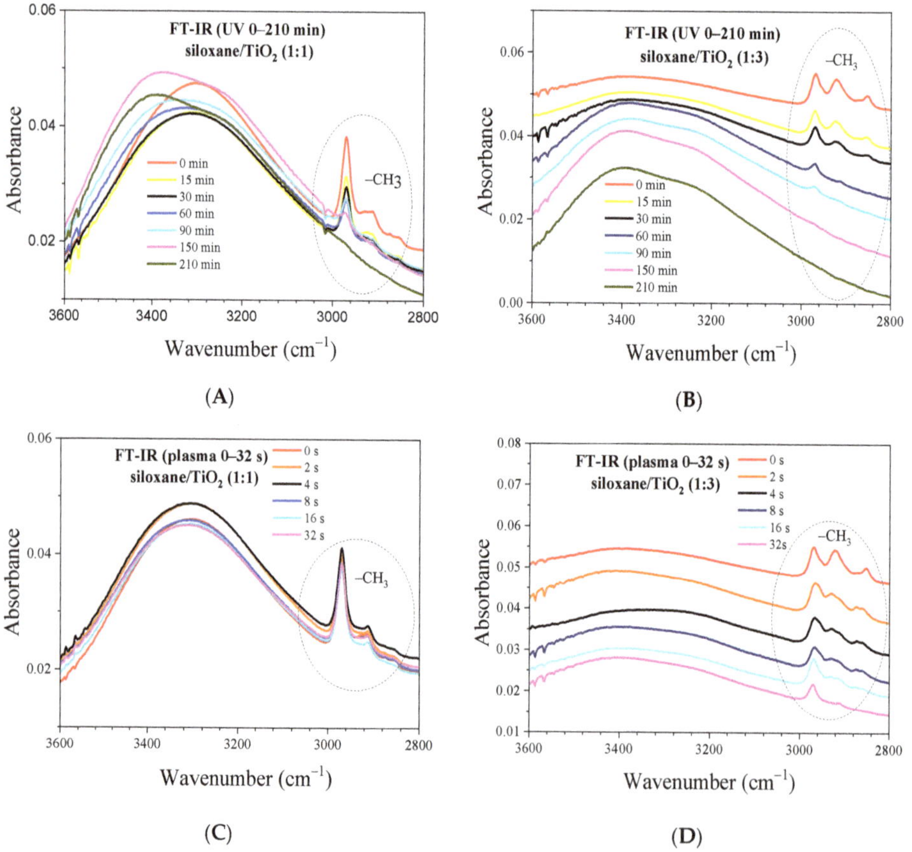

Figure 6. FT-IR measurements of TSC layers. (**A**,**B**) are images for samples treated with UV-irradiation; (**C**,**D**) are images for plasma-treated samples. UV-irradiation has better layer penetration and is therefore effective for layers with a higher concentration of siloxane.

Figure 6C,D shows FT-IR spectra of siloxane/TiO_2 layers prepared by both compositions of ratios 1:1 and 1:3, cured by plasma for 0–32 s. In contrast with the results presented for UV-irradiated layers, the plasma treated layers showed significantly lower efficiency of methyl groups removal. The most profound difference was found for the film of lower porosity (1:1) and higher efficiency was found for the layer of higher porosity. This discrepancy is clearly related to the difference in mechanism of UV-irradiation and plasma treatment. Whereas the dominant energy-transfer mechanism in UV-irradiation is the transport of photons of certain wavelength into the porous film, the plasma treatment works differently. Apart from the UV-irradiation, the plasma generated at atmospheric pressure in ambient air contains various energetic species that can carry energy towards siloxane surface: High-temperature electrons, low-temperature ions, excited species and metastables. These species, however, recombine and extinct in contact with materials surface and thus, the porosity of material plays an important role in the limitation of such method. Since plasma treatment was less efficient for methyl removal at less porous material, it can be concluded that plasma cannot efficiently penetrate into the coating bulk and mineralise it. On the other hand, the coating with high porosity showed better efficiency of methyl removal, and therefore, it is important to find an optimal combination of layer porosity if atmospheric pressure plasma is used for mineralisation of coatings. Although not efficient as UV-irradiation, a clear benefit of plasma treatment is in significantly faster treatment times, in order of ten seconds, which allow to use this method on fast roll-to-roll production lines and employ the coatings in flexible and printed electronics concept.

2.7. Voltammetric Measurements of Nonthermal Curing TSC

Voltammetric measurements can be taken for the photocatalytic activity test. If the treated layer is exposed to UV-irradiation, it generates an electron/hole pair. By applying the external voltage, the electrons can be abstracted from the exciton and the photocurrent is detected in the external circuit. The growth of the photocurrent is related to higher photocatalytic activity [24].

The siloxane/TiO_2 films of ratios 1:1 and 1:3 were exposed to UV-irradiation for 0–120 min and plasma treatment for 0–32 s. Figure 7 shows the voltametric measurement of photocurrent generated by the coating upon exposure to UV. The UV-irradiation pre-treatment for 0–120 min led to a gradual increase of the photocurrent for both coatings (1:1, 1:3) investigated. Higher values of photocurrent were measured for coating with a higher concentration of TiO_2 in siloxane/TiO_2 (ratio 1:3). This is apparently thanks to higher concentration of photo-catalytically active TiO_2 that can not only improve the mineralisation reaction, but also supply more electrons upon UV exposure. Furthermore, the porosity of the coating with siloxane/TiO_2 ratio 1:3 is higher, so the higher photocurrent is most likely an interplay between these factors.

Figure 7C,D shows voltammetric measurements of siloxane/TiO_2 films of ratios 1:1 and 1:3 were exposed to air plasma for 0–32 s. The coating with small porosity yields very small photocurrents, and it seems that plasma treatment had a negative effect on the maximal photocurrent values. On the other hand, the plasma treatment of the coating with higher porosity (1:3) resulted in a gradual increase in photocurrent with plasma-exposure time. The better efficiency in photocurrent generation in the coating of higher porosity is related to higher mineralisation efficiency, as shown in FT-IR results presented in Figure 6D.

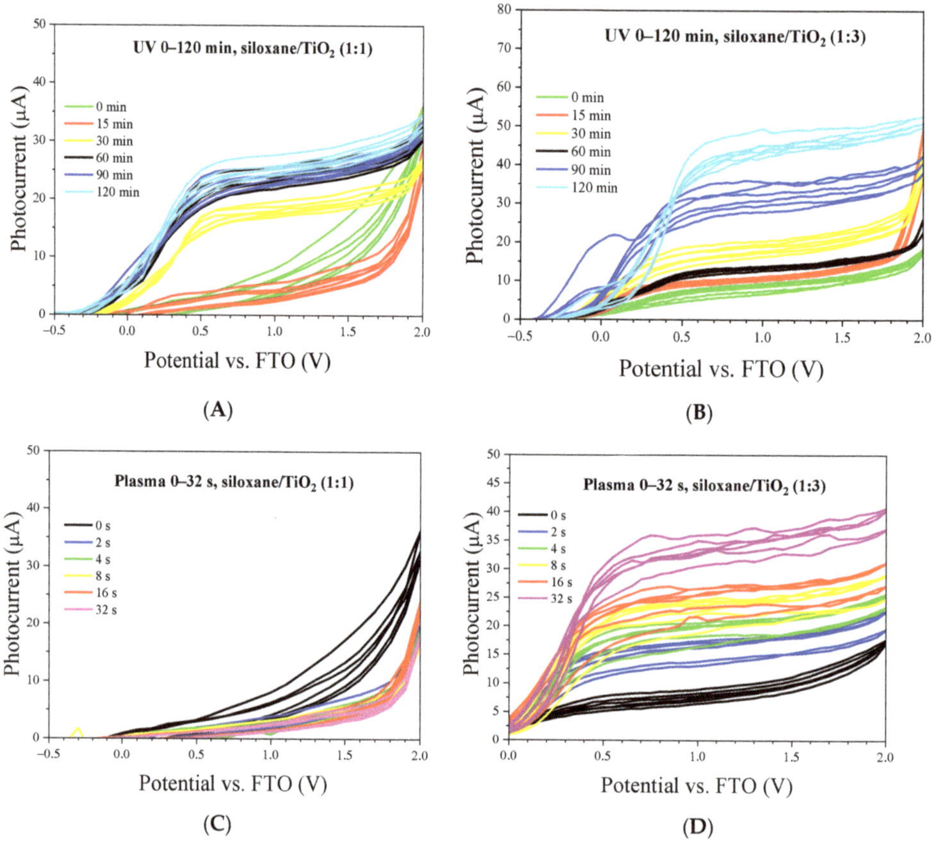

Figure 7. Voltammetric characterisation of TSC printed on FTO-coated glass. (**A**,**B**) are images for samples treated with UV-irradiation; (**C**,**D**) are images for plasma-treated samples. Plasma can not penetrate the layer sufficiently if it contains too much siloxane. In the case of (**C**), the photocurrent is very low to zero.

2.8. Photocatalytic Degradation of Acid Orange 7

Acid Orange 7 (AO7) serves as a model pollutant for the photodegradation test. AO7 turns the aqueous solution to orange and it is a resistant substance that slow degrades. Figure 8 shows the decrease in absorbance signal reflecting the concentration of AO7 upon irradiation of UVA for 1 h. UVA was chosen because radiation with lower wavelengths is less advantageous in terms of energy efficiency. We compared three samples with ratio 1:3: untreated siloxane/TiO_2, UV-irradiated siloxane/TiO_2 for 210 min and plasma-treated siloxane/TiO_2 for 32 s. The slowest degradation of AO7 was observed for untreated siloxane/TiO_2, whereas the fastest degradation was observed for UV-irradiated siloxane/TiO_2 for 210 min. The plasma-treated siloxane/TiO_2 showed significant improvement in AO7 degradation in comparison to the untreated sample, although the efficiency of siloxane/TiO_2 irradiated by UV was better. It should be noted that UVA involved in AO7 degradation test may further contribute to mineralisation of TSC [25].

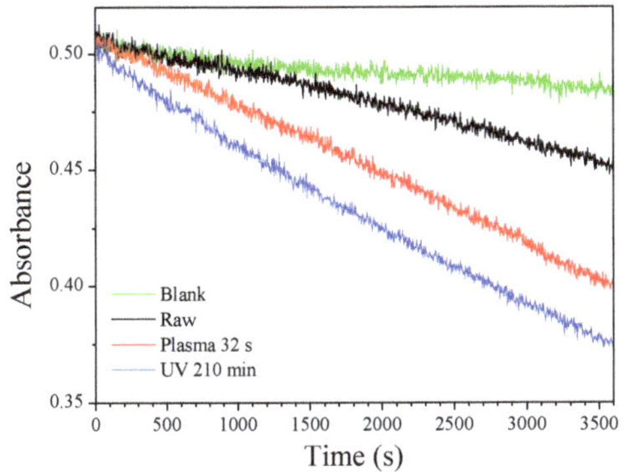

Figure 8. Degradation of 2 mg/L aqueous solution of AO7 with immobilized TSC layers treated by plasma and UV-irradiation.

3. Materials and Methods

3.1. Synthesis of Siloxane

The starting substance for the synthesis of siloxane was methyltriethoxysilane (MTEOS) (Alfa Aesar, 98%, Haverhill, MA, USA). MTEOS was hydrolysed with acidic water. Ethanol was formed during the hydrolysis and subsequently distilled. Siloxane was extracted with diethyl ether (Penta, 99.7%), and it was dissolved in absolute ethanol (Penta, 99.8%, Prague, Czech Republic) after evaporation of the extracting agent. The siloxane solution in ethanol was stored at a temperature below 0 °C [6]. The expected chemical reaction can be summarised by the following equation [26]:

$$R - Si(OC_2H_5)_3 + 3H_2O \rightarrow R - Si(OH)_3 + 3C_2H_5OH \quad (1)$$

$$2R - Si(OH)_3 \leftrightarrow R - Si(OH)_2 - O - Si(OH)_2 - R + H_2O \quad (2)$$

3.2. Preparation of TSC

Two series of samples were prepared for experiments, as shown in Table 3. TiO_2 (P25, Sigma Aldrich, 99.7%, St. Louis, MO, USA, particle size $\leq$ 25 nm, SSA 45–55 m^2/g [27]) was dispersed in dowanol and a siloxane 20% solution was added. The resulting suspension was further diluted with hexanol. The TSC were printed with a Dimatix (DMP-2800) material printer on soda-lime glass and FTO-coated glass.

Table 3. Compositions designed for material printing.

Composition	Siloxane (20% in Ethanol)	TiO_2 (20% P25 in Dowanol)	Hexanol
E38-9AD (1:1)	4 mL	4 mL	20 mL
E38-10AD (1:3)	2 mL	6 mL	20 mL

3.3. Mineralisation of the Printed TSC on the Substrate

The siloxane/TiO_2 coatings were mineralised by two methods: UV-irradiation and plasma treatment. UV-irradiation was generated by a Sylvania UV lamp Figure 9 (mercury, 125 W, Budapest, Hungary). The radiation intensity was set and held at 9 $mW \cdot cm^{-2}$. Samples were placed under 5 mm of distilled water and irradiated for 0, 15, 30, 60, 90, 150, and 210 min.

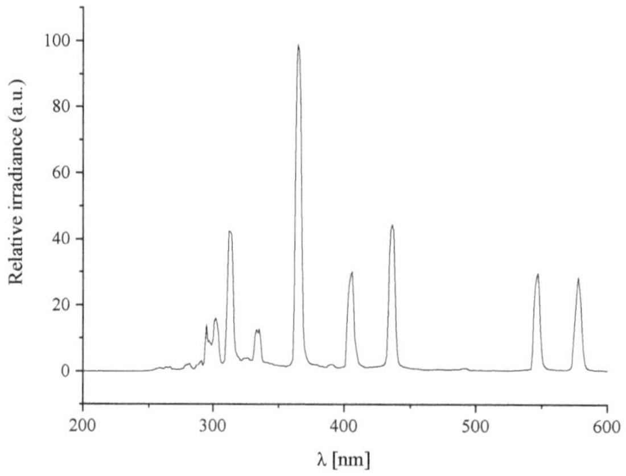

Figure 9. Spectrum of the Sylvania UV lamp (mercury, 125 W).

The plasma treatment was performed by RPS400 (Roplass s.r.o., Brno, Czech Republic) equipped by dielectric barrier discharge (DBD) with a coplanar arrangement of the electrode system: A diffuse coplanar surface barrier discharge (DCSBD) plasma unit. The DCSBD is capable of generating a thin surface plasma of very high-power density up to 100 $W \cdot cm^{-3}$ at very low temperature ~70 °C in open air environment at atmospheric pressure. The details of the DCSBD plasma can be found in Homola et al. [28,29]. The plasma exposure times of siloxane surfaces in this study were 0, 2, 4, 8, 16 and 32 s.

3.4. Characterisation of Siloxane

The properties of siloxane solutions in ethanol were studied by means of various methods. Since the ink-jet printing is compatible only to a certain range of viscosity (10–30 cP) and other rheologic parameters, it is important to determine the viscosity, surface tension, and size of agglomerates, in order to verify if the compositions have the desired properties and do not suffer complications during printing.

The concentration range of siloxane (10, 20, 30, 40, 50%) was used for viscosity measurements carried out on a rotating rheometer AR-G2 (TA Instruments) with a roller sensor in a cylinder at 25 °C. The surface tension was measured by tensiometer KSV Sigma 701 using the *du Noüy ring method* with a platinum ring.

The SEC-MALS-dVI-dRI system was used for GPC measurements. Chromatographic part (Agilent Technologies, Santa Clara, CA, USA): consisting of isocratic pump, degasser, autosampler, column and UV-VIS detector. Detectors (Wyatt Technology, Dernbach, Germany): DAWN HELEOS II: multi-angle light scattering (MALS), VISCOSTAR II: differential viscometer (dVI), OPTILAB T-REX: differential refractometer (dRI). The 10% siloxane solution was further diluted with absolute ethanol 1:1.

The thermal gravimetric analysis was performed with Instrument SDT Q600. The siloxane solution was dried to 100 °C and the solid was exposed to a temperature of up to 1300 °C in the air with a step 10 °C/min. Thermal analysis was also performed to give us insight into changes during the thermal mineralisation of siloxane.

3.5. Characterisation of TSC and Printed Layers on Glass

Brunauer–Emmett–Teller (BET) method was employed to determine the specific surface area (SSA) of TSC. BET was performed on Autosorb iQ. Prior to BET analysis, the samples were dried at 100 °C.

FT-IR Nicolet iS5 was used in order to measure the decrease the methyl groups after non-thermal curing by UV-irradiation and plasma treatment. The TSC samples on soda-lime glass were measured on FT-IR using the Omnic program. The spectral region characteristic of the 2950 cm^{-1} and 2895–2840 cm^{-1} methyl groups was observed [30].

The photo-electrochemical characterization of TSC on the FTO-coated glass was performed by linear sweep voltammetry at room temperature using a two-electrode setup with the 1 cm^2 titania patches. The printed FTO slide was scratched with a diamond knife and thus two isolated FTO strips were created. One strip with the printed titania patch, served as the working electrode and the opposite naked FTO strip as the counter electrode. This setup was fitted into a custom build quartz cuvette. The cuvette was filled with 0.1 M perchloric acid (conductivity 36 $mS \cdot cm^{-1}$) and fitted onto an optical bench equipped with a fluorescent UV-A lamp emitting a broad peak centered at 365 nm (Sylvania Lynx-L 11 W). A magnetic stirrer was placed beneath the cuvette and a magnetic flea inside the cuvette provided efficient electrolyte mixing. The lamp emission was monitored by Gigahertz Optic X97 Irradiance Meter with a UV-3701 probe and the irradiance was set to 2 $mW \cdot cm^{-2}$ by adjusting the lamp-to-cuvette distance. Measurements of generated photocurrents were performer with an electrometer build on the basis of National Instruments Labview platform and supplying a linear voltage gradient of 10 mV/s from −0.5 to 2 V and measuring the generated currents down to submicroampere range [24].

Photocatalytic degradation test of TSC on glass with 12 cm^2 E38-10AD patches was performed by degradation recrystallization AO7 (Centre for Organic Chemistry) solution in water. The quartz cuvette was filled with 2 mg/L aqueous solution of AO7 and fitted onto an optical bench equipped with a fluorescent UV-A lamp emitting a broad peak centered at 365 nm (Sylvania Lynx-L 11 W). The absorbance of the resulting solution at a 4 cm optical path was 0.5 at a wavelength of 480–490 nm. A magnetic stirrer was placed beneath the cuvette and a magnetic flea inside the cuvette provided efficient solution mixing. The lamp emission was monitored by Gigahertz Optic X97 Irradiance Meter with a UV-3701 probe and the irradiance was set to 2 $mWcm^{-2}$ Measurements of decrease in absorbance AO7 were performer with a spectrometer Red Tide USB650 (Ocean Optics, Prague, Czech Republic) and an Ocean View program.

The TSC samples printed on FTO-coated glass were also examined by the TESCAN Mira3 XMH scanning electron microscope. Coatings were printed on FTO, in order to prevent charging and deposition of conductive layer onto TSC.

4. Conclusions

This contribution presents the non-thermal curing of titania siloxane coatings by two atmospheric methods. Titania siloxane coatings are promising alternative for other photo-catalytically active layers on substrates that cannot withstand processing at temperatures >150 °C, e.g., thermally sensitive polymers and paper. This work demonstrates and compares slow and efficient UV-irradiation, and extremely rapid plasma treatment, as methods for low-temperature sintering of titania siloxane photo-catalytically active coatings. It was also investigated that the main mechanism of the non-thermal curing is the removal of the $-CH_3$ methyl groups from the siloxane chain while the bulk structure of the titania content remained unaffected. Although, UV-irradiation showed more efficient and led to the removal of most of the detectable methyls from the siloxane surface, the plasma treatment resulted in much faster processing times in the order of tens of seconds. This, on the other hand, is promising for the implementation of the method into fast roll-to-roll production line. Since both methods, UV and plasma, are scalable to several meters in width, the employment of any of them or even the combination of UV and plasma treatment could be of interest development of new large scale thin-film photocatalysts.

Author Contributions: Conceptualization, T.S. and P.D.; methodology, P.D., T.S.; investigation, T.S.; supervision, P.D. and M.V.; writing—original draft preparation, T.S.; software, T.S.; writing—review and editing, P.D., T.H. and T.S.; data curation, T.S. and R.B.; resources, P.D., M.V. and T.H.; validation, P.D., T.H., R.B. and T.S.; project administration, T.H., M.V. and P.D. All authors have read and agreed to the published version of the manuscript.

Funding: This work has been supported by Czech Science Foundation research project 19-14770Y.

Institutional Review Board Statement: Not applicable.

Informed Consent Statement: Not applicable.

Data Availability Statement: The data presented in this study are available in the article.

Conflicts of Interest: The authors declare that they have no competing interests.

References

1. Levchuk, I.; Sillanpää, M. Chapter 1—Titanium Dioxide–Based Nanomaterials for Photocatalytic Water Treatment. In *Advanced Water Treatment: Advanced Oxidation Processes*; Elsevier: Amsterdam, The Netherlands, 2020; pp. 1–56. ISBN 978-0-12-819225-2.
2. Cha, B.J.; Saqlain, S.; Seo, H.O.; Kim, Y.D. Hydrophilic surface modification of TiO_2 to produce a highly sustainable photocatalyst for outdoor air purification. *Appl. Surf. Sci.* **2019**, *479*, 31–38. [CrossRef]
3. Yoon, K.H.; Noh, J.S.; Kwon, C.H.; Muhammed, M. Photocatalytic behaviour of TiO_2 thin films prepared by sol–gel process. *Mater. Chem. Phys.* **2006**, *95*, 79–83. [CrossRef]
4. Hu, J.; Gao, Q.; Xu, L.; Wang, M.; Zhang, M.; Zhang, K.; Liu, W.; Wu, G. Functionalization of cotton fabrics with highly durable polysiloxane—TiO_2 hybrid layers: Potential applications for photo-induced water–oil separation, UV shielding, and self-cleaning. *J. Mater. Chem. A* **2018**, *6*, 6085–6095. [CrossRef]
5. Wang, M.; Zhang, M.; Zhang, M.; Aizezi, M.; Zhang, Y.; Hu, J.; Wu, G. In-situ mineralized robust polysiloxane–Ag@ZnO on cotton for enhanced photocatalytic and antibacterial activities. *Carbohydr. Polym.* **2019**, *217*, 15–25. [CrossRef]
6. Gregori, D.; Benchenaa, I.; Chaput, F.; Thérias, S.; Gardette, J.-L.; Leonard, D.; Guillard, C.; Parola, S. Mechanically stable and photocatalytically active TiO_2/SiO_2 hybrid films on flexible organic substrates. *J. Mater. Chem. A* **2014**, *2*, 20096–20104. [CrossRef]
7. Shan, A.Y.; Ghazi, T.I.M.; Rashid, S.A. Immobilisation of titanium dioxide onto supporting materials in heterogeneous photocatalysis: A review. *Appl. Catal. A Gen.* **2010**, *389*, 1–8. [CrossRef]
8. Wang, M.; Zhang, M.; Pang, L.; Yang, C.; Zhang, Y.; Hu, J.; Wu, G. Fabrication of highly durable polysiloxane-zinc oxide (ZnO) coated polyethylene terephthalate (PET) fabric with improved ultraviolet resistance, hydrophobicity, and thermal resistance. *J. Colloid Interface Sci.* **2019**, *537*, 91–100. [CrossRef]
9. Zhang, J.; Palaniappan, A.; Su, X.; Tay, F.E. Mesoporous silica thin films prepared by argon plasma treatment of sol–gel-derived precursor. *Appl. Surf. Sci.* **2005**, *245*, 304–309. [CrossRef]
10. Baskaran, S.; Liu, J.; Domansky, K.; Kohler, N.; Li, X.; Coyle, C.; Fryxell, G.E.; Thevuthasan, S.; Williford, R.E. Low Dielectric Constant Mesoporous Silica Films Through Molecularly Templated Synthesis. *Adv. Mater.* **2000**, *14*, 291–294. [CrossRef]
11. Wirnsberger, G.; Yang, P.; Scott, B.J.; Chmelka, B.F.; Stucky, G.D. Mesostructured materials for optical applications: From low-k dielectrics to sensors and lasers. *Spectrochim. Acta Part A Mol. Biomol. Spectrosc.* **2001**, *57*, 2049–2060. [CrossRef]
12. Chen, J.Y.; Pan, F.M.; Chang, L.; Cho, A.T.; Chao, K.J. Thermal stability of trimethylsilylated mesoporous silica thin films as the ultralow-k dielectric for copper interconnects. *J. Vac. Sci. Technol. B Microelectron. Nanometer Struct.* **2005**, *23*, 2034. [CrossRef]
13. Jung, S.-B.; Park, H.-H. Control of surface residual OH polar bonds in SiO2 aerogel film by silylation. *Thin Solid Films* **2002**, *420*, 503–507. [CrossRef]
14. Zhao, C.; Pelaez, M.; Dionysiou, D.D.; Pillai, S.C.; Byrne, J.A.; O'Shea, K.E. UV and visible light activated TiO_2 photocatalysis of 6-hydroxymethyl uracil, a model compound for the potent cyanotoxin cylindrospermopsin. *Catal. Today* **2014**, *224*, 70–76. [CrossRef]
15. Lee, M.; Park, I.J.; Jeong, H.; Kim, B.J.; Yun, Y.; Kim, H.J.; Cho, H.; Lee, S. Thermal-assisted photo-annealed TiO_2 thin films for perovskite solar cells fabricated under ambient air. *Appl. Surf. Sci.* **2020**, *530*, 147221. [CrossRef]
16. Kraus, F.; Cruz, S.; Müller, J. Plasma polymerized silicon organic thin films from HMDSN for capacitive humidity sensors. *Sens. Actuators B Chem.* **2003**, *88*, 300–311. [CrossRef]
17. Gomez-Vega, J.M.; Teshima, K.; Hozumi, A.; Sugimura, H.; Takai, O. Mesoporous silica thin films produced by calcination in oxygen plasma. *Surf. Coat. Technol.* **2003**, *169–170*, 504–507. [CrossRef]
18. Park, H.-H.; Jo, M.; Kim, H.; Hyun, S. Effect of Oxygen Plasma Treatment on SiO_2 Aerogel Films. *J. Mater. Sci. Lett.* **1998**, *17*, 2083–2085. [CrossRef]
19. Huang, J.; Ichinose, I.; Kunitake, T.; Nakao, A. Preparation of Nanoporous Titania Films by Surface Sol−Gel Process Accompanied by Low-Temperature Oxygen Plasma Treatment. *Langmuir* **2002**, *18*, 9048–9053. [CrossRef]
20. Homola, T.; Dzik, P.; Veselý, M.; Kelar, J.; Černák, M.; Weiter, M. Fast and Low-Temperature (70 °C) Mineralization of Inkjet Printed Mesoporous TiO_2 Photoanodes Using Ambient Air Plasma. *ACS Appl. Mater. Interfaces* **2016**, *8*, 33562–33571. [CrossRef]
21. Lidmila, B.; Marie, Š. *Physical Chemistry of Surfaces and Colloidal Systems [In Czech: Fyzikální chemie povrchů a koloidních soustav]*, 4th ed.; VŠCHT: Prague, Czech Republic, 2002; pp. 51–56. ISBN 80-708-0475-0.
22. Dzik, P.; Svoboda, T.; Kaštyl, J.; Veselý, M. Modification of photocatalyst morphology by ball milling and its impact on the physicochemical properties of wet coated layers. *Catal. Today* **2019**, *328*, 65–70. [CrossRef]
23. Ghosh, M.; Lohrasbi, M.; Chuang, S.S.C.; Jana, S.C. Mesoporous Titanium Dioxide Nanofibers with a Significantly Enhanced Photocatalytic Activity. *ChemCatChem* **2016**, *8*, 2525–2535. [CrossRef]

24. Dzik, P.; Veselý, M.; Pachovská, M.; Neumann-Spallart, M.; Buršíková, V.; Homola, T. The influence of curing methods on the physico-chemical properties of printed mesoporous titania patterns reinforced by methylsilica binder. *Catal. Today* **2018**, *313*, 26–32. [CrossRef]
25. Homola, T.; Ďurašová, Z.; Shekargoftar, M.; Souček, P.; Dzik, P. Optimization of TiO_2 Mesoporous Photoanodes Prepared by Inkjet Printing and Low-Temperature Plasma Processing. *Plasma Chem. Plasma Process.* **2020**, *40*, 1311–1330. [CrossRef]
26. Junlabhut, P.; Boonruang, S.; Mekprasart, W.; Pecharapa, W. Ag nanoparticle-doped SiO_2/TiO_2 hybrid optical sensitive thin film for optical element applications. *Surf. Coat. Technol.* **2016**, *306*, 262–266. [CrossRef]
27. Product Specification: TiO_2. Sigma-Aldrich. Available online: https://api.sigmaaldrich.com/deepweb/assets/sigmaaldrich/quality/spec/440/011/637254-BULK_______ALDRICH__.pdf (accessed on 2 December 2020).
28. Homola, T.; Wu, L.Y.L.; Černák, M. Atmospheric Plasma Surface Activation of Poly(Ethylene Terephthalate) Film for Roll-To-Roll Application of Transparent Conductive Coating. *J. Adhes.* **2014**, *90*, 296–309. [CrossRef]
29. Homola, T.; Matoušek, J.; Kormunda, M.; Wu, L.Y.L.; Černák, M. Plasma Treatment of Glass Surfaces Using Diffuse Coplanar Surface Barrier Discharge in Ambient Air. *Plasma Chem. Plasma Process.* **2013**, *33*, 881–894. [CrossRef]
30. George, S. *Infrared and Raman Characteristic Group Frequencies: Tables and Charts*, 3rd ed.; John Wiley & Sons: Hoboken, NJ, USA, 2004; ISBN 978-0-470-09307-8.

catalysts

MDPI

Article

Photocatalytic Removal of Pharmaceuticals from Greywater

Bukola Ojobe [1], Radek Zouzelka [2], Barbora Satkova [1], Magdalena Vagnerova [3], Alzbeta Nemeskalova [3,4], Martin Kuchar [3], Jan Bartacek [1] and Jiri Rathousky [2,*]

1 Department of Water Technology and Environmental Engineering, University of Chemistry and Technology Prague, Technicka 5, 166 28 Prague, Czech Republic; bukola.ojobe@vscht.cz (B.O.); barbora.satkova@vscht.cz (B.S.); jan.bartacek@vscht.cz (J.B.)
2 Centre for Innovations in the Field of Nanomaterials and Nanotechnologies, J. Heyrovsky Institute of Physical Chemistry, Czech Academy of Sciences, Dolejskova 3, 182 23 Prague, Czech Republic; radek.zouzelka@jh-inst.cas.cz
3 Forensic Laboratory of Biologically Active Substances, Department of Chemistry of Natural Compounds, University of Chemistry and Technology Prague, Technicka 5, 166 28 Prague, Czech Republic; magdalena.vagnerova@vscht.cz (M.V.); alzbeta.nemeskalova@vscht.cz (A.N.); kuchara@vscht.cz (M.K.)
4 Department of Analytical Chemistry, University of Chemistry and Technology Prague, Technicka 5, 166 28 Prague, Czech Republic
* Correspondence: jiri.rathousky@jh-inst.cas.cz

Abstract: High concentrations of pharmaceuticals have been detected in greywater effluents treated using up-to-date technologies. Finding a suitable additional treatment before this effluent is reused is urgently needed to ensure the reused water meets quality standards. This paper reports the use of heterogeneous photocatalysis on anatase and rutile nanopowders to remove naproxen, metformin and sulfamethoxazole, at practically relevant concentrations found in membrane bioreactor (MBR)-treated greywater. A low anatase concentration of 400 mg L^{-1} was sufficient to efficiently degrade the pharmaceuticals listed above, with complete degradation observed in 5 h. The effect of background species presented in greywater was, to some extent, comparable to that of the OH-radical scavenger. These results prove that photocatalysis using anatase TiO_2 is a feasible additional treatment for greywater recycling.

Keywords: greywater; photocatalysis; micropollutants; metformin; naproxen; sulfamethoxazole

1. Introduction

In recent years, the rapidly increased water scarcity index has emphasized the importance of recycling and reusing greywater, which can mitigate limited freshwater resources on activities where non-potable water could be utilized. Currently, the biggest obstacle in recycling greywater for irrigation, for instance, is the presence of micropollutants, specifically pharmaceuticals and personal care products [1–3]. Pharmaceuticals are bioactive, persistent and prone to bioaccumulation, posing a high risk when they enter the environment. Diclofenac, for instance, an analgesic, has been linked to the declined population of vultures in India, and carbamazepine, an antiepileptic, has detrimental effects on fish organs [4,5]. Unfortunately, most conventional treatment technologies used for greywater recycling have little or no capacity to remove pharmaceuticals due to various inhibitory factors.

Membrane bioreactors, which make use of the merging of biological degradation and filtration to recycle greywater, produce high-quality effluents. In most cases, regular pollutants (bacteria, inorganic and organic molecules) are entirely removed. However, such as other conventional technologies, this treatment method does not eliminate pharmaceuticals, such as Naproxen (NPX), Metformin (MTF) and Sulfamethoxazole (SMX), effectively. These pharmaceuticals, representing the different groups of Active Pharmaceutical Ingredients (APIs) with different structures and functional groups in play, have been found in concerning concentrations in raw and treated wastewater and even in greywater with low pollution indices. The reasons for this inadequate elimination are:

1. Their resistance to biodegradation;
2. Due to their small molecular sizes, the filtration modules are unable to create a suitable barrier to hold them back.

These sets of pharmaceuticals show various degrees of hydrophilicity with adsorption and biodegradation, a pathway for the elimination of hydrophobic and hydrophilic pharmaceuticals, respectively [6]. The fact is, when considering the effects of greywater reuse, the high removal efficiency of some of these pharmaceuticals is irrelevant due to their bioactive and bioaccumulatory abilities. Any seepage, even at low concentrations, is highly detrimental for reuse [7,8].

To mitigate this situation, various tertiary treatment processes, i.e., additional steps after conventional wastewater treatment, have been used to significantly improve pharmaceuticals' removal. If applied successfully, they can expand the usability of greywater to more than mere flushing. Well-treated effluents can be used for unrestricted irrigation, car washing, fire dousing, laundry, etc. [9,10]. Of these treatment methods, Advanced Oxidation Processes (AOPs), which rely on the in situ production of reactive radical species able to degrade pharmaceuticals in wastewater [11], seem to be the most effective. Among these processes, heterogeneous photocatalysis can be used as a tool to purify pharmaceutically polluted wastewater [12–14].

TiO_2 is the photocatalyst of choice, providing thermodynamically stable, non-toxic material with high photocatalytic activity. In combination with UV light, pharmaceuticals and their transformation products are degraded through a series of reactions [15]. Photocatalysis has proven to be a reliable course of treatment in removing pollutants from wastewater, leading to total degradation [16,17]. This performance is due to the formation of holes (h^+) and reactive oxygen species, such as OH, O_2^-, or HO_2, on the photocatalyst surface [18,19].

In recent times, pharmaceuticals, such as NPX, MTF and SMX, have been successfully degraded using TiO_2 (often commercial Evonik P25 and, in rare cases, anatase TiO_2) as a photocatalyst either in demineralized water solutions or in synthetic and real wastewaters [20–23]. NPX photocatalytic degradation was carried on different scales using its solutions in demineralized water, in river water and artificial wastewater [24–26]. However, the application of the TiO_2 photocatalytic process in degrading NPX in real wastewater to study the effect of scavengers and other factors is very limited. The same holds for other two pharmaceuticals addressed in the present study, i.e., for MTF [27,28] and SMX [29,30].

The main research gap in this subject area is the absence of research extended to real greywater. What is especially missing is the understanding of the effects of different factors on the degradation of these three important pharmaceuticals. Generally, either anatase or anatase/rutile-mixed photocatalysts have been used. We also decided to add rutile, which, owing to different mechanistic pathways in comparison with anatase, may exhibit interesting performance features in the presence of potentially disturbing components in the water to be treated. The pharmaceuticals selected are among the most important present in real and MBR-treated greywater and photocatalysis was applied at environmentally relevant concentrations. Besides the photocatalysis itself, we also focus on other accompanying processes, namely, photolysis, adsorption on the photocatalyst surface and the natural scavenging species. In the long run, this enhanced treatment will help in expanding the usage of greywater and, in doing so, reduce the water scarcity index.

2. Results

2.1. Photocatalysts Used and Their Characterization

Table 1 summarizes the structure and texture characteristics of both TiO_2 powders used. The phase content and the size of coherent regions were determined by the analysis of X-ray diffractograms, the broadening of the diffraction lines being interpreted only with regard to the crystal size and the microscopic deformations being neglected. The data showed that the anatase and rutile samples exclusively contained the relevant phase.

Table 1. Structure and texture characterization of photocatalysts used for the photocatalytic pharmaceuticals degradation.

Photocatalyst	Phase Content [a]/% Anatase	Rutile	Crystallite Size [a]/nm Anatase	Rutile	Particle Size [b]/nm	Particle Size [c]/nm
A30	100	0	16	-	31	20–40
R30	0	100	-	29	100	30–100

[a] from X-ray diffractograms; [b] from N_2 sorption data; [c] from HR-TEM images.

Concerning the particle size, a comparative study using three physically different experimental techniques was carried out, including the X-ray diffraction, nitrogen sorption a HR-TEM microscopy. Surface areas calculated from sorption data using the BET method were converted to mean particle sizes using the spherical model approximation. A comparison of the particle size determined by the techniques mentioned enabled to draw some conclusions concerning the morphology properties of the powders.

Concerning the anatase sample A30, the size of the coherent region determined by X-ray diffraction was approximately half in comparison with the particle size determined by sorption and HR-TEM. Consequently, the extent of agglomeration of this material was limited to the formation of small clumps. Concerning the rutile sample R30, it consisted of crystals ca 30 nm in size which were aggregated to larger particles about three times greater as follows from the comparison of the XRD and sorption data.

2.2. Photocatalytic Removal of Pharmaceuticals in Ultrapure Water and in the Presence of a Radical Scavenger

2.2.1. Naproxen

The photocatalytic elimination of naproxen at an initial concentration of 2 mg L^{-1} was very effective (Figure 1). A low concentration of 400 mg L^{-1} of both allotropes was sufficient to achieve a very high degree of degradation, the complete removal of naproxen being attained within 40 and 180 min for A30 and R30, respectively.

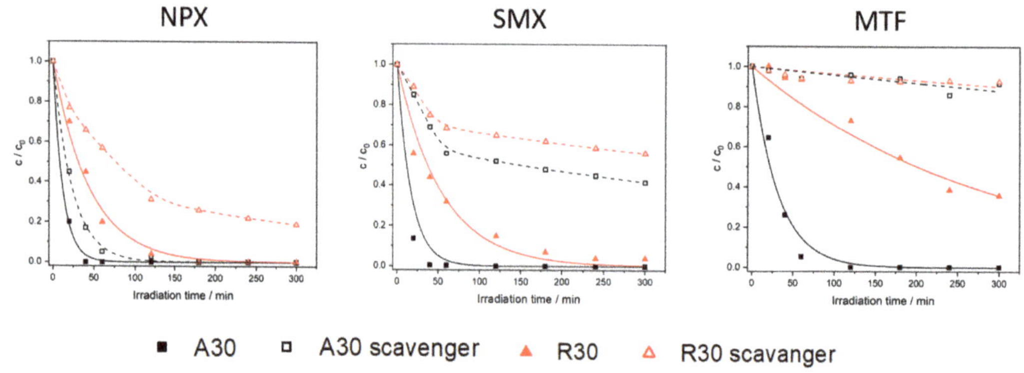

Figure 1. The effect of radical scavenger added on the kinetics of the photocatalytic degradation of naproxen (NPX), sulfamethoxazole (SMX) and metformin (MTF) in the presence of anatase (A30) and rutile (R30).

The two accompanying processes, namely, photolysis and adsorption on the photocatalyst surface, were shown to play some role (Figure 2). Concerning photolysis, naproxen was partially degraded by UV-A irradiation (20%) in the absence of the photocatalyst. This can be connected with the presence of the chromophores. Consequently, the degradation could only occur if the light fell within the absorption spectrum of naproxen.

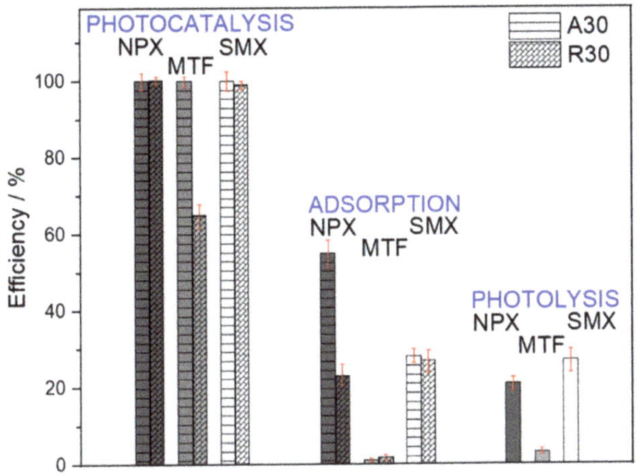

Figure 2. Comparison of the removal efficiencies of photocatalysis, adsorption and photolysis (all after 5 h) for naproxen (NPX), sulfamethoxazole (SMX) and metformin (MTF).

The extent of adsorption on the surface of anatase A30 was significant. After 5 h, when the equilibrium was achieved, the concentration of dissolved naproxen was halved. On rutile this effect was less pronounced, where the naproxen concentration decreased by about 20%. The high sorption on anatase is important from the mechanistic point of view as photocatalytic mechanisms, including adsorbed species, can be expected to play a dominant role.

This conclusion was supported by the outcomes of the scavenger tests (Figure 3). The effect of the radical scavenger was relatively limited, the performance being to a large extent retained. This conclusion was valid especially for anatase A30, whose performance was only slightly affected by the presence of scavengers, complete degradation being achieved within 120 min. For rutile, the performance was slightly worse; however, even here, a high degree of naproxen removal was attained.

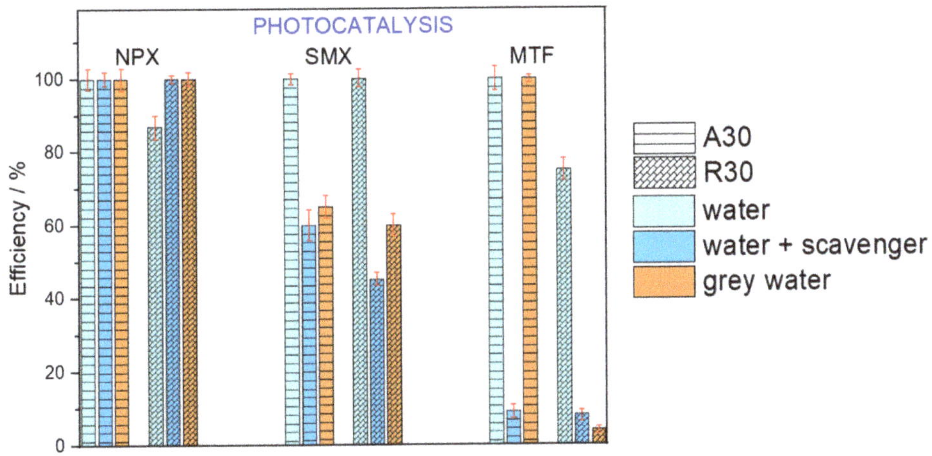

Figure 3. The comparison of the removal efficiencies of photocatalysis for naproxen (NPX), sulfamethoxazole (SMX) and metformin (MET) dissolved in ultrapure water, in ultrapure water with added scavenger and in grey water. The irradiation time equalled 5 h.

2.2.2. Sulfamethoxazole

Interesting was the comparison of data obtained for sulfamethoxazole with those for naproxen (Figure 1). The effect of photocatalysis in ultrapure water was comparable both on anatase and rutile. For anatase, the complete removal was achieved within 60 min for both pollutants.

Concerning the effect of photolysis and adsorption, both these processes play an important role (Figure 2). Adsorption on anatase was, approximately, in comparison with naproxen, while for rutile it was even higher. Additionally, the photolysis of sulfamethoxazole was relatively high, achieving about 25% after 5 h, which was even more than that of naproxen.

The scavenger tests showed a substantial worsening of the performance, both for anatase and rutile (Figure 3). After 5 h, the concentration of sulfamethoxazole for anatase and rutile decreased by only 50 and 40%, respectively. As the effect of radical scavenger was much more pronounced in comparison with naproxen, a different degradation mechanism was to be expected. The role of radicals freed from the photocatalyst surface is clearly more important, which at least for anatase was in agreement with a lower adsorption in comparison with naproxen.

2.2.3. Metformin

The degradation of metformin on anatase in ultrapure water was only slightly slower than that for naproxen and sulfamethoxazole (Figure 1). The performance of rutile was substantially weaker, the decrease in the metformin concentration after 5 h achieving only about 65%. What was, however, especially marked, was the devastating effect of the radical scavenger on the performance of both anatase and rutile. Therefore, a radical mechanism was to be expected for both photocatalysts [22]. This conclusion was in agreement with the adsorption experiments, which showed a practically negligible amount adsorbed. Additionally, the degree of photolysis was very small, which further supports the very low degradation efficiency in the presence of a radical scavenger. Additional kinetic curves are presented in the Supplementary Information (Figures S1 and S2).

2.3. *Photocatalytic Removal of Pharmaceuticals in MBR-Treated Greywater*

The MBR-treated greywater had a low pollutant load, and, hence, was of a high quality (Table 2). The degradation of naproxen and sulfamethoxazole in greywater followed a similar trend as their degradation in ultrapure water with the scavenger (Figure 3). For both water with the scavenger and greywater, complete removal was observed after 5 h of irradiation. Apparently, the significant adsorption of naproxen on the photocatalyst surface enables to preserve high performance, even in the presence of radical scavengers.

Table 2. Characteristics of MBR-treated greywater.

Parameter Range		
Dissolved oxygen/mg L^{-1}		6.7–8.5
Temperature/°C		22.6–26.0
Total nitrogen/mg L^{-1}		0.2–11.7
Total phosphorus/mg L^{-1}		0.1–0.4
Electric conductivity/µS cm^{-1}		412–500
pH		6.5–7.6
Dissolved organic carbon/mg L^{-1}		3.4–5.4
Solids/mg L^{-1}	Total suspended solids	Not detected
	Total dissolved solids	310–358

As for sulfamethoxazole, a substantial decrease in the performance in comparison with ultrapure water was observed. The achieved efficiency, however, was slightly higher in comparison with the experiments with the ultrapure water containing scavenger. Therefore, it can be concluded that also for this pharmaceutical, the effect of background species contained in greywater on its photocatalytic degradation was similar to the role of the radical scavenger.

In the case of metformin, the results were contradictory. For rutile, the removal efficiency in greywater was very low, which corresponded to the effect of the radical scavenger (Figure 3). Surprisingly, in the case of anatase, high performance was observed with approx.100% removal achieved in 5 h, a similarity shared with naproxen.

3. Discussion

This paper showed the potential of TiO_2 photocatalysis for the removal of pharmaceuticals with different chemical structures at practically relevant concentrations found in greywater influent.

Degradation of Pharmaceuticals Using TiO_2

Slow, and for metformin practically negligible photolysis (20%, 0% and 28% for naproxen, metformin and sulfamethoxazole, respectively, after 5 h) clearly showed the degradation was mainly due to the photocatalytic process. The somehow higher degradation rate via photolysis of naproxen and sulfamethoxazole was believed to be due to photooxidation as oxygen was being mixed into the system by intensive stirring, in our case 400 rpm. Alternatively, the photolysis was carried out in the presence of inert SiO_2 particles in suspension, whose particle size similar to those of anatase and rutile allotropes in order to exactly reproduce the light-scattering conditions. However, no difference in the comparison with experiments without particles was observed.

Further, the products of the photolytic and photocatalytic degradation may differ. Published studies show that for, e.g., naproxen, the degree of the degradation due to photocatalysis is much higher than due to photolysis (e.g., the application of only photolysis often leads to the formation of products which are of comparable or even higher toxicity than the untreated water). Therefore, the photocatalysis is a more suitable technology for the purification of water.

In most cases, the adsorption of pharmaceuticals on TiO_2 is viewed as mostly negligible. However, our results showed that it can play an important role in the process of photocatalytic degradation. Especially for naproxen and sulfamethoxazole on the anatase surface, our adsorption experiments conducted separately in the dark demonstrated that adsorption can effectively remove, from the solution, the pharmaceuticals used in this study.

Naproxen was shown to be easily degraded using anatase. As the point of zero charge of TiO_2 was around pH 5–6, its surface was mostly not charged at neutral pH conditions used in our experiments and, also, valid for real greywater. The published studies suggest the carboxylic functional group of naproxen could be responsible for the efficient adsorption by strongly interacting with superficial OH groups leading to complexes. Owing to its large adsorption on the photocatalyst surface, the naproxen molecules can be easily degraded by the direct attack of holes.

This conclusion was supported by the scavenging tests, in which little effect on the degradation of naproxen was observed. In these tests, methanol basically functioned as a radical scavenger, even if some proportion of CH_2OH radicals was formed before the formation of formaldehyde as a final product. Moreover, the adsorption of drugs on the surface could be different in the presence of methanol. Therefore, the OH radicals seemed to not play an important role. Finally, the preliminary TOC measurements showed that naproxen could be effectively mineralized in the range of 60–70%.

The lower adsorption of sulphamethoxazole and of metformin could be due to the very different nature of the molecules with respect to the surface: highly hydrophobic to the drugs, hydrophilic to the surface. Similar to naproxen, sulfamethoxazole was also degraded at least to some extent in the adsorbed state.

Metformin was removed efficiently using anatase, although at a slower reaction rate in comparison with naproxen and sulfamethoxazole. Metformin can gain or lose a proton easily. As it has two distant pK_a values equalling about 3 and about 13 (discussed in [28]), it can form a biprotonated form at a low pH, the monoprotonated form at a medium pH between ca 3 and 13 and, finally, neutral species at the highest pH. At the range of a pH typical for greywater, i.e., 6.5–7.6 (Table 2), metformin was mainly monoprotonated. Its high solubility in water (1000 g/L) and its low K_{ow} led to a practically negligible adsorption on the photocatalyst surface. Due to these features, the degradation mechanism of metformin by a direct attack of holes was less probable. Another issue was the formation of an intermediate product guanylurea, which is also recalcitrant. This leads to active competition for degradation between both species. Consequently, achieving a high degree of metformin degradation is much more difficult in comparison with the other two pharmaceuticals dealt with in this study.

In greywater, both allotropes of TiO_2 were suitable to remove naproxen (20 $\mu g\ L^{-1}$), reaching with anatase a 100% removal efficiency within less than 60 min. For sulfamethoxazole, a similarity was found with the scavenger tests, showing that scavengers in the system slowed down degradation rates, but not enough to stop the reaction. If the experiment was left to run longer (10 h), the complete degradation of sulfamethoxazole was observed.

With metformin (300 $\mu g\ L^{-1}$) in greywater, a surprisingly high efficiency was observed using anatase as the photocatalyst, which was very beneficial from the point of view of the application. This phenomenon could be explained as follows: Dissolved organic matter present in greywater (Table 2) is known to have a higher adsorption affinity onto the TiO_2 surface in comparison with alcohols used as radical scavenges [29]. However, as the adsorption of metformin on the TiO_2 surface is very low, the effect of the preferential adsorption of dissolved organic matter should be negligible, as there is no competition between them. For methanol used as a radical scavenger, a dramatic effect was observed as it efficiently destroyed the OH radicals, which were decisive in the degradation of metformin. Consequently, the effect of these two components, i.e., methanol and dissolved organic matter, was not comparable, which differentiated this pharmaceutical from the other two included in this study.

The very low efficiency of rutile in greywater was in an agreement with the data obtained using a radical scavenger. The practical inability of rutile to degrade metformin in greywater could be due to a number of reasons. The increased ionic strengths expected due to a slightly higher dissolved salt content compared to drinking water can be such a reason why metformin molecules have limited migration power to the rutile surface (this is the primary means of degradation using rutile TiO_2) [30], or there was also a competition between scavengers (methanol or dissolved organic matter) and metformin with scavengers being more mobile, which would be degraded first. Finally, the UV light-induced aggregation of TiO_2 particles, which was much stronger for rutile, might have been the reason of the worse performance of rutile than anatase [31].

4. Materials and Methods

4.1. *Pharmaceuticals*

Metformin hydrochloride, naproxen and sulfamethoxazole (Figure 4) were purchased from Sigma-Aldrich, Taufkirchen, Germany, titanium dioxide nanopowders from U.S. Research Nanomaterials, Houston, TX, USA, methanol and sodium hydroxide from Penta, Czech Republic. Ultrapure water was used for all experiments as a matrix, its purity being tested before each experiment. All apparatuses were washed with distilled water and methanol according to a standard procedure to remove impurities that could contaminate the experiments.

For these experiments, 2 mg L^{-1} of each pharmaceutical was dissolved in water. While metformin hydrochloride and naproxen could be dissolved directly, sulfamethoxazole, which is practically insoluble in water, was alkalified using 0.1 M sodium hydroxide to increase the solution's pH; thus, increasing solubility.

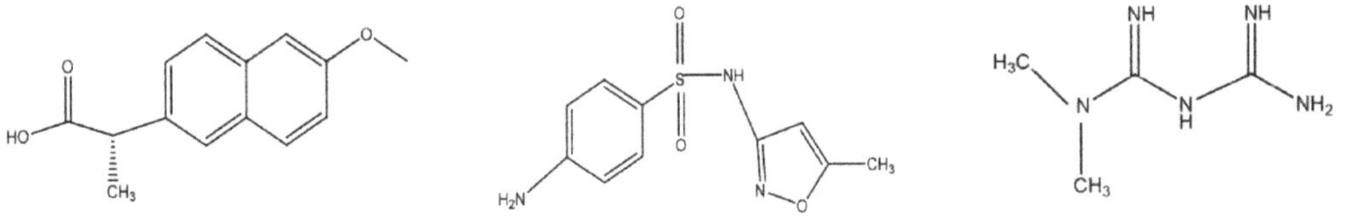

Figure 4. Structural structure of Naproxen (**a**), sulfamethoxazole (**b**) and metformin (**c**).

4.2. TiO_2 Photocatalysts

TiO_2 nanopowders (≥99.5% purity) included two pristine allotropic forms of TiO_2, namely, rutile and anatase, their nominal particle size according to the producer being 30 nm. Structural and textural properties of the nanoparticles were determined by combining several techniques. Surface morphology was followed with scanning electron microscopy (SEM, JEOL JSM 5500LV) combined with high-resolution transmission electron microscopy (HR-TEM, JEOL JEM-2100Plus). Crystallinity (structure and phase composition) of the titania dioxide samples was evaluated by powder X-ray diffraction (pXRD) using PANalytical X'Pert PRO diffractometer equipped with Co anode in the Bragg–Brentano geometry and multichannel X'Celerator detector. The surface area of powders was determined from nitrogen sorption isotherms at ca 77 K (boiling point of liquid nitrogen) using a Micromeritics 3Flex apparatus. Prior to the measurements, the samples were degassed at 250 °C overnight.

4.3. Photocatalytic Experiments

Before the photocatalytic experiments, the photocatalyst samples were cleaned overnight by UV light of a dominant wavelength of 365 nm with an irradiation intensity of 2.0 mW cm^{-2} to decompose any residual organic matter on them. The pharmaceuticals in aqueous solution (2 mg L^{-1}) were photocatalytically degraded at 25 °C using photocatalyst dispersion (400 mg L^{-1}) in a 25 mL quartz cell (Figure 5). As the top of the liquid in the cell was open to air and the solution was intensively stirred (400 rpm), the concentration of dissolved oxygen was constant during the experiment. A Sylvania Lynx-S 11 W BLB lamp irradiated the aqueous dispersion with UV light (365 nm) at a low-power density of 1.0 mW cm^{-2}. Prior to the photocatalytic experiments, the dissolved pharmaceuticals were equilibrated with the photocatalyst surface for 3 h. For each experiment, aliquots each of 100 μL were collected from the solution in the reaction cell every 20 min for the first one hour and once every hour for the next 4 h. The samples were filtered immediately after collection to stop the reaction process from continuing, stored in amber vials to limit light interaction with the analytes, and frozen at −20 °C, to keep the samples from degrading. The samples were defrosted just before they were subjected to HPLC-MS analysis. The pseudo first-order reaction rate constants for the degradation of pharmaceuticals were calculated using nonlinear regression fitting of the kinetic curves.

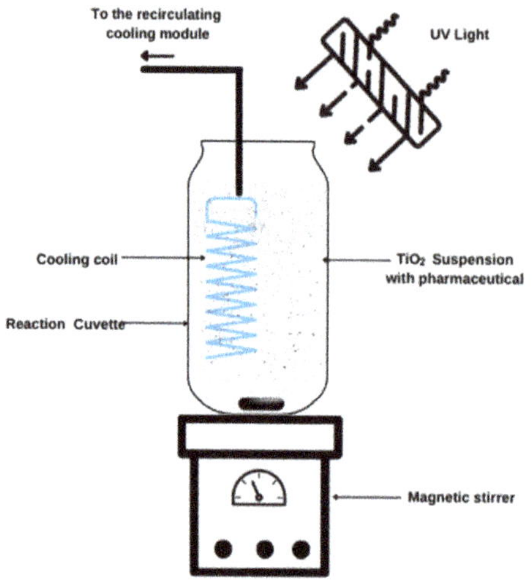

Figure 5. Photocatalytic module for the degradation of pharmaceutical dissolved in greywater.

4.4. *Adsorption, Photolysis and Scavenger Experiments*

To determine the effect of adsorption on the surface of TiO_2 nanoparticles, experiments analogous to the photocatalytic ones were conducted in the dark with the same stirring speed and sampling performed at 0, 3 and 5 h.

The photostability of the pharmaceuticals was tested by subjecting their solutions to UV light without the photocatalysts for 5 h, sampling being conducted at 0, 3 and 5 h.

A solution of 0.1 M methanol was used as a OH radical scavenger, the sampling during the photocatalytic experiments being carried out once every hour as slow degradation rates were expected. The samples obtained were treated analogously to those collected during the photocatalytic degradation experiments.

4.5. *Lab-Scale MBR Treatment of Greywater*

Greywater from an apartment building with 99 flats and approx. 300 inhabitants was collected periodically and treated in a 3 L MBR reactor fitted with two ultrafiltration hollow tube membranes running in series. This lab-sized reactor was run for a year and high-quality effluent was produced throughout the operation of the reactor (Table 2). The hydraulic retention time of the reactor was approx. 24 h at the time of sample collection. HPLC analyses were carried out on raw greywater samples collected and the highest concentrations analysed for naproxen, metformin and sulfamethoxazole were 20, 300 and 50 $\mu g\ L^{-1}$, respectively.

4.6. *Analytical Methods*

The separation of analytes was carried out using HPLC (Agilent Infinity 1290 HPLC, Agilent Technologies, Santa Clara, CA, USA) equipped with a Poroshell 120 EC-C18 (100 mm × 2.1 mm, 2.7 µm particle size) column with a guard column (both by Agilent Technologies, Santa Clara, CA, USA). The column temperature was set at 40 °C, and the injection volume was 1 µL. The mobile phases consisted of 0.5 mM ammonium fluoride in water (solvent A) and methanol (solvent B). The flow rate was 0.4 mL/min. The mobile phase composition started at 20% B and was linearly increased to 100% B in 5 min. After each run, the column was re-equilibrated for 1 min. The 6460 QqQ Triple Quad mass spectrometer (Agilent Technologies, Santa Clara, CA, USA) equipped with Agilent Jet Stream technology in positive electrospray (ESI+) ionization mode was used for the analyte detection. The quantitative analysis was performed using dynamic Multiple Reaction Monitoring (dMRM). The following MRM transitions were monitored:

metformin—130→85, 71; naproxen—231→170, 153, 141; sulphamethoxazole—254→155.8, 107.9, 92. The capillary voltage was set at 3500 V. The sheath gas flow and temperature were set at 11.0 L min^{-1} and 250 °C, respectively. The drying gas flow and temperature were set at 5 L min^{-1} and 300 °C, respectively. The nebulizer pressure was 45 psi. The quantification was performed using external calibration. Agilent Mass Hunter software was used for data collection and processing.

5. Conclusions

We demonstrated the high efficiency of anatase TiO_2 in degrading photocatalytically naproxen, metformin and sulfamethoxazole in greywater. Even if their chemical compositions and physico-chemical properties substantially differ, a high degree of their removal by photocatalysis in greywater was achieved. Despite the vastly differing sorption properties of individual pharmaceuticals on the photocatalyst surface, this property did not constitute a critical parameter in the degradation process. This observation has a major importance in a real application of the photocatalytic technology in greywater treatment, as the content of the background components may be vastly varied.

As our results showed a high potential of the photocatalysis application, further work will be focused on more recalcitrant pharmaceuticals, which may be contained in other real greywater samples, and on immobilised forms of TiO_2. Immobilized TiO_2 opens a new way for the implementation of photocatalytic technology in continuous water treatment set-ups.

Supplementary Materials: The following are available online at https://www.mdpi.com/2073-4344/11/9/1125/s1, Figure S1: Kinetics of photocatalytic degradation of metformin (MTF), naproxene (NPX) and sulfomethaxazole (SMX) carried out in the suspension of anatase and rutile 30 nm nanoparticles; Figure S2: UV-A photolysis of metformin (MTF), naproxene (NPX) and sulfomethaxazole (SMX).

Author Contributions: Conceptualization, J.R. and J.B.; methodology, J.R. and R.Z.; validation, B.O., M.V. and A.N.; investigation, B.O., B.S., M.V., A.N. and M.K.; data curation, B.O., J.R., M.V., A.N. and R.Z.; writing—original draft preparation, B.O.; writing—review and editing, J.R., R.Z. and J.B.; supervision, J.R., M.K. and J.B.; project administration, J.R. and J.B.; funding acquisition, J.R. and J.B. All authors have read and agreed to the published version of the manuscript.

Funding: This research was funded by the Czech Science Foundation (GACR, grant no. 19-21801S) and the Technology Agency of the Czech Republic (TACR, grant no. TH03030408).

Institutional Review Board Statement: Not applicable.

Informed Consent Statement: Not applicable.

Data Availability Statement: Not applicable.

Acknowledgments: This work was also supported by the Ministry of Education, Youth and Sports of the Czech Republic and The European Union—European Structural and Investments Funds in the frame of Operational Programme Research, Development and Education—project Pro-NanoEnviCz (project no. CZ.02.1.01/0.0/0.0/16_013/0001821), which provided access to a Micromeritics 3Flex apparatus, Nicolet FTIR 6700 spectrophotometer and Beckman Coulter Optima XPN-100 ultracentrifuge. Further, we thank the University of Chemistry and Technology, Prague for the financial support from specific university research (MSMT no 21-SVV/2019), the Academy of Sciences of the Czech Republic for the financial support (projects no. MSM200402101 and Strategy AV21, 16/2021) and Ministry of Education, Youth and Sports of the Czech Republic (Barrande programme no. 8J20FR012).

Conflicts of Interest: The authors declare no conflict of interest.

References

1. Zhang, T.; Li, B. Occurrence, transformation, and fate of antibiotics in municipal wastewater treatment plants. *Crit. Rev. Environ. Sci. Technol.* **2011**, *41*, 951–998. [CrossRef]

2. Benotti, M.J.; Trenholm, R.A.; Vanderford, B.J.; Holady, J.C.; Stanford, B.D.; Snyder, S. Pharmaceuticals and endocrine disrupting compounds in U.S. drinking water. *Environ. Sci. Technol.* **2008**, *43*, 597–603. [CrossRef]
3. Luo, Y.; Guo, W.; Ngo, H.H.; Nghiem, L.D.; Hai, F.; Zhang, J.; Liang, S.; Wang, X. A review on the occurrence of micropollutants in the aquatic environment and their fate and removal during wastewater treatment. *Sci. Total Environ.* **2014**, *473–474*, 619–641. [CrossRef]
4. Green, R.E.; Newton, I.; Shultz, S.; Cunningham, A.A.; Gilbert, M.; Pain, D.J.; Prakash, V. Diclofenac poisoning as a cause of vulture population declines across the Indian subcontinent. *J. Appl. Ecol.* **2004**, *41*, 793–800. [CrossRef]
5. Weber, F.-A.; der Beek, T.A.; Carius, A.; Grüttner, G.; Hickmann, S.; Ebert, I.; Hein, A.; Küster, A.; Rose, J.; Koch-Jugl, J. *Pharmaceuticals in the Environment—The Global Perspective Occurrence, Effects, and Potential Cooperative Action under SAICM*; German Environment Agency: Dessau-Roßlau, Germany, 2014.
6. Bolong, N.; Ismail, A.; Salim, M.R.; Matsuura, T. A review of the effects of emerging contaminants in wastewater and options for their removal. *Desalination* **2009**, *239*, 229–246. [CrossRef]
7. Boyjoo, Y.; Pareek, V.K.; Ang, M. A review of greywater characteristics and treatment processes. *Water Sci. Technol.* **2013**, *67*, 1403–1424. [CrossRef] [PubMed]
8. Liu, S.; Butler, D.; Memon, F.; Makropoulos, C.; Avery, L.; Jefferson, B. Impacts of residence time during storage on potential of water saving for grey water recycling system. *Water Res.* **2010**, *44*, 267–277. [CrossRef] [PubMed]
9. Etchepare, R.; van der Hoek, J.P. Health risk assessment of organic micropollutants in greywater for potable reuse. *Water Res.* **2015**, *72*, 186–198. [CrossRef]
10. Van Der Hoek, J.P.; Tenorio, J.L.I.; Hellinga, C.; Van Lier, J.B.; Van Wijk, A.J.M. Green village delft—Integration of an autarkic water supply in a local sustainable energy system. *J. Water Reuse Desalination* **2014**, *4*, 154–163. [CrossRef]
11. Butkovskyi, A.; Sevenou, L.; Meulepas, R.J.W.; Leal, L.H.; Zeeman, G.; Rijnaarts, H.H.M. Micropollutant removal from black water and grey water sludge in a UASB-GAC reactor. *Water Sci. Technol.* **2017**, *77*, 1137–1148. [CrossRef]
12. Govindan, K.; Angelin, A.; Kalpana, M.; Rangarajan, M.; Shankar, P.; Jang, A. Electrocoagulants characteristics and application of electrocoagulation for micropollutant removal and transformation mechanism. *ACS Appl. Mater. Interfaces* **2019**, *12*, 1775–1788. [CrossRef] [PubMed]
13. Maher, E.K.; O'Malley, K.N.; Dollhopf, M.E.; Mayer, B.K.; McNamara, P.J. Removal of estrogenic compounds from water via energy efficient sequential electrocoagulation-electrooxidation. *Environ. Eng. Sci.* **2020**, *37*, 99–108. [CrossRef]
14. Lee, C.M.; Palaniandy, P.; Dahlan, I. Pharmaceutical residues in aquatic environment and water remediation by TiO_2 heterogeneous photocatalysis: A review. *Environ. Earth Sci.* **2017**, *76*, 611. [CrossRef]
15. Foteinis, S.; Borthwick, A.G.; Frontistis, Z.; Mantzavinos, D.; Chatzisymeon, E. Environmental sustainability of light-driven processes for wastewater treatment applications. *J. Clean. Prod.* **2018**, *182*, 8–15. [CrossRef]
16. Gimeno, O.; García-Araya, J.; Beltrán, F.; Rivas, F.; Espejo, A. Removal of emerging contaminants from a primary effluent of municipal wastewater by means of sequential biological degradation-solar photocatalytic oxidation processes. *Chem. Eng. J.* **2016**, *290*, 12–20. [CrossRef]
17. Gassie, L.W.; Englehardt, J.D. Advanced oxidation and disinfection processes for onsite net-zero greywater reuse: A review. *Water Res.* **2017**, *125*, 384–399. [CrossRef]
18. Miranda-García, N.; Suárez, S.; Sanchez, B.; Coronado, J.; Malato, S.; Maldonado, M.I. Photocatalytic degradation of emerging contaminants in municipal wastewater treatment plant effluents using immobilized TiO_2 in a solar pilot plant. *Appl. Catal. B Environ.* **2011**, *103*, 294–301. [CrossRef]
19. Pirilä, M.; Saouabe, M.; Ojala, S.; Rathnayake, B.; Drault, F.; Valtanen, A.; Huuhtanen, M.; Brahmi, R.; Keiski, R.L. Photocatalytic degradation of organic pollutants in wastewater. *Top. Catal.* **2015**, *58*, 1085–1099. [CrossRef]
20. Du Pisani, P.L. Direct reclamation of potable water at Windhoek's Goreangab reclamation plant. *Desalination* **2006**, *188*, 79–88. [CrossRef]
21. Chen, X.; Richard, J.; Liu, Y.; Dopp, E.; Tuerk, J.; Bester, K. Ozonation products of triclosan in advanced wastewater treatment. *Water Res.* **2012**, *46*, 2247–2256. [CrossRef]
22. Prashanth, V.; Priyanka, K.; Remya, N. Solar photocatalytic degradation of metformin by TiO_2 synthesized using Calotropis gigantea leaf extract. *Water Sci. Technol.* **2021**, *83*, 1072–1084. [CrossRef]
23. Jallouli, N.; Elghniji, K.; Hentati, O.; Ribeiro, A.R.; Silva, A.; Ksibi, M. UV and solar photo-degradation of naproxen: TiO_2 catalyst effect, reaction kinetics, products identification and toxicity assessment. *J. Hazard. Mater.* **2016**, *304*, 329–336. [CrossRef] [PubMed]
24. Kanakaraju, D.; Motti, C.; Glass, B.D.; Oelgemöller, M. Solar photolysis versus TiO_2-mediated solar photocatalysis: A kinetic study of the degradation of naproxen and diclofenac in various water matrices. *Environ. Sci. Pollut. Res.* **2016**, *23*, 17437–17448. [CrossRef] [PubMed]
25. Roncaroli, F.; Blesa, M.A. Kinetics of adsorption of carboxylic acids onto titanium dioxide. *Phys. Chem. Chem. Phys.* **2010**, *12*, 9938–9944. [CrossRef] [PubMed]
26. Dobson, K.D.; McQuillan, A. In situ infrared spectroscopic analysis of the adsorption of aliphatic carboxylic acids to TiO_2, ZrO_2, Al_2O_3, and Ta_2O_5 from aqueous solutions. *Spectrochim. Acta Part A Mol. Biomol. Spectrosc.* **1999**, *55*, 1395–1405. [CrossRef]
27. Molinari, A.; Bruni, M.; Maldotti, A. Chemo- and Regio-Selectivity in the TiO_2-Mediated Photooxidation of 1,n-diols. *J. Adv. Oxid. Technol.* **2008**, *11*, 143–148. [CrossRef]

28. Hernández, B.; Pflüger, F.; Kruglik, S.G.; Cohen, R.; Ghomi, M. Protonation–deprotonation and structural dynamics of antidiabetic drug metformin. *J. Pharm. Biomed. Anal.* **2015**, *114*, 42–48. [CrossRef]
29. Brame, J.; Long, M.; Li, Q.; Alvarez, P. Inhibitory effect of natural organic matter or other background constituents on photocatalytic advanced oxidation processes: Mechanistic model development and validation. *Water Res.* **2015**, *84*, 362–371. [CrossRef]
30. Samayamanthula, D.R.; Sabarathinam, C.; Bhandary, H. Treatment and effective utilization of greywater. *Appl. Water Sci.* **2019**, *9*, 90. [CrossRef]
31. Zhou, C.; Bashirzadeh, Y.; Bernadowski, T.A.; Zhang, X. UV Light-Induced Aggregation of Titania Submicron Particles. *Micromachines* **2016**, *7*, 203. [CrossRef]

catalysts

MDPI

Article

A Facile Synthesis of $Bi_2O_3/CoFe_2O_4$ Nanocomposite with Improved Synergistic Photocatalytic Potential for Dye Degradation

Abdul Basit Naveed, Fakhira Riaz, Azhar Mahmood *, Ammara Shahid and Saman Aqeel

School of Natural Sciences, National University of Sciences and Technology, H-12, Islamabad 44000, Pakistan; baasitlali@gmail.com (A.B.N.); fakhirariaz22@gmail.com (F.R.); ammarashahid04@gmail.com (A.S.); samanaqeel8910@gmail.com (S.A.)

* Correspondence: dr.azhar@sns.nust.edu.pk; Tel.: +92-51-9085-5574

Abstract: Semiconductor-based photocatalysis is a probable approach to overcoming many pollution problems and eradicating toxic organic materials from wastewater. This research endeavor aimed to explore the synergistic potential of different semiconductor nanocomposites for photocatalytic degradation of organic pollutants in contaminated water. A facile hydrothermal approach was employed to synthesize bismuth oxide and cobalt ferrite nanoparticles from their precursors—bismuth nitrate pentahydrate, ferric chloride hexahydrate and cobalt chloride hexahydrate—with various concentrations and conditions to optimize the product. Subsequently, nanocomposites of bismuth oxide and cobalt ferrite were prepared by solid-state mixing in varying concentrations followed by calcination. UV/visible diffuse reflectance spectroscopy, X-ray diffraction, scanning electron microscopy and elemental dispersive X-ray spectroscopic techniques have corroborated the successful synthesis of nanocomposites. The energy gaps of bismuth oxide and cobalt ferrite nanocomposites were computed in the range of 1.58–1.62 eV by Tauc plots. These nanocomposite materials were ascertained for photocatalytic potential to degrade methyl orange organic dye in water. A nanocomposite with equiquantic proportions has shown the best photocatalytic degradation activity, which may be attributed to the type-II band configuration and a synergistic effect, because Bi_2O_3 acts as an electron sink. This synergism has reduced the cogent band gap, hindered electron hole recombination and increased electron hole availabilities for photodegradation reactions, thus ensuing an efficient photodegradation co-work of $Bi_2O_3/CoFe_2O_4$ nanocomposites.

Keywords: photocatalyst; nanocomposite; energy gap; dye degradation; hydrothermal synthesis

1. Introduction

Water pollution is a major environmental problem worldwide. Wastewater contains numerous hazardous and toxic substances such as dyes, resins, heavy metals, phenolic compounds, pesticides and herbicides etc. [1–3]. Dye pollutants are carcinogenic and mutagenic; they can lead to central nervous system dysfunction and may cause morbidity in human as well as in aquatic life [4,5]. Conventional wastewater treatment pathways are not effective because of the contumacious nature of synthetic dyes and the high salinity of wastewater [6,7]. Ozonation and chlorination are also quite incapable because of their high operating costs [8,9]. The conventional physical methods, such as adsorption via activated carbon, ion exchange through synthetic adsorbent resins, reverse osmosis, ultrafiltration, coagulation by chemical agents etc., have been employed for the removal of toxic dyes from water [10,11]. These procedures are successful for transferring organic pollutants from water to another form, thereby producing secondary pollution which needs more treatment for removal of solid wastes and regeneration of the adsorbent and, thus, raises the cost of the process [12–14]. Among these contaminants, dyes, phenolics and pesticides have been of major concern because of their harmfulness to the environment. Advanced oxidation processes have potential for the removal of water polluted with hard degradable organic compounds. Moreover, hybrid processes combining membrane separation and heterogeneous photocatalysis, called photocatalytic membrane reactors, offer an amazing solution by complementing the advantages and overcoming the challenges of each other.

Photocatalytic oxidation involves advanced oxidation techniques for the degradation of dyes and many other toxic organic pollutants [15–17]. Nowadays, semiconductors are used as photocatalysts as they are cost-effective and have furnished good results. The most effective photocatalysts found in the literature are selenides, sulfides, metal oxides and the compounds with a modest band gap energy of 1.1–3.8 eV, such as TiO_2, ZnO and WO_3. However, these metal sulfides and selenides have issues, such as instability, photo anodic corrosion and toxicity [18–22].

Recently, cobalt ferrite has hugely attracted researchers for its utilization in photocatalytic wastewater treatment [23–29]. $CoFe_2O_4$ and its composites are low-cost and possess extraordinary magnetic properties, which results in higher adsorption ability as compared to other photocatalysts. Their photocatalytic activity improves owing to their narrow band gap, rapid transfer capability for charge carriers, higher visible light response, chemical functionalities, morphologies and annealing temperature. Dutta et al. reviewed the photocatalytic activity of $CoFe_2O_4$ against noxious organic pollutants in water and pointed out its limitations, such as aggregation, the recombination rate of electron holes and photocorrosion [25]. Their study concluded that the formation of $CoFe_2O_4$-based heterojunctions is seemingly acceptable for the improved photocatalytic performance, because the combination of $CoFe_2O_4$ with other photocatalysts that have various band gap positions would exhibit synergistic effects, which permits more pollutant compounds to be accessible for photodegradation and enhances the rate of charge separation, thus reducing electron hole recombination. Habibi and Parhizkar prepared spinel phase $CoFe_2O_4$ nanocomposites through a hydrothermal method using $CoFe_2O_4$ and iron nitrate as precursors and ethylene glycol as a chelating agent [26]. Their product has demonstrated an energy gap of 1.3 eV and 68.0% breakdown of Reactive Red 4 (azo dye) within 100 min. They mainly attribute this enhanced activity to the shift of electrons and holes to the surface of nanocomposites, which hinders electron hole recombination. Cubic ferro spinel nanostructures of $Co_{0.2}Zn_{0.8}Fe_2O_4$ were prepared by A. Boudjemaa et al. via coprecipitation [23]. These structures were n-type semiconductors with a lower band gap which dramatically enhanced photocatalytic activity under visible light irradiation. Z. Yamani studied photocatalytic degradation performance of magnetically separable $CoFe_2O_4$/BiOCl composites against Rhodamine B under UV light irradiation [29]. Yamani concluded that the adsorption of RhB molecules was favored on the surface, caused by the development of heterojunctions on the interface between BiOCl and ferrites. Y. Deng et al. successfully synthesized a recyclable photo-Fenton catalyst, of composition $CoFe_2O_4$@polypyrrole, by simple in situ Fenton oxidization polymerization [24]. This catalyst system has exhibited 90% discoloration in up to five cycles against methyl orange, methylene blue and rhodamine B. The degradation measurement in the presence of reactive species scavengers has revealed that the holes and the hydroxyl radical were the major reactive oxygen species in the $CoFe_2O_4$@PPy system. Y. Deng et al. has attributed this performance enhancement to the combination of enhanced Fenton response by coordinated PPy and Fe^{2+} redox pairs and photocatalytic performance by improved photo-led charge separation and adsorption. A boost in catalytic performance of pristine $CoFe_2O_4$ was observed by R. Sharma when yttrium was doped into a cobalt ferrite single-phase cubic spinel structure [27]. This increased BET surface area values and a reduced band gap. The good recyclability of prepared photocatalysts has proven the good stability of the materials used in wastewater treatment. A $CoFe_2O_4$/PMS (peroxymonosulphate) system, prepared by sol-gel combustion method, has demonstrated good photodegradation of methyl orange and phenol [28]. Surface hydroxyls of Co^{2+}/PMS have played a significant role in the generation of sulphate and hydroxyl free radicals in neutral medium, producing more surface active sites.

Among numerous semiconductors, Bi_2O_3 and $CoFe_2O_4$ are potent candidates for photoactive nanocomposite preparations due to their distinctive structures and physical aspects, such as band gap, thermal stability, high oxygen ion conductivity and high refractive index. However, bare Bi_2O_3 and $CoFe_2O_4$ show less photo catalytic activity, owing to photo corrosion, rapid charge carrier recombination and structural conversions [30–33]. Different methods can be utilized to increase the photocatalytic activity of Bi_2O_3 and $CoFe_2O_4$, such as doping of transition metal oxides or rare earth metals and formation of nanocomposites in order to sensitize the photocatalyst to photo energy and to keep electron hole pairs separate. One of the most efficient ways to resolve this issue is to prepare nanocomposites of compatible semiconductor materials with a synergistic effect that will help charge migration [34–47].

The present study has explored the compatibility of Bi_2O_3 and $CoFe_2O_4$ for the preparation of photoactive nanocomposites with good photodegradation potential against methyl orange dye. To our knowledge, no earlier research has been reported in the literature regarding the synthesis of $Bi_2O_3/CoFe_2O_4$ nanocomposites and their dye degradation potential.

2. Results

This study aimed to develop different nanocomposites of bismuth oxide and cobalt ferrite semiconductor materials that work synergistically to improve photocatalytic activity for efficient photodegradation of methyl orange dye. As-prepared nanocomposite materials were characterized by X-ray diffraction (XRD), scanning electron microscopy (SEM) and energy dispersive X-ray (EDX) spectroscopic techniques while their energy gaps were gauged by UV/Vis diffuse reflectance spectroscopy (DRS) via Tauc plots. Photodegradation experiment results were employed to study photocatalytic efficiency.

2.1. Phase Analysis

XRD patterns of prepared samples were scanned in the range of 10° to 70° by X-ray diffractometer (D8 advance, BRUKER, Camarillo, CA, USA) with Cu-Kα radiation (k = 1.5406 Å) at room temperature. All the peaks in Figure 1a were found to be consistent with the available standard JCPDS 01-077-0426 and can be indexed to a single-phase cubic $CoFe_2O_4$ structure. Reflection (220), exhibited at 30.12° exclusively, depends on the Co^{2+} cations occupying the cubic sites. No additional and intermediate phases were observed within the sensitivity of the experimental measurements. The crystallite size of as-prepared $CoFe_2O_4$ particles was computed by the classical Scherrer formula [48]. $Dhkl = k\lambda/\beta\cos\theta$, where Dhkl is the crystallite size derived from the (311) peak of the XRD spectrum, k the sphere shape factor (0.89), θ the angle of diffraction, β the difference between the full width at half-maximum (FWHM) of the sample peak and the standard SiO_2 used to calibrate the intrinsic width associated with the instruments and λ the X-ray wavelength (1.5406 Å). The obtained crystallite size of as-prepared $CoFe_2O_4$ particles was about 26.6 nm (Table 1). The XRD pattern of Bi_2O_3 (Figure 1e) was in agreement with the available standard JCPDS 00-022-0515. The crystal system of Bi_2O_3 was a tetragonal form which is a metastable two-dimensional superstructure of β-Bi_2O_3 [49]. Diffraction lines of Bi_2O_3 patterns have indicated that the powder has a good degree of crystallinity (Figure 1e). The crystallite size of Bi_2O_3 was computed from the full width at half-maximum intensity of the (201) X-ray line diffraction and was found to be approximately 32 nm.

With the increase in Bi_2O_3 content in $Bi_2O_3/CoFe_2O_4$ nanocomposites, the intensity of the (201) peak at 27.42°, which is identified as the main peak of the Bi_2O_3 phase, gradually increased, whereas that of the (311) peak at 35.16°, inherent from the $CoFe_2O_4$ phase, was decreased. It was evident from Figure 1b–d that individual peaks of Bi_2O_3 and $CoFe_2O_4$ varied directly with their proportions in as-prepared nanocomposite materials, thus depicting no chemical reaction between the constituent chemicals. Furthermore, the purity of the samples is also apparent from the lack of any extra peak in the diffraction patterns of all prepared photocatalysts.

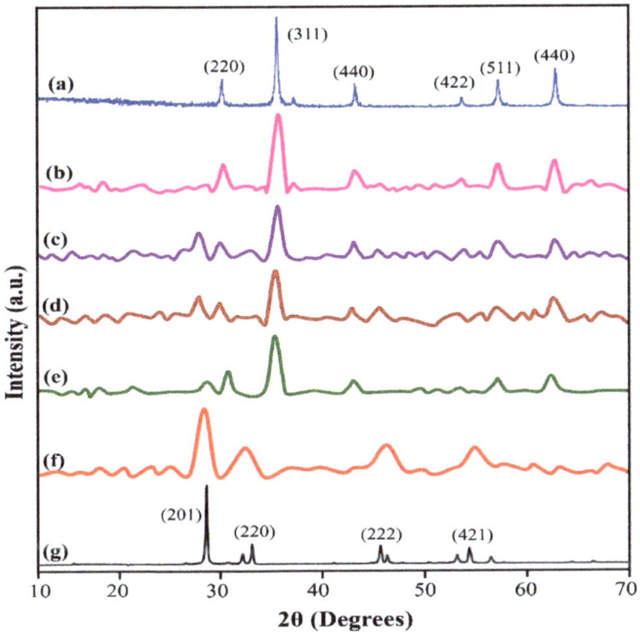

Figure 1. XRD patterns of (**a**) $CoFe_2O_4$ [32]; (**b**) $CoFe_2O_4$ (as-prepared); (**c**) NC-3; (**d**) NC-2; (**e**) NC-1; (**f**) Bi_2O_3 (as-prepared); (**g**) Bi_2O_3 [43].

Table 1. Crystallite size of prepared samples.

Sample	Average Crystallite Size (nm)
Bi_2O_3	32
$CoFe_2O_4$	26.6
NC-1	19
NC-2	21
NC-3	30

2.2. Elemental Analysis

EDX analysis was performed by a VEGA3 machine (TESCAN, Kohoutovice, Czech Republic) to determine the elemental composition of prepared samples. Each element showed its specific peaks in the graph, signifying its quantified presence in the sample. All peaks corresponded to $CoFe_2O_4$ with no extra peak showing the purity of the sample. The spectrum depicted signals of cobalt, iron and oxygen with 17.43%, 30.48% and 52.09% by weight, respectively. The composition, determined by energy-dispersive spectroscopy, showed the stoichiometry of $CoFe_2O_4$. Since the atomic numbers of cobalt and iron are closer, the ratio of the X-ray intensities from these elements generally associated with their composition. The intensity ratio of Co:Fe was about 1:2, as determined by the EDX. In this case, X-ray intensities do not need to be compensated for absorption and fluorescence effects, as these particles are smaller than the free path required for X-ray transmission through solids, i.e., 100 nm. The EDX pattern of Bi_2O_3 has shown no extra peak, which indicates that the sample was pure. The spectrum has shown signals of O and Bi with 23.06% and 76.94% by weight, respectively (Table 2).

Table 2. EDX elemental data of as-prepared samples.

Sample	Elemental Weight Percentage			
	Co	Fe	Bi	O
$CoFe_2O_4$	17.43	30.48	-	52.09
Bi_2O_3	-	-	23.06	76.94
NC-1	21.20	44.19	19.97	14.63
NC-2	13.13	28.14	25.42	33.31
NC-3	10.73	17.74	47.89	23.63

Qualitative and quantitative EDX results from the nanocomposites agreed with as-prepared compositions, which indicates that no element was lost due to any volatile product formation by heating during nanocomposite calcination.

2.3. Morphological Analysis

A field-emission scanning electron microscope (VEGA3 TESCAN) was used for morphological studies of as-prepared Bi_2O_3, $CoFe_2O_4$ and their nanocomposites. SEM micrographs (Figure 2a–e) were recorded at 200 and 500 nm scales to examine the morphology of the prepared materials. The textures of pure $CoFe_2O_4$ and Bi_2O_3 were observed as granular nanoparticles (Figure 2a,b).

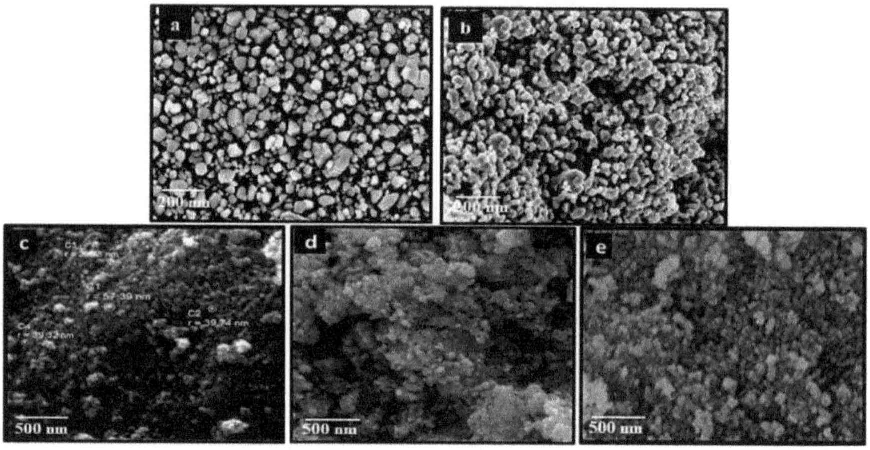

Figure 2. SEM images of (**a**) Bi_2O_3; (**b**) $CoFe_2O_4$; (**c**) NC-1; (**d**) NC-2; (**e**) NC-3.

Scanning electron micrographs of as-prepared nanocomposites have shown irregular granular microstructures with slight agglomeration despite the dipolar interaction between the particles. For compositional uniformity from particle to particle, it was demonstrated that both Bi_2O_3 and $CoFe_2O_4$ phases were positioned on the grain surface side-by-side. Hence, the prepared $Bi_2O_3/CoFe_2O_4$ nanocomposites were considered to be a tightly contacting heterojunction structure between Bi_2O_3 and $CoFe_2O_4$, formed in nano-size level. SEM images of the $Bi_2O_3/CoFe_2O_4$ nanocomposites have shown no significant change in the morphology of the constituent nanoparticles. However, it was established that as Bi_2O_3 contents were increased, the grains in the nanocomposite matrix became larger and more agglomerated, while the shape of the grains was not changed (Figure 2c–e). SEM micrographs have shown that the average particle size of prepared nanoparticles and their composites was below 100 nm, which was also in agreement with those computed from the X-ray powder diffraction peaks.

2.4. Diffuse Reflectance Spectroscopy and Tauc Plots

DRS spectra were employed to construct Tauc plots of the prepared catalysts. Tauc plots were plotted between photon energy and $(\alpha h\nu)^2$ to calculate band gaps. Bi_2O_3 has shown a single hump at 521 nm. Extrapolation of the linear region of the Tauc plot (inset Figure 3a) has furnished a band gap of 2.55 eV for hydrothermally prepared Bi_2O_3. $CoFe_2O_4$ has an intrinsic black color, so it has shown absorption in a wide range of wavelengths [50]. The absorption spectra of cobalt ferrite exhibited a single hump at 751 nm. As cobalt ferrite is a direct semiconductor, the band gap value was 1.63 eV, computed from the Tauc plot (inset Figure 3b). The band gaps of the nanocomposites, namely NC-1, NC-2 and NC-3, were calculated as 1.62, 1.58 and 1.61 eV, respectively, by Tauc plots (inset Figure 3c–e). For the $Bi_2O_3/CoFe_2O_4$ nanocomposites, absorbance reduced as the loading of Bi_2O_3 increased. However, those values of absorbance could be useful in predicting the relative amount of $CoFe_2O_4$ and Bi_2O_3 in the synthesized composites by using the Beer–Lambert law.

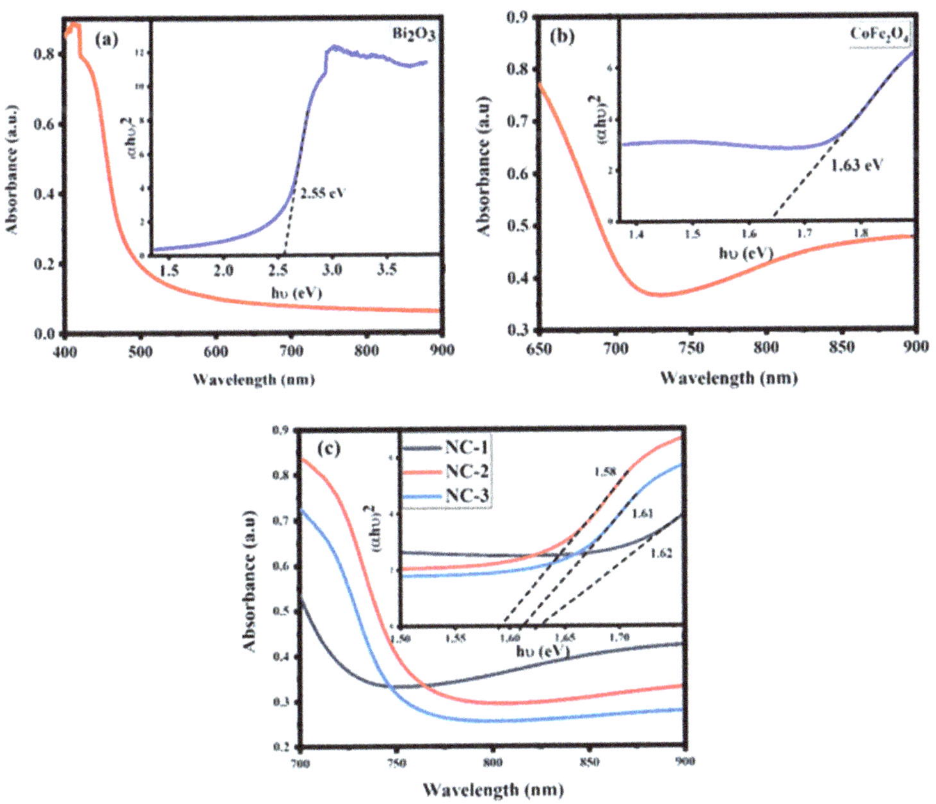

Figure 3. DRS spectra and Tauc plots of (**a**) Bi_2O_3; (**b**) $CoFe_2O_4$; (**c**) NC-1, NC-2, NC-3.

2.5. Photodegradation Activity

The degradation of methyl orange dye was studied by logging changes in the absorption value of a charged dye solution at its $\lambda_{max} \approx 462$ nm for different hourly time intervals of a light irradiation experiment (Table 3). Figure 4 shows degradation spectra of methyl orange for each prepared catalyst along with a blank sample. The photodegradation activities of all prepared nanocatalysts were plotted in terms of percentage degradation versus passage of time (Figure 5). NC-1 (1:3 $Bi_2O_3/CoFe_2O_4$) has also shown good degradation results. It degraded up to 82% of the dye in 5 h of light exposure. NC-2 (1:1 $Bi_2O_3/CoFe_2O_4$) has shown the best degradation results. It degraded up to 92% of the dye in 5 h of the light irradiation experiment, as evidenced from the associated absorbance spectra taken during photocatalytic bleaching. NC-3 (3:1 $Bi_2O_3/CoFe_2O_4$) has shown moderate degradation results. It degraded up to 78% of the dye in 5 h of light illumination (Figure 6). These results have shown that all prepared nanocomposites exhibited good photodegradation activities as compared to bare materials. The blank with no nanocatalyst has exhibited a minimal decrease, i.e., about 2%, in absorbance when subjected to light irradiation for the period of 5 h witnessed in Figure 4. This shows an evident synergism phenomenon operating between the constituents of the nanocomposites.

Table 3. Absorbance values of dye solutions charged by various as-prepared photocatalysts during photodegradation experiment in comparison to blank.

Sample	Absorbance					
Time (h)	0	1	2	3	4	5
MO-Blank	1.00	0.998	0.995	0.991	0.986	0.980
$CoFe_2O_3$	0.99	0.97	0.94	0.86	0.67	0.37
Bi_2O_3	0.95	0.91	0.89	0.82	0.53	0.24
NC-I	0.85	0.79	0.76	0.63	0.32	0.15
NC-II	0.82	0.77	0.74	0.58	0.34	0.06
NC-III	0.88	0.84	0.82	0.67	0.48	0.19

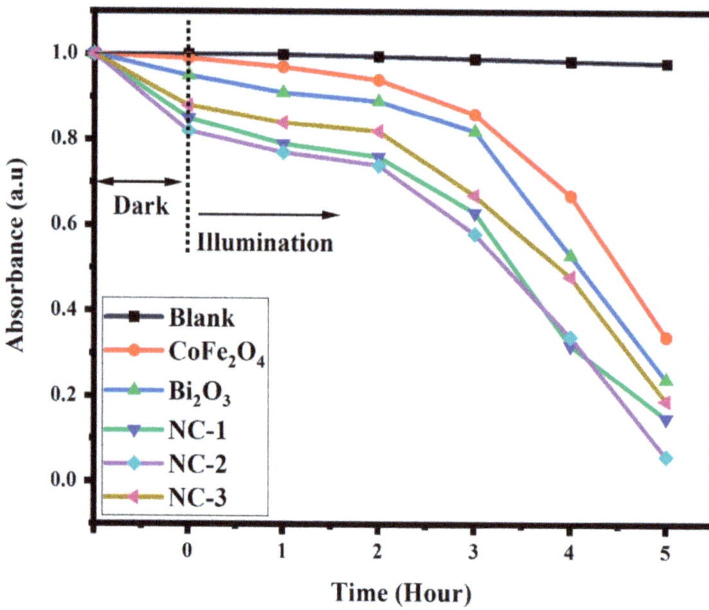

Figure 4. Absorbance of dye solutions charged by various as-prepared photocatalysts during photodegradation experiment.

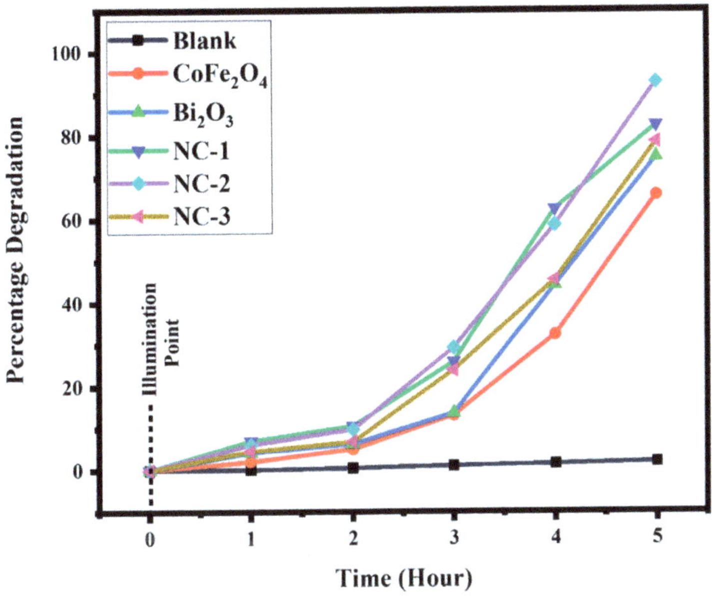

Figure 5. Photodegradation activities of as-prepared photocatalysts against methyl orange dye solution.

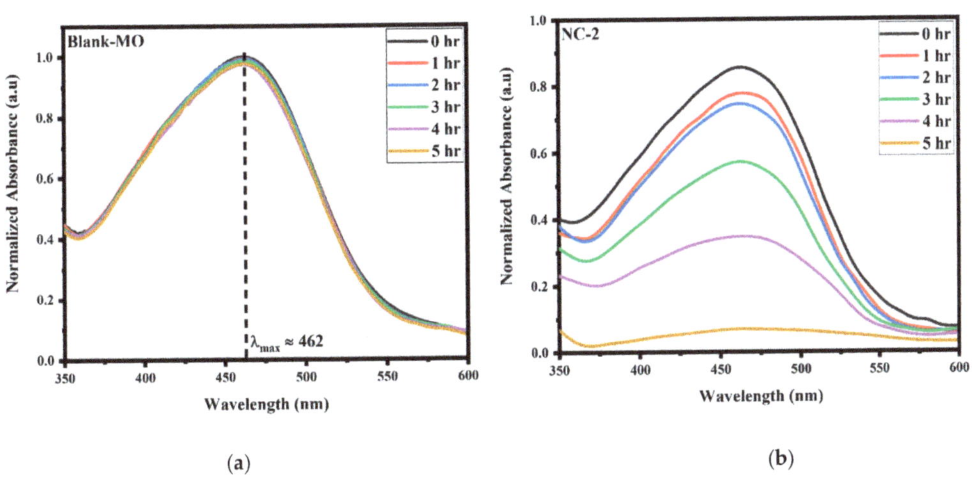

Figure 6. UV/visible spectral information for (**a**) Blank and (**b**) NC-2 under illumination.

Photodegradation efficiency of all prepared catalysts against methyl orange was calculated by the formula:

$$\text{Percentage Efficiency} = [(A_o - A_t)/A_o] \times 100$$

where A_o is absorbance at time 0 min and A_t is absorbance after time "t" min of the photodegradation experiment. A_o and A_t were recorded at λ_{max} of the dye [51]. Figure 7 shows the comparison of the percentage efficiency of all prepared catalysts against methyl orange. The highest efficiency was calculated for NC-2, which is due to the optimum constitution of Bi_2O_3 and $CoFe_2O_4$ (1:1) as well as the effective migration of exciton pairs in the type-II band configuration between the constituent materials, resulting in the suppression of charge recombination.

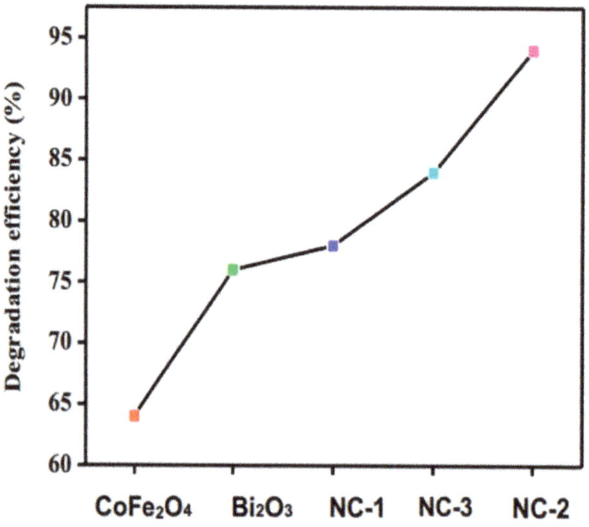

Figure 7. Comparison of percentage efficiency of all prepared photocatalysts for methyl orange degradation.

2.6. Proposed Photodegradation Scheme

According to the aforementioned results, a speculative photocatalytic mechanism based on bibliographic data is proposed in Figure 8 [39,52–54]. Under continuous light irradiation, electrons become excited from valence bands of $CoFe_2O_4$ and Bi_2O_3 to their conduction bands, meanwhile producing holes in their respective valence bands. As both the valence band and conduction band of the $CoFe_2O_4$ are lower in energy than those of Bi_2O_3, cobalt ferrite acts as a light energy trapper. Subsequently, these electrons and holes are rearranged and enriched into the conduction band of Bi_2O_3 and the valence band of $CoFe_2O_4$, respectively. A band-to-band transition bridge is established between Bi_2O_3 and cobalt ferrite in which the energy levels are mismatched between each other. Since the conduction band potential of $CoFe_2O_4$ is more negative than that of Bi_2O_3, excited electrons are constantly shifted to the conduction band of Bi_2O_3. Meanwhile, as the valence band of $CoFe_2O_4$ is less positive than that of the Bi_2O_3, holes in the VB of Bi_2O_3 will transfer continuously to the VB of cobalt ferrite. This synergism will reduce the cogent band gap, hinder electron hole recombination and will increase electron hole availabilities for photodegradation reactions, thus ensuring efficient photodegradation co-work of $Bi_2O_3/CoFe_2O_4$ nanocomposites.

These holes in the VB of cobalt ferrite will react to H_2O/OH^- and produce highly reactive $OH^\bullet$ radicals to start the process of degradation. Simultaneously, electrons gathered in the conduction band of Bi_2O_3 react with adsorbed oxygen and form oxidants such as superoxide ions. These active species such as $OH^\bullet$ and $O_2^{\bullet-}$ radicals react to degrade molecules of the dye into H_2O, CO_2 or other small molecular products. Chemical equations of this speculative photocatalytic mechanism based on bibliographic data are illustrated as:

$$CoFe_2O_4 + Bi_2O_3 + 2h\nu \rightarrow CoFe_2O_4\ (e^- + h^+) + Bi_2O_3\ (e^- + h^+) \quad (1)$$

$$CoFe_2O_4\ (e^- + h^+) + Bi_2O_3\ (e^- + h^+) \rightarrow CoFe_2O_4\ (2h^+) + Bi_2O_3\ (2e^-) \quad (2)$$

$$Bi_2O_3\ (2e^-) + 2O_2 \rightarrow 2O_2^{\bullet -} + Bi_2O_3 \quad (3)$$

$$CoFe_2O_4\ (2h^+) + 2OH^- \rightarrow CoFe_2O_4 + 2OH^{\bullet} \quad (4)$$

$$Dye + O_2^{\bullet -} + OH^{\bullet} \rightarrow CO_2 + H_2O + \text{other by-product} \quad (5)$$

The $Bi_2O_3/CoFe_2O_4$ heterojunction designed in this study is not similar to the type-I heterojunction structure. It is considered for the $Bi_2O_3/CoFe_2O_4$ that $CoFe_2O_4$ plays its role as the main photocatalyst, while the Bi_2O_3 works as a photosensitizer [14,55,56]. The VB edge of Bi_2O_3 is lower than that of $CoFe_2O_4$; hence, this system is a type-II heterojunction [57,58]. Since these heterojunctions are tightly bounded at the nanoscale, the hole transfers through the junction are efficacious. Furthermore, the $CoFe_2O_4$, working as the main photocatalyst, is found side-by-side in the $Bi_2O_3/CoFe_2O_4$ nanocomposite. Hence, formation of a heterojunction structure does not shield the active sites of catalysts.

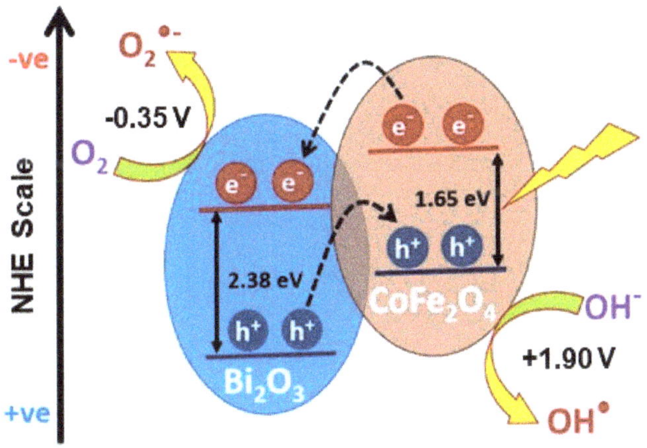

Figure 8. Schematic illustration of energy level alignment and photodegradation action of Bi_2O_3/$CoFe_2O_4$ nanocomposites.

3. Materials and Methods

All the materials, viz., bismuth (III) nitrate pentahydrate, cobalt chloride hexahydrate, ferric chloride hexahydrate, polyvinyl pyrrolidone (MW 40,000), sodium hydroxide, nitric acid and glycerol were purchased from Daejung, Korea and used without further purification. Ethanol and deionized water were used for washing and preparation of solutions. Various concentrations of precursor solution and conditions of reactions were explored for material synthesis based on reported methods. However, the following optimized methods proved to avoid agglomeration of individual products and composites.

3.1. Synthesis of Bi_2O_3 Nanoparticles

Bismuth nitrate pentahydrate (0.97 g) was dissolved in 10 mL of 1.12 M HNO_3 along with vigorous stirring in order to avoid hydrolyzation of Bi^{3+} ions. Subsequently, 0.072 g of polyvinylpyrrolidone (PVP) was added as surfactant into the above mixture and stirred by manual stirring for about 15–20 min. Later, 4 M NaOH solution was gradually added with constant stirring until the pH of the solution turned basic (pH = 11) and white precipitates appeared. After stirring for 10 min, the resultant suspension was placed into a Teflon-lined autoclave (KH100, Guangzhou Aolantec, China) at 90 °C for 1 h. The suspension was allowed to cool at room temperature and its white color turned to yellow. The resultant yellow precipitates were centrifuged in a centrifuge (Hermle, Germany) and washed with distilled water/ethanol several times. The resultant product was dried in a vacuum oven (Memmert, Germany) for 2 h at 80 °C. This furnished yellow, solid particles of bismuth oxide which were ground into a fine powder. This powder was further calcined at 400 °C for 4 h [42,43].

3.2. Synthesis of $CoFe_2O_4$ Nanoparticles

Magnetic cobalt ferrite nanoparticles were prepared via a simple, one-pot hydrothermal process. An aqueous solution containing 4 mmol $FeCl_3{\cdot}6H_2O$ was mixed with 0.078 M glycerol (surfactant) under constant stirring. Afterward, 2 mmol $CoCl_2{\cdot}6H_2O$ solution was added slowly into the above mixture to prepare a uniform solution. Highly alkaline conditions (pH = 13) of the reaction mixture were maintained by dropwise addition of 6 M NaOH solution. This resulted in a brownish-black precipitate mixture. Subsequently, the reaction mixture was transferred into a Teflon-lined autoclave and kept in an oven for 6 h at 200 °C. The resultant suspension was centrifuged and washed with distilled water several times. Later, the precipitates were dried at 100 °C for 6 h in a vacuum oven. The resultant cobalt ferrite product was ground into a powder [44,45].

3.3. Synthesis of $Bi_2O_3/CoFe_2O_4$ Nanocomposites

A physical method was opted for in order to prepare the nanocomposites. This involved simple, solid-state mixing of already synthesized bismuth oxide and cobalt ferrite nanoparticles, followed by 3 h grinding via pestle and mortar. Resultant mixtures were calcined at 400 °C for 4 h. Three nanocomposites with different proportions of $CoFe_2O_4$ and Bi_2O_3 were prepared (Table 4).

Table 4. Details of $Bi_2O_3/CoFe_2O_4$ nanocomposites.

	Wt.% of Bi_2O_3	**Wt.% of $CoFe_2O_4$**
NC-1	25	75
NC-2	50	50
NC-3	75	25

3.4. Degradation Experiment

Photodegradation studies were carried out against methyl orange dye to explore the photocatalytic properties of as-prepared nanocomposites. A 300 W xenon arc lamp was used as the light source for degradation studies and to irradiate beakers charged with dye stock solution and catalyst. A stock solution of 30 ppm methyl orange dye was prepared in distilled water. For each photodegradation run, 50 mL stock solution was poured into beakers and 10 mg of each prepared catalyst nanocomposite was added into these separately. These solutions were kept in the dark with constant stirring for 2 h to attain adsorption–desorption equilibrium. Afterward, samples were kept in a photoreactor chamber and light was shone on the system. After different predetermined intervals of time, 5 mL of each solution was stripped out and centrifuged for 5 min at 10,000 rpm to let the catalyst settle down. The solutions obtained were examined by UV–visible spectrophotometer (UV/Vis Lambda 365, Perkin Elmer, Akron, OH, USA) to ascertain the residual dye contents in the solution in order to study the photocatalytic activity and the level of methyl orange degradation [46,47].

4. Conclusions

$Bi_2O_3/CoFe_2O_4$ photocatalysts were successfully prepared by a facile agitation and the influence of different proportions of constituents was also investigated systematically. The results of XRD, SEM, EDX and UV–visible DRS spectroscopy have corroborated the synthesis of nanocomposites without any further chemical reaction or by-products, nor any change in morphology of the constituent chemicals. $Bi_2O_3/CoFe_2O_4$ heterojunctions have shown a high efficiency for decomposing organic dye under light irradiation as compared to both of the individual components. Furthermore, $Bi_2O_3/CoFe_2O_4$ induced mineralization by utilizing the electrons and the holes that were enriched in the conduction band of Bi_2O_3 and the valence band of $CoFe_2O_4$, respectively. The results suggested that $CoFe_2O_4$ acts as the main photocatalyst, while Bi_2O_3 works as a photosensitizer. The nanocomposite with 1:1 $Bi_2O_3/CoFe_2O_4$ composition showed the highest photodegradation efficiency, i.e., 92% for the degradation of methyl orange dye. The superior photocatalytic performance of the nanocomposite may be credited to its type-II band configuration, which results in the suppression of charge recombination because Bi_2O_3 acts as an electron sink. As per our research, no prior study has been reported in the literature regarding the synthesis of a $Bi_2O_3/CoFe_2O_4$ nanocomposite and its dye degradation potential. Therefore, good photocatalytic degradation activities, demonstrated by these nanocomposites, for contaminated aqueous solutions make these materials a potent candidate to develop robust photocatalysts for practical applications in water decontamination technology. The $Bi_2O_3/CoFe_2O_4$ is environmentally friendly, the synthesis is simple, and, if it were scaled up, the manufacturing cost would be low. This also leaves scope for researchers to further explore its photocatalytic activities against other dye/drug pollutants and its water splitting potential.

Author Contributions: Conceptualization, A.B.N., A.M. and A.S.; methodology A.B.N., F.R. and S.A.; formal analysis, F.R. and A.S.; writing—original draft preparation A.B.N., A.S. and S.A.; writing—review and editing, A.M.; supervision, A.M. All authors have read and agreed to the published version of the manuscript.

Funding: This research received no external funding.

Acknowledgments: The authors are thankful to Institute of Environmental Sciences & Engineering, School of Chemical & Materials Engineering, U.S.–Pakistan Center for Advanced Studies in Energy of NUST and the National Centre for Physics of QAU for extending their technical support to this research work.

Conflicts of Interest: The authors declare no conflict of interest.

References

1. Gao, Q.; Xu, J.; Bu, X.-H. Recent advances about metal–organic frameworks in the removal of pollutants from wastewater. *Coord. Chem. Rev.* **2019**, *378*, 17–31. [CrossRef]
2. Chong, M.N.; Jin, B.; Chow, C.W.; Saint, C. Recent developments in photocatalytic water treatment technology: A review. *Water Res.* **2010**, *44*, 2997–3027. [CrossRef] [PubMed]
3. Mohammed, R.R. Removal of heavy metals from wastewater using black tea waste. *Arab. J. Sci. Eng.* **2012**, *37*, 1505–1520. [CrossRef]
4. Manaa, Z.; Chebli, D.; Bouguettoucha, A.; Atout, H.; Amrane, A. Low-Cost Photo-Fenton-Like Process for the Removal of Synthetic Dye in Aqueous Solution at Circumneutral pH. *Arab. J. Sci. Eng.* **2019**, *44*, 9859–9867. [CrossRef]
5. Dellamatrice, P.M.; Silva-Stenico, M.E.; de Moraes, L.A.B.; Fiore, M.F.; Monteiro, R.T.R. Degradation of textile dyes by cyanobacteria. *Braz. J. Microbiol.* **2017**, *48*, 25–31. [CrossRef] [PubMed]
6. Kooh, M.R.R.; Dahri, M.K.; Lim, L.B.; Lim, L.H. Batch adsorption studies on the removal of acid blue 25 from aqueous solution using Azolla pinnata and soya bean waste. *Arab. J. Sci. Eng.* **2016**, *41*, 2453–2464. [CrossRef]
7. Gupta, V.K.; Ali, I.; Saleh, T.A.; Nayak, A.; Agarwal, S. Chemical treatment technologies for waste-water recycling—An overview. *Rsc. Adv.* **2012**, *2*, 6380–6388. [CrossRef]
8. Esplugas, S.; Bila, D.M.; Krause, L.G.T.; Dezotti, M. Ozonation and advanced oxidation technologies to remove endocrine disrupting chemicals (EDCs) and pharmaceuticals and personal care products (PPCPs) in water effluents. *J. Hazard. Mater.* **2007**, *149*, 631–642. [CrossRef]
9. Moo-Young, H.K. Pulp and paper effluent management. *JSTOR* **2007**, *79*, 1733–1741. [CrossRef]
10. Maleš, L.; Fakin, D.; Bračič, M.; Gorgieva, S. Efficiency of Differently Processed Membranes Based on Cellulose as Cationic Dye Adsorbents. *Nanomaterials* **2020**, *10*, 642. [CrossRef]
11. Gogate, P.R.; Pandit, A.B. A review of imperative technologies for wastewater treatment II: Hybrid methods. *Adv. Environ. Res.* **2004**, *8*, 553–597. [CrossRef]
12. Rauwel, P.; Uhl, W.; Rauwel, E. *Editorial for the Special Issue on 'Application and Behavior of Nanomaterials in Water Treatment'*; Multidisciplinary Digital Publishing Institute: Basel, Switzerland, 2019.
13. Coleman, H.M.; Eggins, B.R.; Byrne, J.A.; Palmer, F.L.; King, E. Photocatalytic degradation of 17-β-oestradiol on immobilised TiO_2. *Appl. Catal. B Environ.* **2000**, *24*, L1–L5. [CrossRef]
14. Akpan, U.G.; Hameed, B.H. Parameters affecting the photocatalytic degradation of dyes using TiO_2-based photocatalysts: A review. *J. Hazard. Mater.* **2009**, *170*, 520–529. [CrossRef]

15. Chauhan, N.; Singh, V.; Kumar, S.; Dhiman, R. Influence of Nickel, Silver, and Sulphur Doping on the Photocatalytic Efficiency of Mesoporous ZnO Nanoparticles. *Arab. J. Sci. Eng.* **2020**, *45*, 249–259. [CrossRef]
16. Kabra, K.; Chaudhary, R.; Sawhney, R.L. Treatment of hazardous organic and inorganic compounds through aqueous-phase photocatalysis: A review. *Ind. Eng. Chem. Res.* **2004**, *43*, 7683–7696. [CrossRef]
17. Hoffmann, M.R.; Martin, S.T.; Choi, W.; Bahnemann, D.W. Environmental applications of semiconductor photocatalysis. *Chem. Rev.* **1995**, *95*, 69–96. [CrossRef]
18. Ismail, N.J.; Othman, M.H.D.; Kamaludin, R.; Esham, M.I.M.; Ali, N.A.; Rahman, M.A.; Jaafar, J.; Bakar, S.A. Characterization of Bauxite as a Potential Natural Photocatalyst for Photodegradation of Textile Dye. *Arab. J. Sci. Eng.* **2019**, *44*, 10031–10040. [CrossRef]
19. Khan, M.M.; Adil, S.F.; Al-Mayouf, A. *Metal Oxides as Photocatalysts*; Elsevier: Amsterdam, The Netherlands, 2015.
20. Zhou, H.; Qu, Y.; Zeid, T.; Duan, X. Towards highly efficient photocatalysts using semiconductor nanoarchitectures. *Energy Environ. Sci.* **2012**, *5*, 6732–6743. [CrossRef]
21. Saravanan, R.; Gupta, V.K.; Narayanan, V.; Stephen, A. Comparative study on photocatalytic activity of ZnO prepared by different methods. *J. Mol. Liq.* **2013**, *181*, 133–141. [CrossRef]
22. Khan, M.M.; Ansari, S.A.; Pradhan, D.; Ansari, M.O.; Lee, J.; Cho, M.H. Band gap engineered TiO_2 nanoparticles for visible light induced photoelectrochemical and photocatalytic studies. *J. Mater. Chem. A* **2014**, *2*, 637–644. [CrossRef]
23. Boudjemaa, A.; Popescu, I.; Juzsakova, T.; Kebir, M.; Helaili, N.; Bachari, K.; Marcu, I.C. M-Substituted (M=Co, Ni and Cu) Zinc Ferrite Photo-Catalysts for Hydrogen Production by Water Photo-Reduction. *Int. J. Hydrog. Energy* **2016**, *41*, 11108–11118. [CrossRef]
24. Deng, Y.; Zhao, X.; Luo, J.; Wang, Z.; Tang, J. Magnetic Recyclable $CoFe_2O_4$@ Ppy Prepared by in Situ Fenton Oxidization Polymerization with Advanced Photo-Fenton Performance. *RSC Adv. Tang* **2020**, *10*, 1858–1869. [CrossRef]
25. Dutta, V.; Sheetal, S.; Pankaj, R.; Ahmad, H.B.; Vinod, K.G.; Pardeep, J. Review on Augmentation in Photocatalytic Activity of $CoFe_2O_4$ Via Heterojunction Formation for Photocatalysis of Organic Pollutants in Water. *J. Saudi Chem. Soc. Singh* **2019**, *23*, 1119–1136.
26. Habibi, M.H.; Janan, J. Cobalt Ferrite Nano-Composite Coated on Glass by Doctor Blade Method for Photo-Catalytic Degradation of an Azo Textile Dye Reactive Red 4: Xrd, Fesem and Drs Investigations. *Spectrochim. Acta Part A Mol. Parhizkar Biomol. Spectrosc.* **2015**, *150*, 879–885. [CrossRef]
27. Sharma, R.; Kumar, V.; Bansal, S.; Singhal, S. Boosting the Catalytic Performance of Pristine $CoFe_2O_4$ with Yttrium (Y^{3+}) Inclusion in the Spinel Structure. *Mater. Res. Bull. Singhal* **2017**, *90*, 94–103. [CrossRef]
28. Devi, L.G.; Srinivas, M. Hydrothermal synthesis of reduced graphene oxide-CoFe2O4 heteroarchitecture for high visible light photocatalytic activity: Exploration of efficiency, stability and mechanistic pathways. *J. Environ. Chem. Eng.* **2017**, *5*, 3243–3255. [CrossRef]
29. Yamani, Z. Magnetic Properties and Photocatalytic Degradation Performance of MFe_2O_4 (M=Co, Ni)/Biocl Composites Catalysts under Uv Light Irradiation. *Arab. J. Sci. Eng.* **2018**, *43*, 383–388. [CrossRef]
30. Fan, H.; Pan, S.; Teng, X.; Ye, C.; Li, G.; Zhang, L. δ-Bi_2O_3 thin films prepared by reactive sputtering: Fabrication and characterization. *Thin Solid Film.* **2006**, *513*, 142–147. [CrossRef]
31. Shan, D.; Zhang, J.; Xue, H.-G.; Zhang, Y.-C.; Cosnier, S.; Ding, S.-N. Polycrystalline bismuth oxide films for development of amperometric biosensor for phenolic compounds. *Biosens. Bioelectron.* **2009**, *24*, 3671–3676. [CrossRef]
32. Shafi, K.V.; Gedanken, A.; Prozorov, R.; Balogh, J. Sonochemical preparation and size-dependent properties of nanostructured $CoFe_2O_4$ particles. *Chem. Mater.* **1998**, *10*, 3445–3450. [CrossRef]
33. Zi, Z.; Sun, Y.; Zhu, X.; Yang, Z.; Dai, J.; Song, W. Synthesis and magnetic properties of $CoFe_2O_4$ ferrite nanoparticles. *J. Magn. Magn. Mater.* **2009**, *321*, 1251–1255. [CrossRef]
34. Ullah, I.; Ali, S.; Hanif, M.A.; Shahid, S.A. Nanoscience for environmental remediation: A review. *Int. J. Chem. Biochem. Sci.* **2012**, *2*, 60–77.
35. Dong, F.; Guo, S.; Wang, H.; Li, X.; Wu, Z. Enhancement of the visible light photocatalytic activity of C-doped TiO_2 nanomaterials prepared by a green synthetic approach. *J. Phys. Chem. C* **2011**, *115*, 13285–13292. [CrossRef]
36. Narayana, R.L.; Matheswaran, M.; Aziz, A.A.; Saravanan, P. Photocatalytic decolourization of basic green dye by pure and Fe, Co doped TiO_2 under daylight illumination. *Desalination* **2011**, *269*, 249–253. [CrossRef]
37. Dippong, T.; Levei, E.A.; Deac, I.G.; Neag, E.; Cadar, O. Influence of Cu^{2+}, Ni^{2+}, and Zn^{2+} Ions Doping on the Structure, Morphology, and Magnetic Properties of Co-Ferrite Embedded in SiO_2 Matrix Obtained by an Innovative Sol-Gel Route. *Nanomaterials* **2020**, *10*, 580. [CrossRef] [PubMed]
38. Serpone, N.; Emeline, A. *Semiconductor Photocatalysis-Past, Present, and Future Outlook*; ACS Publications: Washington, DC, USA, 2012.
39. Chen, S.; Hu, Y.; Ji, L.; Jiang, X.; Fu, X. Preparation and characterization of direct Z-scheme photocatalyst $Bi_2O_3/NaNbO_3$ and its reaction mechanism. *Appl. Surf. Sci.* **2014**, *292*, 357–366. [CrossRef]
40. Kang, I.-C.; Zhang, Q.; Yin, S.; Sato, T.; Saito, F. Improvement in photocatalytic activity of TiO_2 under visible irradiation through addition of N-TiO_2. *Environ. Sci. Technol.* **2008**, *42*, 3622–3626. [CrossRef] [PubMed]
41. Bai, S.; Liu, H.; Sun, J.; Tian, Y.; Chen, S.; Song, J.; Luo, R.; Li, D.; Chen, A.; Liu, C.-C. Improvement of TiO2 photocatalytic properties under visible light by WO_3/TiO_2 and MoO_3/TiO_2 composites. *Appl. Surf. Sci.* **2015**, *338*, 61–68. [CrossRef]

42. Oudghiri-Hassani, H.; Rakass, S.; Al Wadaani, F.T.; Al-Ghamdi, K.J.; Omer, A.; Messali, M.; Abboudi, M. Synthesis, characterization and photocatalytic activity of α-Bi_2O_3 nanoparticles. *J. Taibah Univ. Sci.* **2015**, *9*, 508–512. [CrossRef]
43. Mallahi, M.; Shokuhfar, A.; Vaezi, M.; Esmaeilirad, A.; Mazinani, V. Synthesis and characterization of bismuth oxide nanoparticles via sol-gel method. *AJER* **2014**, *3*, 162–165.
44. Cote, L.J.; Teja, A.S.; Wilkinson, A.P.; Zhang, Z.J. Continuous hydrothermal synthesis of $CoFe_2O_4$ nanoparticles. *Fluid Phase Equilibria* **2003**, *210*, 307–317. [CrossRef]
45. Maaz, K.; Mumtaz, A.; Hasanain, S.; Ceylan, A. Synthesis and magnetic properties of cobalt ferrite ($CoFe_2O_4$) nanoparticles prepared by wet chemical route. *J. Magn. Magn. Mater.* **2007**, *308*, 289–295. [CrossRef]
46. Li, Y.; Li, X.; Li, J.; Yin, J. Photocatalytic degradation of methyl orange by TiO_2-coated activated carbon and kinetic study. *Water Res.* **2006**, *40*, 1119–1126. [CrossRef] [PubMed]
47. Wang, C.; Wang, X.; Xu, B.-Q.; Zhao, J.; Mai, B.; Sheng, G.; Fu, J. Enhanced photocatalytic performance of nanosized coupled ZnO/SnO_2 photocatalysts for methyl orange degradation. *J. Photochem. Photobiol. A Chem.* **2004**, *168*, 47–52. [CrossRef]
48. Mendelson, M.I. Average grain size in polycrystalline ceramics. *J. Am. Ceram. Soc.* **1969**, *52*, 443–446. [CrossRef]
49. Medernach, J.W.; Snyder, R.L. Powder diffraction patterns and structures of the bismuth oxides. *J. Am. Ceram. Soc.* **1978**, *61*, 494–497. [CrossRef]
50. Khan, M.S.; Ashiq, M.N.; Ehsan, M.F.; He, T.; Ijaz, S. Controlled synthesis of cobalt telluride superstructures for the visible light photo-conversion of carbon dioxide into methane. *Appl. Catal. A Gen.* **2014**, *487*, 202–209. [CrossRef]
51. Bera, K.K.; Majumdar, R.; Chakraborty, M.; Bhattacharya, S.K. Phase control synthesis of α, β and α/β Bi_2O_3 hetero-junction with enhanced and synergistic photocatalytic activity on degradation of toxic dye, Rhodamine-B under natural sunlight. *J. Hazard. Mater.* **2018**, *352*, 182–191. [CrossRef]
52. Cong, Y.; Zhang, J.; Chen, F.; Anpo, M.; He, D. Preparation, photocatalytic activity, and mechanism of nano-TiO_2 co-doped with nitrogen and iron (III). *J. Phys. Chem. C* **2007**, *111*, 10618–10623. [CrossRef]
53. Huang, S.; Xu, Y.; Xie, M.; Xu, H.; He, M.; Xia, J.; Huang, L.; Li, H. Synthesis of magnetic $CoFe_2O_4/g$-C_3N_4 composite and its enhancement of photocatalytic ability under visible-light. *Colloids Surf. A Physicochem. Eng. Asp.* **2015**, *478*, 71–80. [CrossRef]
54. Fu, J.; Tian, Y.; Chang, B.; Xi, F.; Dong, X. BiOBr–carbon nitride heterojunctions: Synthesis, enhanced activity and photocatalytic mechanism. *J. Mater. Chem.* **2012**, *22*, 21159–21166. [CrossRef]
55. Konstantinou, I.K.; Albanis, T.A. TiO_2-assisted photocatalytic degradation of azo dyes in aqueous solution: Kinetic and mechanistic investigations: A review. *Appl. Catal. B Environ.* **2004**, *49*, 1–14. [CrossRef]
56. Thiruvenkatachari, R.; Vigneswaran, S.; Moon, I.S. A review on UV/TiO_2 photocatalytic oxidation process (Journal Review). *Korean J. Chem. Eng.* **2008**, *25*, 64–72. [CrossRef]
57. He, G.; Ding, J.; Zhang, J.; Hao, Q.; Chen, H. One-step ball-milling preparation of highly photocatalytic active $CoFe_2O_4$–reduced graphene oxide heterojunctions for organic dye removal. *Ind. Eng. Chem. Res.* **2015**, *54*, 2862–2867. [CrossRef]
58. Kumar, A.; Kumar, A.; Sharma, G.; Ala'a, H.; Naushad, M.; Ghfar, A.A.; Guo, C.; Stadler, F.J. Biochar-templated g-$C_3N_4/Bi_2O_2CO_3/CoFe_2O_4$ nano-assembly for visible and solar assisted photo-degradation of paraquat, nitrophenol reduction and CO_2 conversion. *Chem. Eng. J.* **2018**, *339*, 393–410. [CrossRef]

catalysts

Article

Facile Synthesis of Various ZrO_2 Phases and ZrO_2-MO_2 (M = Ti, Hf) by Thermal Decomposition of a Single UiO-66 Precursor for Photodegradation of Methyl Orange

Ira Nur Arba'atul Jannah [1], Hanu Fiorena Sekarsari [1], Sri Mulijani [2], Karna Wijaya [3], Arief Cahyo Wibowo [4,*,†] and Aep Patah [1,*]

[1] Division of Inorganic and Physical Chemistry, Faculty of Mathematics and Natural Sciences, Institut Teknologi Bandung, Jl. Ganesha 10, Bandung 40132, Indonesia; iranurarbaatuljannah@gmail.com (I.N.A.J.); hanufiorena@gmail.com (H.F.S.)
[2] Department of Chemistry, Faculty of Mathematics and Natural Sciences, Institut Pertanian Bogor, Campus of IPB Dramaga, Bogor 16680, Indonesia; srimulijani@apps.ipb.ac.id
[3] Department of Chemistry, Faculty of Mathematics and Natural Sciences, Universitas Gadjah Mada, Bulaksumur Bls. 21, Yogyakarta 55281, Indonesia; karnawijaya@ugm.ac.id
[4] School of Advanced Technology and Multidiscipline, Universitas Airlangga, Campus C Mulyorejo, Surabaya 60115, Indonesia
* Correspondence: arief.wibowo@adu.ac.ae (A.C.W.); aep@itb.ac.id (A.P.)
† Present address: Department of Applied Sciences, College of Arts and Sciences, Abu Dhabi University, Abu Dhabi 59911, United Arab Emirates.

Abstract: A zirconia-based catalyst with controlled crystalline phases is synthesized through a simple thermal decomposition of a parent UiO-66 single precursor. The introduction of Ti(IV) and Hf(IV) cation into the Zr(IV) framework has been successfully obtained to tune the photocatalytic activity over methyl orange (MO) solution. Their resulting crystalline phases, morphologies, elemental analysis, band gap values, surface area, and photocatalytic degradation study over MO dye are presented and discussed. The tetragonal zirconia (t-ZrO_2) catalyst exhibits the highest photocatalytic activity with 89% decoloration efficiency under UV irradiation (λ = 254 nm) for 300 min compared to m-ZrO_2 (67%), the mixed phases (t-ZrO_2 and m-ZrO_2), as well as the synthesized mixed oxides ZrO_2-MO_2 (M = Ti or Hf), where the photocatalytic activities are 74% and 63%, respectively. This result is on par with commercially available anatase TiO_2 and other reported t-ZrO_2 catalysts.

Keywords: photodegradation; methyl orange; UiO-66; post-synthetic exchange; thermal decomposition

1. Introduction

Zirconium(IV) oxide (ZrO_2) is a very interesting semiconductor with a myriad of cutting-edge applications, such as a catalyst [1], a sensor of chemical species [2], corrosion-resistant coatings [3], and solid oxide fuel cells [4]. ZrO_2 exhibits three different phases: monoclinic (m-ZrO_2), tetragonal (t-ZrO_2), and cubic (c-ZrO_2). The crystalline phases of interest in catalysis are the crystalline phases of m-ZrO_2, low temperature stabilized t-ZrO_2, and a mix of monoclinic and tetragonal phases [5]. Having a wide band gap value and a high negative value of the conduction band potential, ZrO_2 has been widely studied as a photocatalyst in different chemical reactions [6]. The reported band gap energy of ZrO_2 ranges between 3.25 and 5.1 eV, depending on the preparation technique of the sample [7].

Methyl orange (MO), as one of the widespread carcinogenic water pollutants, is one of the very common water-soluble azo dyes that is extensively used in several industries such as textiles, paper, printing, and food [8–10]. One simple method to mitigate such water pollutants is photocatalytic degradation using photoactive catalysts, such as zirconia (ZrO_2). Based on the existing literature, we note that the photocatalytic degradation activity of zirconia depends on its adopted synthetic route and the resulting crystalline phase(s).

Basahel et al. [11] reported nanosized ZrO_2 powders with near pure monoclinic (m-ZrO_2), tetragonal (t-ZrO_2), and cubic (c-ZrO_2) structures synthesized by varied methods, using different precursors. In their paper, MO photodegradation reached 99% within 110 min using m-ZrO_2 nanoparticles, while the t-ZrO_2 and c-ZrO_2 nanocatalysts showed 90% and 80% degradation, respectively. Reddy et al. [12] showcased that photodecomposition of MO decreased with increasing irradiation time using t-ZrO_2. On the other hand, a low percentage (37%) of MO removal was presented by Mansouri et al. [13], where the ZrO_2 nanoparticle was synthesized by the sol–gel method.

Another widely adopted route for synthesizing nano oxide catalysts is using Metal–Organic Frameworks (MOFs) as a sacrificial template. The morphology of the parent MOF can be retained after a certain degree of controlled calcination. Using the optimized condition, specific crystalline phase(s) can also be easily isolated [14]. MOFs can easily be modified in three different ways to achieve high photocatalytic activity and stability, such as (i) through the introduction of metal sites; (ii) through an indirect route (post-synthetic modification), e.g., as a sacrificial template to make nano oxide catalysts; and (iii) through the creation of defects [15]. There are some reports showing nano ZrO_2-derived MOFs for photocatalytic degradation [8,10,16]. The 50% TiO_2-ZrO_2 binary oxide catalyst exhibited a high phenol degradation (99%) and a high reaction rate (0.73 mg L^{-1} min^{-1}), as reported by Kambur et al. [17]. Gota and Suresh obtained the Zr/Ti ratio (6.3), which degraded 88.72% of reactive red dye after 70 min [18]. Porous TiO_2/ZrO_2 obtained by calcining TiO_2/UiO-66 nanocomposites, reported by Abdi [8], displayed the excellent RhB photocatalytic degradation performance of 95% after 180 min.

Apart from the studies mentioned above, research on isolating crystalline phases of ZrO_2 and their effect on photocatalytic degradation of dyes is scarce. Herein, we report a simple route for isolating highly photoactive t-ZrO_2, m-ZrO_2, mixed t- and m-ZrO_2 with and without MO_2 (M = Ti or Hf) using UiO-66 as a single sacrificial template, without the use of additional precursors. All synthesized catalysts are photocatalytic active, and t-ZrO_2 showcases the highest photocatalytic degradation of MO, comparable to commercially available anatase TiO_2 and other reported t-ZrO_2 catalysts.

2. Results

A phase-pure, zirconium-based MOF, UiO-66, was obtained through the solvothermal method. The process of dissolving $ZrCl_4$ with DMF needs to be carried out under relatively inert conditions to avoid the form of irreversibly zirconyl chloride ($Zr(OH)_2(H_2O)_4Cl$). At the same time, DMF deprotonates H_2O, accelerating the formation of UiO-66 [19]. Apart from DMF, acetic acid was employed to increase the solubility of $ZrCl_4$ in DMF and functioned as a modulator [20]. The as-synthesized UiO-66 displays typical UiO-66 peaks at 7.32° and 8.46° 2θ without any unidentified peak, confirming the successful formation of phase-pure UiO-66, Figure S1a. The expected morphology of UiO-66 is shown in Figure S1b with a particle size of 420 nm, as observed using ImageJ® software.

2.1. Tetragonal and Monoclinic ZrO_2

The synthesis of t-ZrO_2, m-ZrO_2, mixed phases (t- and m-ZrO_2), and ZrO_2-MO_2 (M = Ti or Hf) was successfully carried out through a facile optimized calcination process using UiO-66 as a sacrificial template. Overall, calcination at 500 °C for 2 h obtained the nearly pure tetragonal ZrO_2 phase (t-ZrO_2). Elevating the time (and temperature) of calcination increases monoclinic phase composition. The close-to-pure m-ZrO_2 is obtained at 800 °C for 24 h. Therefore, the mixed-phase of t- and m-ZrO_2 can be easily obtained by using conditions between the two aforementioned conditions.

To find the optimum calcination condition referenced above for isolating near-pure t- and m-ZrO_2, we conducted experiments at two calcination temperatures of 800 °C (for 2, 8, and 24 h) and 500 °C (for 2, 6, and 8 h). The synthesized ZrO_2 is denoted as ZrO_2-A-B, where A indicates the calcination temperature and B represents the calcination time. The patterns of t-ZrO_2 have many peaks at 30.2°, 34.6°, 35.3°, 50.2°, 50.7°, and 60.2°, which were assigned to the (101), (002), (110), (112), (200), and (211) planes; thus, *t*-ZrO_2 has three facets of {101}, {002}, and {112}. Meanwhile, *m*-ZrO_2 has typical peaks at 28.3°, 31.6°, 34.3°, 35.3°, 40.9°, and 44.8°, attributed to the $(11\bar{1})$, (111), (020), (200), $(21\bar{1})$, and (112) planes, with three crystal facets of {111}, {020}, and {112}. As shown in Figure 1, a mixture of monoclinic dominant phase (JCPDS 78-1807) and tetragonal (JCPDS 79-1769) ZrO_2 is already formed even in the shortest calcination time of 2 h when calcining at 800 °C. Elevating the calcination time at 8 and 24 h shifts the mixture of tetragonal and monoclinic phases to nearly pure monoclinic, as indicated by the disappearing 30.2° 2θ peak. Furthermore, increasing the calcination time increases the peak intensity, indicating an increase in crystallinity and particle size of ZrO_2. Thus, close-to-pure *m*-ZrO_2 is obtained at 800 °C calcination for 24 h. Therefore, the dominant tetragonal phase would be observed below 800 °C, as reported by Basahel et al. [11] and Li et al. [21]. Lowering the calcination temperature to 500 °C isolates more *t*-ZrO_2, with the near-pure pattern showing a 2 h calcination time. To obtain the ratio of the tetragonal-to-monoclinic phase, we use Toraya's method, which calculated the volume ratio by X-ray diffraction [22]. Table 1 showcases the yield of *t*-ZrO_2 and *m*-ZrO_2 for each sample obtained from the varying calcination temperatures and times above. The table shows that different calcination temperatures and time variations can gradually transform the ZrO_2 phase.

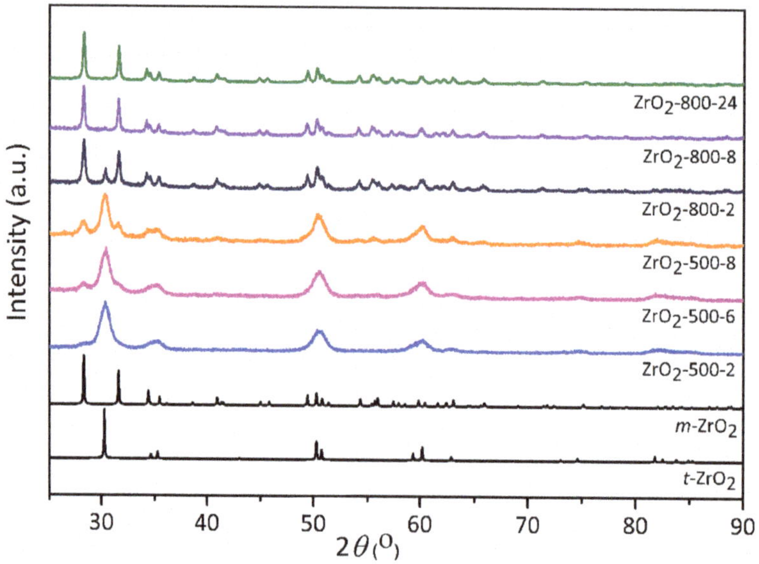

Figure 1. PXRD patterns of ZrO_2 at different calcination temperatures and times.

Table 1. Phase ratio of ZrO_2.

Sample ZrO_2	**% Phase**	
	***t*-ZrO_2**	***m*-ZrO_2**
ZrO_2-800-24	3.90	96.1
ZrO_2-800-8	6.90	93.1
ZrO_2-800-2	15.7	84.3
ZrO_2-500-8	60.0	40.0
ZrO_2-500-6	71.1	28.9
ZrO_2-500-2	84.4	15.6

In general, the morphology of the UiO-66 particle after calcination did not differ from that of pristine UiO-66, yet appeared denser and agglomerated, with a lighter color due to the expected change in chemical composition, giving a solid ZrO_2. It is observed that the ZrO_2 (the calcined UiO-66) size was much smaller than that of the UiO-66, Figure 2, as expected, due to organic linker decompositions. The approximate particle size of ZrO_2 was only about 116 nm, compared to parent UiO-66 420 nm in size.

Figure 2. SEM images of UiO-66 particle after calcination at 500 °C.

2.2. ZrO_2-MO_2 (M = Ti or Hf) Composites

Aside from investigating ZrO_2 phases against MO photocatalytic degradation, we also studied the composite of ZrO_2 with other oxides of the same group (Ti and Hf). HfO_2 has not been reported as a photocatalytic active material due to having a wide band gap; however, mixing HfO_2 with ZrO_2 has mixed results in a favorable photocatalytic band gap. Further, tuning the ZrO_2 band gap can also be completed by adding notorious photoactive TiO_2 by making a ZrO_2-TiO_2 composite. We have successfully synthesized these composites, ZrO_2-MO_2 (M = Ti or Hf), by a simple thermal decomposition of Ti- or Hf-substituted Zr-UiO-66 as a single precursor.

The process of forming the composites starts with partially replacing the Zr(IV) cations with Ti(IV) in a post-synthetic exchange (PSE) reaction of UiO-66 with $TiBr_4$ at 85 °C for 1, 3, and 5 days. The post-synthetic cation exchange process of Ti^{4+} to UiO-66 generates a proportion of Ti^{4+} to Zr^{4+} of around 2:1. This process did not cause the changes, even for the structure in which Hf $^{4+}$ exists. There is no change in either PXRD patterns or morphology of UiO-66(Zr/Ti) compared to those of UiO-66, indicating a successful partial substitution of Zr(IV) by Ti(IV), Figures S2–S4. The Ti(IV) cation amount was 0.04, 0.05, and 2 % atom for 1, 3, and 5 days, respectively, as observed by SEM-EDS.

ZrO_2-MO_2 composites are shown in Figure 3, denoted as (Zr/M)-Xd-A-B, where X represents the reaction time. From that figure, it is known that anatase TiO_2 starts to form under the decomposition process at 500 °C for 2 h, with a sharp anatase TiO_2 peak (JCPDS 21-1272) appearing at 25.4°, with 2θ characteristics to {100} facet, observed for the sample with PSE reaction for 5 d. The amount of TiO_2 increases as we increase reaction times (1, 3, and 5 d) during UiO-66(Zr/Ti), Table 2. The highest t/m-ZrO_2 ratio with the highest percentage of TiO_2 is 1.58, observed in the (Zr/Ti)-5d-500-2 sample. It is noteworthy from our results that anatase TiO_2 can provide stability to the tetragonal structure of ZrO_2 as shown in Figure 4, which is susceptible to forming a monoclinic phase. We observe such a phase of pure t-ZrO_2 formation in the presence of anatase TiO_2 in our (Zr/Ti)-5d-500-2 sample, Table 2. Re-heating the sample further to 800 °C for 24 h did not transform the sample into pure m-ZrO_2, confirming our hypothesis above on the role of TiO_2 in stabilizing the tetragonal phase of ZrO_2.

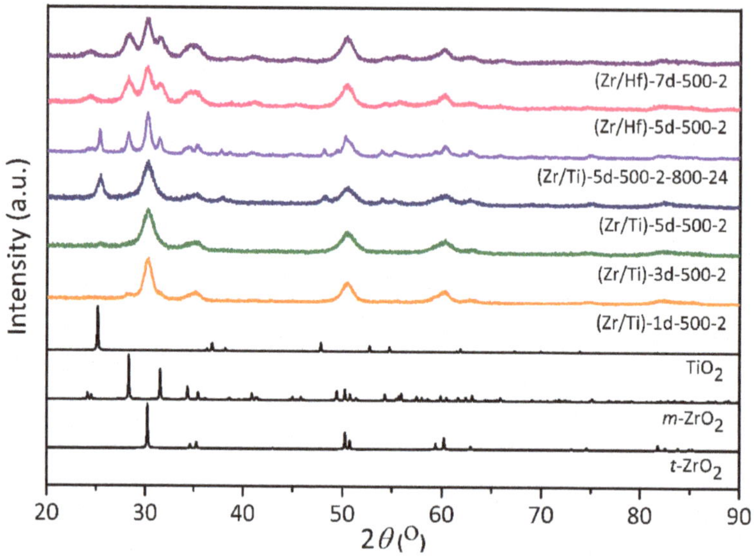

Figure 3. PXRD patterns of ZrO_2-TiO_2 and ZrO_2-HfO_2 composites.

Table 2. The volume ratio of binary oxide catalysts.

ZrO_2-MO_2	% Phase		
	t-ZrO_2	*m*-ZrO_2	TiO_2 Anatase
(Zr/Hf)-7d-500-2	37.9	62.1	-
(Zr/Hf)-5d-500-2	34.6	65.4	-
(Zr/Ti)-5d-500-2-800-24	29.7	54.6	15.7
(Zr/Ti)-5d-500-2	44.2	28.0	27.8
(Zr/Ti)-3d-500-2	81.0	-	19.0
(Zr/Ti)-1d-500-2	78.9	21.1	-

Figure 4. SEM image of ZrO_2-TiO_2 composite for (Zr/Ti)-5d-500-2.

Like Ti(IV) substitutions, Hf(IV) doping does not significantly affect the diffraction pattern of UiO-66, as shown in Figure S5. The Hf(IV) cation exchange lasted for 5 and 7 days. The binary oxide ZrO_2-HfO_2 composites were obtained by calcining UiO-66(Zr/Hf) at 500 °C for 2 h, featuring overlapping patterns between monoclinic and tetragonal phases for both ZrO_2 and HfO_2. The observed broad peaks resulted from mixing two metal oxides (ZrO_2 and HfO_2) with the same phases with similar diffraction patterns, Figure 3. In Table 2, the ratios of *t*/*m*-ZrO_2 are 0.53 and 0.61 in (Zr/Hf)-5d-500-2 and -7d-500-2, respectively.

Shifting peaks in PXRD patterns were observed for different metal dopants, Figure 3, as expected in accordance with the atomic radius of the dopant. With the same PSE reaction time for 5 days, a slight change of the {101} facet from 35.3° to 35.2° was observed for Ti(IV) doping. Hf(IV) doping showed a different character, where there was a left shift from 35.3° to 35.0° for {101} facet. According to Takeuchi and Inoue [23], due to a larger ionic radius, Hf(IV) doping tends to expand the crystallite size diameter of the {101} facet more significantly than Ti(IV) doping, as observed in our result.

2.3. Band Gap Characterization

The estimated band gap values for different catalysts and anatase TiO_2 are shown in Figure 5. The pristine TiO_2 anatase has a band gap of 3.16 eV. The ZrO_2-TiO_2 composite derived from UiO-66 (Zr/Ti) reduces the band gap of ZrO_2 from 4.81–5.01 eV to 3.01–3.14 eV, Figure 5 (purple line). Despite having a more significant concentration of electron density on oxygen atoms than that of ZrO_2, which would be advantageous for photocatalysis purposes because such defect levels would serve as electron-trapping sites for accelerated transfer of photogenerated electrons, the band gap of HfO_2 is generally too wide, 5.68 eV, to be an excellent photocatalyst [24]. Alloying with a lower band gap could lower the composite band gap to become photoactive. However, the introduction of Hf to form the ZrO_2-HfO_2 composite did not change the band gap of ZrO_2 significantly (from 5.68 eV to 5.01 eV). We conclude from the above band gap values that all catalysts are photoactive in the UV region. Therefore, we proceed with their measurements in the photocatalytic degradation of MO. The band absorption of all *t*-ZrO_2-containing compounds is less sharp than those of *m*-ZrO_2-containing compounds. This may indicate the existence of an indirect band gap within all *t*-ZrO_2-containing combinations.

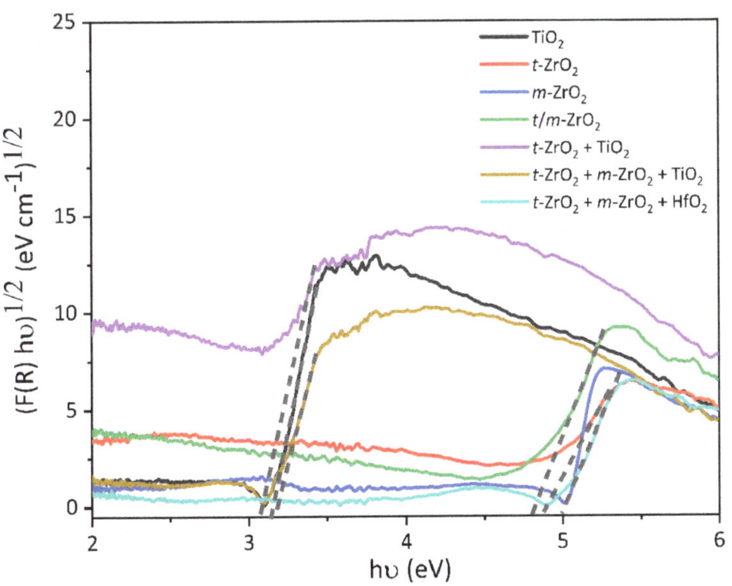

Figure 5. The plot of the Jacobian corrected Kubelka-Munk function for all catalysts.

2.4. *Photocatalysis Activity of ZrO_2, ZrO_2-TiO_2, and ZrO_2-HfO_2*

The performance of different catalysts in the photocatalytic degradation of MO is summarized in Figure 6. MO has two characteristics of the UV-visible absorption spectrum, i.e., the peaks at 464 nm and 272 nm. The first maximum peak of 464 nm indicates the azo groups (-N=N-), while the lower-intensity peak at 272 nm represents aromatic substances from benzene rings, Figure 6a. Typical catalysts are more likely to destroy the azo groups than benzene rings. TiO_2 and *t*-ZrO_2 catalysts can simultaneously break both the azo and aromatic rings [25].

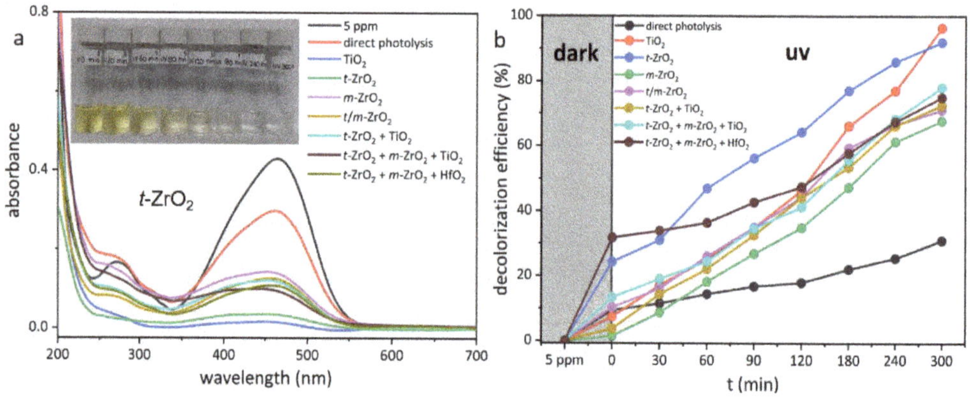

Figure 6. UV-vis spectra (**a**) and decoloration efficiency (**b**) of photocatalytic degradation MO with different catalysts.

The decoloration efficiency observed in Figure 6b is divided into two regions, i.e., before the UV light is on (dark condition) and after the UV light is on. Each catalyst has a different decoloration value during dark conditions due to different catalyst adsorption capacities. The highest adsorption percentage, above 20%, occurred for (*t*-ZrO_2 + *m*-ZrO_2 + HfO_2) and *t*-ZrO_2. This may be due to a larger surface area and pore volume, as observed for *t*-ZrO_2, than for *m*-ZrO_2 and pristine TiO_2 [26], as summarized in Table 3. After the first photocatalytic test, the *t*-ZrO_2 catalyst was re-characterized by PXRD to investigate its phase stability and reusability, Figure S6. The PXRD pattern remained unchanged with no additional peak, which supports the stability of the *t*-ZrO_2 catalyst.

Table 3. Surface area, pore radius, and pore volume of ZrO_2.

Sample	S_{BET} (m^2 g^{-1})	Pore Radius (nm)	Pore Volume (cm^3 g^{-1})
TiO_2	6.40	-	-
t-ZrO_2	57.388	152.45	0.0942
m-ZrO_2	3.656	154.32	0.0623

3. Discussion

The photoactivity of *t*-ZrO_2 in degrading MO is eminently higher than other catalysts, reaching a maximum decoloration efficiency of 89% within 300 min. The concentration of MO decreases rapidly with the reaction rate constant of 0.0063 min^{-1} for *t*-ZrO_2. Other catalysts exhibit moderate MO decoloration efficiency (63–74%), with reaction rate constants ranging from 0.031 to 0.042 min^{-1}, Figures 6 and 7 and Table 4. The photocatalytic degradation reactions of all catalysts follow pseudo-first-order kinetics, as pointed out in Figure 7b.

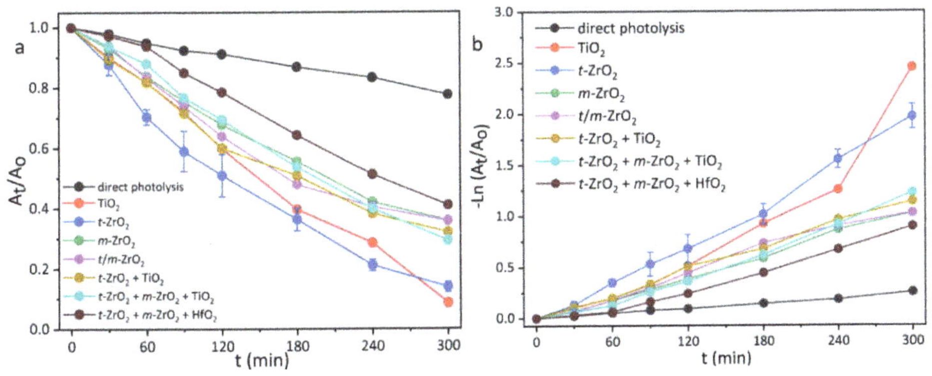

Figure 7. The concentration changes over measuring time (**a**) and the kinetics of photocatalytic degradation MO with different catalysts (**b**).

Table 4. Properties of catalysts and their photocatalytic degradation of MO.

Sample	*t*/*m*-ZrO_2 Ratio	Crystallite Size (nm)	Band Gap (eV)	Photo-degradation (%)	Rate Constant (min^{-1})
TiO_2	-	41.7	3.16	96.14	0.0074
t-ZrO_2	5.42	7.60	4.81	89.42	0.0063
m-ZrO_2	0.04	32.9	5.03	67.23	0.0035
t/*m*-ZrO_2	1.50	20.2	4.83	67.72	0.0037
t-ZrO_2 + TiO_2	1.58	8.40	3.01	71.42	0.0039
t-ZrO_2 + *m*-ZrO_2 + TiO_2	0.54	23.7	3.14	74.62	0.0042
t-ZrO_2 + *m*-ZrO_2 + HfO_2	0.53	6.60	5.01	63.29	0.0031

Table 4 summarizes the properties of each catalyst and their MO photodegradation performance. The catalyst with the largest-to-smallest rate constant is TiO_2 > *t*-ZrO_2 > *t*-ZrO_2 + *m*-ZrO_2 + TiO_2 > *t*-ZrO_2+TiO_2 > *t*/*m*-ZrO_2 > *m*-ZrO_2 > *t*-ZrO_2 + *m*-ZrO_2 + HfO_2. Basahel et al. [11] reported that a photocatalyst with a high surface area, pore volume, and pore size enhances dye adsorption and subsequent photocatalytic activity. This may partly explain the highest MO photocatalytic degradation observed for *t*-ZrO_2. We further observe that within the same *t*/*m* ratio, the substitutions of Ti(IV) to Zr(IV) to form ZrO_2+TiO_2 composites improve, in general, the photodegradation activity of *t*/*m*-ZrO_2. Moderate photocatalytic degradation activity is observed for the ZrO_2-HfO_2 composite, following its relatively wide band gap (5.1 eV).

Sakfali et al. [27], Kumar and Ojha [28], and Yu et al. [29] reported the correlation between calcination temperature, the resulting ZrO_2 phases, and their corresponding midgap states and oxygen vacancies, as well as their photocatalytic properties in the degradation of various compounds under UV irradiation. Such studies stressed the significance of ZrO_2 phases to photodegrade a compound under UV irradiation, similar to our study. They used the photoluminescence (PL) technique to determine the charge carrier separation and recombination efficiency correlated to the photodegradation activity of ZrO_2 phases. More investigations for the PL spectrum analysis will be conducted in the near future to gain a deeper insight into the recombined efficiency contribution of conduction band electrons and valence band holes to the photocatalytic degradation activity of our ZrO_2 phases and ZrO_2-MO_2 composites.

We further compared our highest photocatalytic-active *t*-ZrO_2 with other studies for MO photodegradation and found that our result is on par with other reported *t*-ZrO_2 and other catalysts, Table 5. Because we only completed a simple thermal decomposition of a bulk UiO-66, we believe that we can further enhance the photocatalytic activity by using a variety of nano UiO-66 shapes as sacrificial templates. Such investigation of resulting nano oxides morphologies vs. photocatalytic activity will be conducted in the near future in our lab.

Table 5. Comparison of the present study with other reported photodegradations of methyl orange.

Catalyst	[MO] (mg L^{-1})	Catalyst Dosage (g/L)	UV Source (nm)	Time (min)	Photo-Degradation (%)	Ref.
Cu/ZnO	20	1.0	UV light	240	88	[30]
t-ZrO_2 *m*-ZrO_2 *c*-ZrO_2	10	1.0	350-400	120	92 95 85	[11]
TiO_2-ZrO_2 hollow spheres	50	1.0	313.2	60	8	[31]
VS_4/CP	10	1.0	Xe lamp	30	67	[32]
Cu-BDC UiO-66	20	0.5	UV light	40	3.3 36.4	[33]
TiO_2 nanocellulose	40	1.0	365	30	99.7	[34]
t-ZrO_2	10	0.1	Florescent lamp (20 W)	60	18 (pH 7) 28 (pH 3)	[35]
t-ZrO_2	5	1.0	254	120	53	this work

4. Materials and Methods

Methyl orange solutions 100 ppm were made by mixing 25 mg of powdered methyl orange with demineralized aqua in a 250 mL volumetric flask. The MO solutions of 2, 4, 5, 6, 8, and 10 ppm were diluted from 100 ppm. The 10% acetic acid solution was also made by diluting 10 mL of acetic acid glacial in a 100 mL volumetric flask. Other substances were zirconium(IV) chloride, titanium(IV) bromide, and hafnium(IV) chloride, bought from the Aldrich company. At the same time, terephthalic (1,4-benzene dicarboxylate) acid 98%, titanium(IV) oxide, dimethylformamide, chloroform, acetic acid glacial, and methyl orange were purchased from the Merck company.

Synthesis of ZrO_2 was carried out through UiO-66 as a sacrificial template. The synthesis of UiO-66 was conducted according to Cavka et al. with some modifications [36]. A total of 7 mmol of $ZrCl_4$ was dissolved in 70 mL of DMF inside a Glove Box. Then, 7 mmol of H_2BDC (terephthalic acid) and 0.7 mL of acetic acid were added. The mixture was dispersed by sonication for 45 min, then heated at 120 °C for 24 h under a solvothermal setup. The resulting solids were washed with chloroform. The white powders then dried to produce UiO-66. The *m*-ZrO_2 was obtained by calcination of UiO-66 at 800 °C for 24 h, labeled as UiO-66-800-24. At the same time, *t*-ZrO_2 was obtained by calcination at 500 °C for 2 h, named UiO-66-500-2. The mixed phases were obtained within the time variations above.

The binary oxide was obtained from UiO-66, doped by Ti(IV) and Hf(IV). UiO-66(Zr/Ti) was synthesized by dissolving 0.3 mmol of $TiBr_4$ salts with 6 mL of DMF in a Glove Box. The dispersion process was assisted through sonication, and UiO-66 was added to the solution and then heated at 85 °C for 1, 3, and 5 days. The mixture was centrifuged and washed with 3 × 10 mL of DMF and 10 mL of methanol. The solids were left to immerse in methanol for 3 days and exchanged with fresh methanol [37]. The solids were then dried under a vacuum at 50 °C for 2 h. The UiO-66(Zr/Hf) was produced like UiO-66(Zr/Ti) but using $HfCl_4$ and different reaction times of 5 and 7 days. The powder of UiO-66(Zr/Ti) and UiO-66(Zr/Hf) was then calcined at the optimum temperature to obtain photocatalyst ZrO_2-TiO_2 and ZrO_2-HfO_2 composites. Labeling examples of UiO-66 precursor to making ZrO_2-MO_2 (M = Ti or Hf) composites: (Zr/Ti)-1d-500-2 indicates UiO-66 underwent PSE with $TiBr_4$ for 1 d, followed by calcination at 500 °C for 2 h; (Zr/Hf)-5d-500-2 indicates UiO-66 underwent PSE with $HfCl_4$ for 5 d, followed by calcination at 500 °C for 2 h.

The crystalline phase identification of as-synthesized materials of UiO-66 (Zr), UiO-66 (Zr/Ti), UiO-66 (Zr/Hf), ZrO_2, ZrO_2-TiO_2, and ZrO_2-HfO_2 was carried out using powder X-ray diffractometer, a Rigaku Miniflex 600 instrument with Cu-K_α radiation (λ = 0.1540593 nm). The morphological analysis of the product was observed using Scanning Electron Microscopy (SEM) measurement SU3500 JEOL/EO version 1.0, and the element composition analysis was observed using EDS mapping. Band gap measurements of photocatalyst materials were completed using a Thermo Scientific Evolution 220 UV-vis spectrophotometer. The nitrogen sorption method performed surface area and pore size measurements using a Quantachrome Instruments Nova 3200e. The multipoint Brunauer–Emmett–Teller (BET) method was employed to analyze the surface area of samples.

The diffraction pattern of different samples was measured using Toraya's method [22]. The integrated intensity ratio X_m can be expressed by Equation (1) as

$$X_m = \frac{I_m(\bar{1}11) + I_m(111)}{I_m(\bar{1}11) + I_m(111) + I_t(101)} \quad (1)$$

where I_m and I_t represent the intensity of monoclinic and tetragonal phases on each plane. Intensity was measured using OriginLab software. The volume fraction, V_m, can be obtained by Equation (2) [38]. The volume fraction of TiO_2 and HfO_2 was calculated on the plane with the highest diffraction pattern.

$$V_m = \frac{1.311X_m}{1 + 0.311X_m} \quad (2)$$

Each photocatalyst was dissolved in 90 mL methyl orange solution using ultrasonication for 10 min with a 1:1 (m/v) ratio and then stirred for 30 min before the UV light was turned on. This procedure was according to Zhang et al. [31]. The degradation experiment was recorded with a UV lamp (λ_{max} = 254 nm). At 0, 30, 60, 90, 120, 180, 240, and 300 min, the solution was taken out, centrifuged, and analyzed further using a Thermo Scientific Genesys 10S UV-Vis spectrophotometer to see how the photocatalyst performed either on UV or visible light. The measured samples are TiO_2 pristine, *t*-ZrO_2 (UiO-66-500-2), *m*-ZrO_2 (UiO-66-800-24), *t*/*m*-ZrO_2 (UiO-66-500-8), the binary oxide *t*-ZrO_2+TiO_2 ((Zr/Ti)-5d-500-2), and the mixing phase of binary oxide from *t*-ZrO_2+*m*-ZrO_2+TiO_2 ((Zr/Ti)-5d-500-2-800-24) and *t*-ZrO_2+*m*-ZrO_2+HfO_2 ((Zr/Hf)-5d-500-2).

5. Conclusions

A simple synthetic route for making a ZrO_2 catalyst comprising nearly phase-pure, tetragonal, monoclinic, mixed *t*/*m*-ZrO_2, and ZrO_2-MO_2 (M = Ti or Hf) composites has been effectively prepared by using only a single precursor, namely UiO-66, as a sacrificial template. All the synthesized catalysts are photoactive against MO dye degradation, with *t*-ZrO_2 showcasing the highest photodegradation activity, with 89% decoloration efficiency under UV irradiation for 300 min with a reaction rate constant of 0.0063 min^{-1}. Decoloration efficiencies for the *m*-ZrO_2 phase and ZrO_2-MO_2 (M = Ti or Hf) are 67%, 74%, and 63%, respectively. This result places this catalyst among the best photocatalysts for photocatalytic degradation activity.

Supplementary Materials: The following supporting information can be downloaded at: https://www.mdpi.com/article/10.3390/catal12060609/s1, Figure S1: PXRD pattern (a) and SEM images (b) of UiO-66; Figure S2: PXRD patterns of UiO-66(Zr/Ti) 1, 3, and 5d; Figure S3: SEM images of UiO-66(Zr/Ti) 1, 3, and 5d; Figure S4: Elemental mapping and EDS spectrum of UiO-66(Zr/Ti) 1, 3, and 5d; Figure S5: PXRD patterns of UiO-66 (Zr/Hf) 5 and 7d; Figure S6: PXRD patterns of t-ZrO2 before and after successive catalytic cycle; Table S1: SEM-EDS analysis of % atomic of UiO-66(Zr/Ti).

Author Contributions: Writing—original draft preparation, formal analysis, investigation, visualization, validation, I.N.A.J. and H.F.S.; supervision, project administration, S.M. and K.W.; conceptualization, methodology, resources, supervision, writing—review and editing, funding acquisition, project administration, A.P. and A.C.W. All authors have read and agreed to the published version of the manuscript.

Funding: This research was funded by Institut Teknologi Bandung and Universitas Airlangga, grant number 165/UN3.15/PT/2021.

Acknowledgments: The authors acknowledge Riset Kolaborasi Indonesia 2021 (Institut Teknologi Bandung, Universitas Airlangga, Institut Pertanian Bogor, and Universitas Gadjah Mada) for supporting this research. A.P. acknowledges P2MI FMIPA ITB FY 2022 for financial support. The authors are grateful to Yessi Permana and Laela Mukaromah for PXRD measurements.

Conflicts of Interest: The authors declare no conflict of interest.

References

1. Kaddouri, A.; Mazzocchia, C.; Tempesti, E.; Anouchinsky, R. On the Activity of ZrO_2 Prepared by Different Methods. *J. Therm. Anal.* **1998**, *53*, 97–109. [CrossRef]
2. Gannavarapu, K.P.; Ganesh, V.; Thakkar, M.; Mitra, S.; Dandamudi, R.B. Nanostructured Diatom-ZrO_2 Composite as a Selective and Highly Sensitive Enzyme Free Electrochemical Sensor for Detection of Methyl Parathion. *Sens. Actuators B Chem.* **2019**, *288*, 611–617. [CrossRef] [PubMed]
3. Yu, J.; Ji, G.; Shi, Z.; Wang, X. Corrosion Resistance of ZrO_2 Films under Different Humidity Coal Gas Conditions at High Temperature. *J. Alloys Compd.* **2019**, *783*, 371–378. [CrossRef]
4. Cha, S.W.; Cho, G.Y.; Lee, Y.; Park, T.; Kim, Y.; Lee, J. Effects of Carbon Contaminations on Y_2O_3-Stabilized ZrO_2 Thin Film Electrolyte Prepared by Atomic Layer Deposition for Thin Film Solid Oxide Fuel Cells. *CIRP Ann. Manuf. Technol.* **2016**, *65*, 515–518. [CrossRef]
5. Campa, M.C.; Ferraris, G.; Gazzoli, D.; Pettiti, I.; Pietrogiacomi, D. Rhodium Supported on Tetragonal or Monoclinic ZrO_2 as Catalyst for the Partial Oxidation of Methane. *Appl. Catal. B Environ.* **2013**, *142*, 423–431. [CrossRef]
6. Navío, J.A.; Hidalgo, M.C.; Colón, G.; Botta, S.G.; Litter, M.I. Preparation and Physicochemical Properties of ZrO_2 and Fe/ZrO_2 Prepared by a Sol-Gel Technique. *Langmuir* **2001**, *17*, 202–210. [CrossRef]
7. Botta, S.G.; Navío, J.A.; Hidalgo, M.C.; Restrepo, G.M.; Litter, M.I. Photocatalytic Properties of ZrO_2 and Fe/ZrO_2 Semiconductors Prepared by a Sol-Gel Technique. *J. Photochem. Photobiol. A Chem.* **1999**, *129*, 89–99. [CrossRef]
8. Abdi, J.; Yahyanezhad, M.; Sakhaie, S.; Vossoughi, M.; Alemzadeh, I. Synthesis of Porous TiO_2/ZrO_2 Photocatalyst Derived from Zirconium Metal Organic Framework for Degradation of Organic Pollutants under Visible Light Irradiation. *J. Environ. Chem. Eng.* **2019**, *7*, 103096. [CrossRef]
9. Butova, V.V.; Soldatov, M.A.; Guda, A.A.; Lomachenko, K.A.; Lamberti, C. Metal-organic Frameworks: Structure, properties, methods of synthesis and characterization. *Russ. Chem. Rev.* **2016**, *85*, 280–307. [CrossRef]
10. Zhang, X.; Zhang, M.; Zhang, J.; Zhang, Q.; Tsubaki, N.; Tan, Y.; Han, Y. Methane Decomposition and Carbon Deposition over Ni/ZrO_2 Catalysts: Comparison of Amorphous, Tetragonal, and Monoclinic Zirconia Phase. *Int. J. Hydrogen Energy* **2019**, *44*, 17887–17899. [CrossRef]
11. Basahel, S.N.; Ali, T.T.; Mokhtar, M.; Narasimharao, K. Influence of Crystal Structure of Nanosized ZrO_2 on Photocatalytic Degradation of Methyl Orange. *Nanoscale Res. Lett.* **2015**, *10*, 1–13. [CrossRef] [PubMed]
12. Reddy, C.V.; Babu, B.; Reddy, I.N.; Shim, J. Synthesis and Characterization of Pure Tetragonal ZrO_2 Nanoparticles with Enhanced Photocatalytic Activity. *Ceram. Int.* **2018**, *44*, 6940–6948. [CrossRef]
13. Mansouri, M.; Mozafari, N.; Bayati, B.; Setareshenas, N. Photo-Catalytic Dye Degradation of Methyl Orange Using Zirconia–Zeolite Nanoparticles. *Bull. Mater. Sci.* **2019**, *42*, 230. [CrossRef]
14. Yan, X.; Lu, N.; Fan, B.; Bao, J.; Pan, D.; Wang, M.; Li, R. Synthesis of Mesoporous and Tetragonal Zirconia with Inherited Morphology from Metal–Organic Frameworks. *Cryst. Eng. Comm.* **2015**, *17*, 6426–6433. [CrossRef]
15. Bavykina, A.; Kolobov, N.; Khan, I.S.; Bau, J.A.; Ramirez, A.; Gascon, J. Metal–Organic Frameworks in Heterogeneous Catalysis: Recent Progress, New Trends, and Future Perspectives. *Chem. Rev.* **2020**, *120*, 8468–8535. [CrossRef] [PubMed]
16. Butova, V.V.; Budnyk, A.P.; Charykov, K.M.; Vetlitsyna-Novikova, K.S.; Lamberti, C.; Soldatov, A.V. Water as a Structure-Driving Agent between the UiO-66 and MIL-140A Metal–Organic Frameworks. *Chem. Commun.* **2019**, *55*, 901–904. [CrossRef]
17. Kambur, A.; Pozan, G.S.; Boz, I. Preparation, Characterization and Photocatalytic Activity of TiO_2-ZrO_2 Binary Oxide Nanoparticles. *Appl. Catal. B Environ.* **2012**, *115–116*, 149–158. [CrossRef]
18. Gota, K.R.; Suresh, S. Preparation and Its Application of TiO_2/ZrO_2 and TiO_2/Fe Photocatalysts: A Perspective Study. *Asian J. Chem.* **2014**, *26*, 7087–7101. [CrossRef]
19. Greer, H.F.; Liu, Y.; Greenaway, A.; Wright, P.A.; Zhou, W. Synthesis and Formation Mechanism of Textured MOF-5. *Cryst. Growth Des.* **2016**, *16*, 2104–2111. [CrossRef]
20. Katz, M.J.; Brown, Z.J.; Colón, Y.J.; Siu, P.W.; Scheidt, K.A.; Snurr, R.Q.; Hupp, J.T.; Farha, O.K. A Facile Synthesis of UiO-66, UiO-67 and Their Derivatives. *Chem. Commun.* **2013**, *49*, 9449–9451. [CrossRef]
21. Li, W.; Wang, K.; Huang, J.; Liu, X.; Fu, D.; Huang, J.; Li, Q.; Zhan, G. M x O y –ZrO_2 (M = Zn, Co, Cu) Solid Solutions Derived from Schiff Base-Bridged UiO-66 Composites as High-Performance Catalysts for CO_2 Hydrogenation. *ACS Appl. Mater. Interfaces* **2019**, *11*, 33263–33272. [CrossRef] [PubMed]
22. Toraya, H.; Yoshimura, M.; Somiya, S. Calibration Curve for Quantitative Analysis of the Monoclinic-Tetragonal ZrO_2 System by X-Ray Diffraction. *J. Am. Ceram. Soc.* **1984**, *67*, C119–C121. [CrossRef]

23. Takeuchi, A.; Inoue, A. Classification of Bulk Metallic Glasses by Atomic Size Difference, Heat of Mixing and Period of Constituent Elements and Its Application to Characterization of the Main Alloying Element. *Mater. Trans.* **2005**, *46*, 2817–2829. [CrossRef]
24. Balog, M.; Schieber, M.; Michman, M.; Patai, S. The Chemical Vapour Deposition and Characterization of ZrO_2 Films from Organometallic Compounds. *Thin Solid Film* **1977**, *47*, 109–120. [CrossRef]
25. Yang, S.; Jin, R.; He, Z.; Qiao, Y.; Shi, S.; Kong, W.; Wang, Y.; Liu, X. An Experimental Study on the Degradation of Methyl Orange by Combining Hydrodynamic Cavitation and Chlorine Dioxide Treatments. *Chem. Eng. Trans.* **2017**, *59*, 289–294. [CrossRef]
26. Nainani, R.; Thakur, P.; Chaskar, M. Synthesis of Silver Doped TiO_2 Nanoparticles for the Improved Photocatalytic Degradation of Methyl Orange. *J. Mater. Sci. Eng. B* **2012**, *2*, 52–58.
27. Sakfali, J.; Ben Chaabene, S.; Akkari, R.; Dappozze, F.; Berhault, G.; Guillard, C.; Saïd Zina, M. High Photocatalytic Activity of Aerogel Tetragonal and Monoclinic ZrO_2 Samples. *J. Photochem. Photobiol. A Chem.* **2022**, *430*, 113970. [CrossRef]
28. Kumar, S.; Ojha, A.K. Oxygen Vacancy Induced Photoluminescence Properties and Enhanced Photocatalytic Activity of Ferromagnetic ZrO_2 Nanostructures on Methylene Blue Dye under Ultra-Violet Radiation. *J. Alloys Compd.* **2015**, *644*, 654–662. [CrossRef]
29. Yu, D.; Li, L.; Wu, M.; Crittenden, J.C. Enhanced Photocatalytic Ozonation of Organic Pollutants Using an Iron-Based Metal-Organic Framework. *Appl. Catal. B Environ.* **2019**, *251*, 66–75. [CrossRef]
30. Fu, M.; Li, Y.; Wu, S.; Lu, P.; Liu, J.; Dong, F. Sol-Gel Preparation and Enhanced Photocatalytic Performance of Cu-Doped ZnO Nanoparticles. *Appl. Surf. Sci.* **2011**, *258*, 1587–1591. [CrossRef]
31. Zhang, J.; Li, L.; Xiao, Z.; Liu, D.; Wang, S.; Zhang, J.; Hao, Y.; Zhang, W. Hollow Sphere TiO_2-ZrO_2 Prepared by Self-Assembly with Polystyrene Colloidal Template for Both Photocatalytic Degradation and H_2 Evolution from Water Splitting. *ACS Sustain. Chem. Eng.* **2016**, *4*, 2037–2046. [CrossRef]
32. Cai, R.; Zhang, B.; Shi, J.; Li, M.; He, Z. Rapid Photocatalytic Decolorization of Methyl Orange under Visible Light Using VS4/Carbon Powder Nanocomposites. *ACS Sustain. Chem. Eng.* **2017**, *5*, 7690–7699. [CrossRef]
33. Haryanto, A.; Mukaromah, L.; Permana, Y.; Patah, A. Photocatalytic Activity of CuBDC and UiO-66 MOFs for Methyl Orange Degradation. *J. Chem. Technol. Metall.* **2021**, *56*, 791–795.
34. Liu, G.; Pan, X.; Li, J.; Li, C.; Ji, C. Facile Preparation and Characterization of Anatase TiO_2/Nanocellulose Composite for Photocatalytic Degradation of Methyl Orange. *J. Saudi Chem. Soc.* **2021**, *25*, 101383. [CrossRef]
35. Farhan Hanafi, M.; Sapawe, N. Remarkable Degradation of Methyl Orange by Tetragonal Zirconia Catalyst. *Mater. Today Proc.* **2018**, *5*, 21849–21852. [CrossRef]
36. Cavka, J.H.; Jakobsen, S.; Olsbye, U.; Guillou, N.; Lamberti, C.; Bordiga, S.; Lillerud, K.P. A New Zirconium Inorganic Building Brick Forming Metal Organic Frameworks with Exceptional Stability. *J. Am. Chem. Soc.* **2008**, *130*, 13850–13851. [CrossRef] [PubMed]
37. Kim, M.; Cahill, J.F.; Fei, H.; Prather, K.A.; Cohen, S.M. Postsynthetic Ligand and Cation Exchange in Robust Metal–Organic Frameworks. *J. Am. Chem. Soc.* **2012**, *134*, 18082–18088. [CrossRef]
38. Aboulayt, A.; Onfroy, T.; Travert, A.; Clet, G.; Maugé, F. Relationship between Phosphate Structure and Acid-Base Properties of Phosphate-Modified Zirconia—Application to Alcohol Dehydration. *Appl. Catal. A Gen.* **2017**, *530*, 193–202. [CrossRef]

catalysts

MDPI

Article

Abatement of Naphthalene by Persulfate Activated by Goethite and Visible LED Light at Neutral pH: Effect of Common Ions and Organic Matter

Andrés Sánchez-Yepes, David Lorenzo, Patricia Sáez, Arturo Romero and Aurora Santos *

Chemical Engineering and Materials Department, Complutense University of Madrid, 28040 Madrid, Spain; andsan28@ucm.es (A.S.-Y.); dlorenzo@ucm.es (D.L.); patrisae@ucm.es (P.S.); aromeros@ucm.es (A.R.)
* Correspondence: aursan@ucm.es

Abstract: Naphthalene (NAP) has received particular attention due to its impact on the environment and human health, mandating its removal from water systems. In this work, the abatement of NAP in the aqueous phase was achieved using persulfate (PS) activated by Fe (III) and monochromatic LED light at a natural pH. The reaction was carried out in a slurry batch reactor using goethite as the Fe (III) source. The influence of the PS concentration, goethite concentration, irradiance, temperature and presence of organic matter, chloride, and bicarbonate on the abatement of NAP was studied. These variables were shown to have a different effect on NAP removal. The irradiance showed a maximum at 0.18 W·cm^{-2} where the photonic efficiency was the highest. As for the concentration of goethite and PS, the influence of the first one was negligible, whereas for PS, the best results were reached at 1.2 mM due to a self-inhibitory effect at higher concentrations. The temperature effect was also negative in the PS consumption. Regarding the effect of ions, chloride had no influence on NAP conversion but carbonates and humic acids were affected. Lastly, this treatment to remove NAP has proved to be an effective technique since minimum conversions of 0.92 at 180 min of reaction time were reached. Additionally, the toxicity of the final samples was decreased.

Keywords: goethite; LED light; naphthalene; persulfate; PAHs

1. Introduction

Rapid industrialisation and urbanisation have led to anthropogenic activities that discharge various environmental pollutants such as polycyclic aromatic hydrocarbons (PAHs) [1]. PAHs are composed of two or more fused aromatic benzene rings and due to their inherent properties, they are persistent pollutants having a wide range of biological toxicity. So, although there are more than 100 different types of PAHs, the United States Environmental Protection Agency (USEPA) has categorised 16 PAHs on the priority list due to their adverse impact on the environment and human health [2,3]. The list includes naphthalene, acenaphthylene, acenaphthene, fluorene, phenanthrene, anthracene, fluoranthene, pyrene, chrysene, benzo (a) anthracene, benzo (b) fluoranthene, benzo (k) fluoranthene, benzo (a) pyrene, indeno123-cd perylene, dibenzo (ah) anthracene, and benzo (ghi) perylene. Among these, naphthalene (NAP) is one of the most common PAHs in the environment. NAP is mainly found in moth repellents, deodorants, and cigarettes. Additionally, it is widely used in the chemical industry as a raw material for the production of dyes and pesticides, among others [4]. Unfortunately, it is reported to be a carcinogenic and mutagenic compound that has been detected in rivers, industrial effluents, and sewage wastewaters, among others [1,5,6]. It is hydrophobic, flammable, highly volatile, and chemically stable in water media and its removal from water is necessary [4]. Several technologies have been proposed for its removal, from physical technologies such as adsorption [4,7] to chemical technologies such as advanced oxidation processes (AOPs) [8] using H_2O_2 and sodium persulfate common oxidants. These oxidants must be activated to produce radical species able to degrade rapidly persistent pollutants to increase their oxidising potential. The degradation of NAP has been studied in the literature using H_2O_2 and PS activated by iron through different methods, for example, using Fe (II) (Fenton reaction) [9,10], ultraviolet and visible light (photo-Fenton) [11], power (electro Fenton) [12], and persulfate activated by iron [13–15] or by using both oxidants [16]. Despite the removal of NAP reported in those works often reaching 90%, there are several drawbacks, such as the need to operate at an acidic pH and the generation of iron in the solution, which must be removed in additional treatments (for example, neutralisation, separation, and management of the iron hydroxide sludge generated). These drawbacks can be overcome if a stable solid catalyst is used. This catalyst has to be cheap, environmentally friendly, easily separable from the solution, and with depictable leaching of the active phase (iron) [17]. Naturally occurring iron minerals, whose main characteristics are abundance and affordability, have been studied in the heterogenous process, including hematite, pyrite, goethite, and magnetite, which have been successfully applied in AOPs [18]. Heterogeneous Fenton has been applied in NAP removal with success without light [14] using siderite as the catalyst and with visible light [11]. However, the visible light used was source Hg or Xe lamps, which need complicated cooling and are toxic. For this reason, commercial visible light-emitting diode lamps (VIS LED) have been recently proposed as an innovative light source for the removal of organic pollutants from water at a neutral pH using natural iron minerals and hydrogen peroxide as the oxidant [19]. However, the efficiency of H_2O_2 decreased in the presence of high carbonate contents in water due to non-productive reactions [20–23], and the use of PS is recently gaining attention. PS shows high aqueous solubility, high stability, is relatively inexpensive, and generates benign end products [24–27]. It can be activated with a transition metal complex, using light or heat, adding hydroxide peroxide, or using an alkali (pH > 10) [23,28–30] producing sulfate radicals. Sulfate radicals are strong oxidants, less selective, and more stable on a broader pH interval than hydroxyl radicals.

This work aims to remove NAP in the aqueous phase using PS activated by VIS LED as a light source and goethite as a heterogeneous catalyst. To our knowledge, this system has not been previously studied. The influence of different variables has been addressed, such as PS and catalyst concentrations, the irradiance of the light source used, and the temperature and presence of other ions and natural organic matter in the solution. In addition, the toxicity of the water obtained after the treatment has been studied.

2. Results and Discussion

The conversion of NAP and PS obtained in runs summarised in Table 1 has been calculated with the reaction time by Equation (1) and Equation (2), respectively.

$$X_{NAP} = 1 - \frac{C_{NAP}}{C_{NAP,0}} \tag{1}$$

$$X_{PS} = 1 - \frac{C_{PS}}{C_{PS,0}} \tag{2}$$

X_{NAP} *and* X_{PS} are the NAP and PS conversions, respectively; C_{NAP} *and* C_{PS} are the NAP and PS concentrations at a reaction time t, respectively; and $C_{NAP,0}$ *and* $C_{PS,0}$ are the naphthalene and persulfate initial concentrations, respectively.

Blank runs B1, B2, and B3 in Table 1 give NAP conversion lower than 10% at 180 min.

The complete profile of NAP and PS concentrations with time has been studied in R1 (Table 1) using $C_{NAP,0} = 0.1$ mM, $C_{PS,0} = 1.2$ mM, $I = 0.18\ W{\cdot}cm^{-2}$, and $C_{GOE} = 0.1\ g{\cdot}L^{-1}$ and 25 °C. The reaction was carried out for a total time of 360 min. The measured NAP and PS conversions with reaction times in R1 are shown in Figure S2 of the Supplementary Materials. As shown in Figure S2, the conversion of NAP was 0.79 in the first 60 min. After this time, depletion of this conversion was observed, reaching a conversion of 0.96 at 180 min. After this time, the conversion was kept almost constant until 360 min. The same trend was observed in the PS conversion.

The PS consumption was 0.086 in the first hour, then increased to 0.19 at 180 min and remained constant during the next 180 min. For these reasons, runs R2 to R9 and S1 to S3 were maintained for 180 min. In these runs, samples were taken at 30 min and 180 min to study the effects of the variables on NAP oxidation.

Table 1. Experimental conditions of experiments. Initial NAP concentration 0.1 mM.

Experiment	T (°C)	C_{PS} (mM)	C_{GOE} ($g{\cdot}L^{-1}$)	*I* ($W{\cdot}cm^{-2}$)	C_{Cl^-} (ppm)	C_{HCO3^-} (ppm)	$C_{H.Ac}$ (ppm)	t (min)
B1	25	1.2	0	0.24	0	0	0	180
B2	25	0	0.1	0.24	0	0	0	180
B3	25	1.2	0.1	0	0	0	0	180
R1	25	1.2	0.1	0.18	0	0	0	360
R2	25	1.2	0.1	0.1	0	0	0	180
R3	25	1.2	0.1	0.18	0	0	0	180
R4	25	1.2	0.1	0.24	0	0	0	180
R5	25	1.2	0.2	0.18	0	0	0	180
R6	25	0.6	0.1	0.18	0	0	0	180
R7	25	2.4	0.1	0.18	0	0	0	180
R8	25	4.8	0.1	0.18	0	0	0	180
R9	40	1.2	0.1	0.18	0	0	0	180
S1	25	1.2	0.1	0.18	500	0	0	180
S2	25	1.2	0.1	0.18	0	600	0	180
S3	25	1.2	0.1	0.18	0	0	50	180

2.1. Effect of Irradiance

The influence of the absolute irradiance in NAP reduction was evaluated by varying the nominal power of the LED light source irradiated over the reactor window.

Figure 1a,b show the NAP and PS conversions, respectively, obtained at 30 and 180 min in runs B3, R2, R3, and R4, carried out at several irradiances (within the range 0 to 0.24 $W{\cdot}cm^{-2}$). As can be seen, run B3 was carried out in the absence of LED light, and the NAP conversion was close to 0.10 at 180 min. This conversion could be explained by the loss of NAP from the medium by evaporation or adsorption [13]. At the same time, the production of sulfate radicals was not observed since no PS conversion was experimentally obtained in the absence of LED light. On the contrary, the LED light enhances the reduction of Fe (III) to Fe (II), promoting PS activation and sulfate radical production, as indicated in Equations (3) and (4)

$$Fe(III) - goethite \xrightarrow{light} Fe(II) - goethite \quad (3)$$

$$Fe(II) - goethite + S_2O_8^{2-} \rightarrow Fe(III) - goethite + SO_4^{-2} + SO_4^{-}\cdot \quad (4)$$

When the GOE surface is irradiated (runs R2, R3, and R4) with the monochromatic LED, sulfate radicals are produced and oxidise NAP. As shown in Figure 1, the higher the irradiance, the higher the NAP and PS conversions after 30 min of reaction time. However, the difference between the NAP conversion values using irradiance of 0.18 and 0.24 $W{\cdot}cm^{-2}$ was lower than 5%. This fact has been explained in the literature, considering that the number of photons that can trigger more electron holes on the GOE surface increases with the lamp irradiance until a specific irradiance value. Above that value, the effect of lamp irradiance on the pollutant conversion is insignificant [19,31]. Under experimental conditions in R2, R3, and R4, an almost total conversion of NAP was achieved at 180 min.

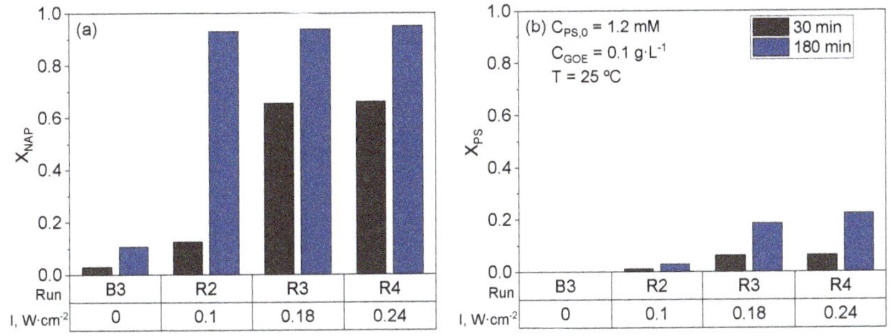

Figure 1. Effect of irradiance in (**a**) conversion of NAP; (**b**) conversion of PS at 30 min and 180 min. $C_{NAP,0} = 0.1$ mM, $C_{PS,0} = 1.2$ mM, $C_{GOE} = 0.1$ g·L^{-1} and 25 °C.

In Table S1, the initial and final pHs measured at the end of the reactions (180 min) are summarised. As shown in Table S1, final pHs are close to 3, except for B3 which has a final pH of 5.44. The acidification of the reaction media can be attributed to PS decomposition and the formation of acid intermediates during naphthalene oxidation.

The photodegradation of NAP under different operating conditions of irradiance was analysed using the photonic efficiency η_p $\left(\text{mol}_{\text{NAP}}\cdot\text{Einstein}^{-1}\right)$ defined in Equation (5). This photonic efficiency is the ratio between the number of moles of pollutant converted and the number of photons reaching the reactor window over a defined range of wavelengths (350–600 nm).

$$\eta_p = \frac{C_{NAP,0}\cdot X_{NAP}\cdot V_R}{F_E\cdot\left(t_f - t_0\right)} \tag{5}$$

where $C_{NAP,0}$ is the initial concentration of NAP (mol·L^{-1}) and X_{NAP} is the conversion of NAP achieved at the time t_f (30 min of reaction). V_R refers to the reactor volume (0.1 L) and F_E is the flow of photons (Einstein s^{-1}) irradiated over the reactor window.

The photon flow that reached the reactor window can be estimated by applying Equation (6) measuring the absolute irradiance (I_a) in three different positions of the reactor window.

$$F_E = \int_0^A \int_{\lambda_0}^{\lambda_f} \frac{I_a}{h\cdot\frac{c}{\lambda}}\cdot d\lambda\cdot dA \tag{6}$$

I_a is the average of absolute irradiance measured with the spectrometer coupled with an optical fibre and a cosine corrector in W·cm^{-2}·nm^{-1}; λ is the wavelength, and h and c are the Plank constant at light speed, respectively. A is the total evaluated area of the reactor window (cm^2).

The photon flow was calculated from the average values of absolute irradiance in Figure S3 and integrating Equation (6). In Table 2, the photonic efficiencies are summarised. These values were calculated using the experimental results of the NAP conversion after 30 min at different irradiance values (Figure 1).

Table 2. Photonic efficiencies at different ca irradiances, calculated from the experimental values of NAP conversion in Figure 1 and the photons flow irradiated over the reactor window.

I (W·cm^{-2})	0.1	0.18	0.24
η_p $\left(\text{mol}_{\text{NAP}}\cdot\text{Einstein}^{-1}\right)$	1.82×10^{-4}	5.36×10^{-4}	4.03×10^{-4}

As can be seen in Table 2, the η_p presented a maximum value when the reactor was irradiated with $I = 0.18$ W·cm^{-2}. This value inferred the use of a higher irradiance value performing an unnecessary waste of photon flow that was not used to activate the PS in the NAP abatement. On the other hand, irradiance values lower than 0.18 W·cm^{-2} were not considered since the photonic efficiency decreased, meaning lower NAP oxidation.

2.2. Effect of Goethite Concentration

The decrease of NAP was studied in the absence of the catalyst and using 0.1 and 0.2 g·L^{-1} of GOE with 1.2 mM of PS and 0.18 W·cm^{-2} of lamp irradiance. The conversion of NAP obtained and the PS consumption are plotted in Figure 2. As can be seen, significant NAP conversion is only noticed when the catalyst is added to the reaction medium. The wavelength used (470 nm) was selected attending to the optical properties of GOE [19], promoting the reduction of Fe (III) in the GOE surface to Fe (II) (Equation (3)), and yielding sulfate radicals (Equation (4)) that caused the NAP oxidation.

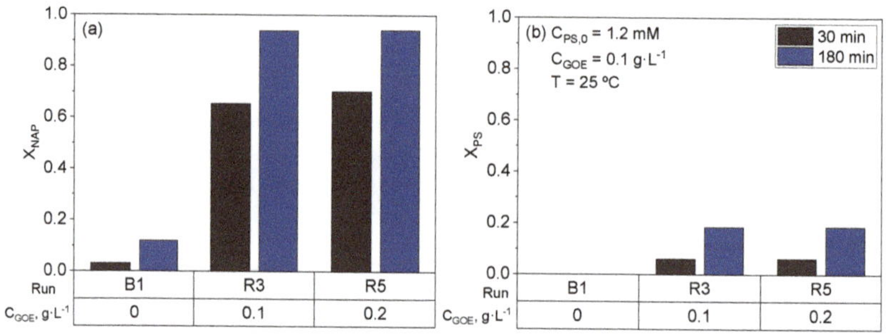

Figure 2. Effect of goethite concentration in (**a**) conversion of NAP; (**b**) conversion of PS at 30 min and 180 min. $C_{NAP,0} = 0.1$ mM, $C_{PS,0} = 1.2$ mM, $I = 0.18$ W·cm^{-2} and 25 °C.

As shown in Figure 2a, negligible differences in the conversion of NAP and PS were found in the GOE concentration range from 0.10 to 0.20 g·L^{-1} at 30 min of reaction time and using the same irradiance and initial oxidant concentration. This low difference of conversions observed in runs R3 and R5 was in agreement with that observed in a previous work where the increase in the catalyst concentration did not increase the reaction rate [32]. The minor differences observed using a twofold catalyst concentration can be attributed to the reaction taking place on the catalyst surface. In a well-mixed slurry reactor, the solid can form agglomerates reducing the active surface area of the catalyst. The latter affected the conversions of NAP and PS in two ways, promoting (i) a defective liquid–solid contact, meaning a lower contact between PS and the active sites and (ii) the scattering-out of the light, yielding a decrease in the irradiated area of the catalyst reducing the reaction of Fe (III) to Fe (II) [32].

The iron leaching from the solid catalyst to the aqueous phase was studied, measuring the total iron in the solution at 180 min of reaction time by atomic emission spectroscopy. The iron concentration in all runs was lower than 0.03 mg·L^{-1}. The low content of Fe in the aqueous phase proves that the reaction occurs only on the solid catalyst surface.

Regarding the pH, except for B1 ($pH_f = 5.01$), the final pH values reached were 3.20 (Table S1), with a similar conversion of PS at 180 min in R3 and R5

2.3. Effect of Temperature

The decrease of NAP was investigated under two different temperatures 25 °C and 40 °C. The NAP and PS conversions obtained at the same initial concentrations of NAP, PS, and irradiance are summarised in Figure 3. As shown, the temperature slightly enhanced the reduction of NAP at 30 min, with a NAP conversion of 0.65 at 25 °C and 0.75 at 40 °C. However, after 180 min of reaction time, the NAP conversion obtained at both temperatures reached similar values (close to unity). However, the consumption of PS was higher at 40 °C than at 25 °C at 30 min and 180 min (Figure 3b).

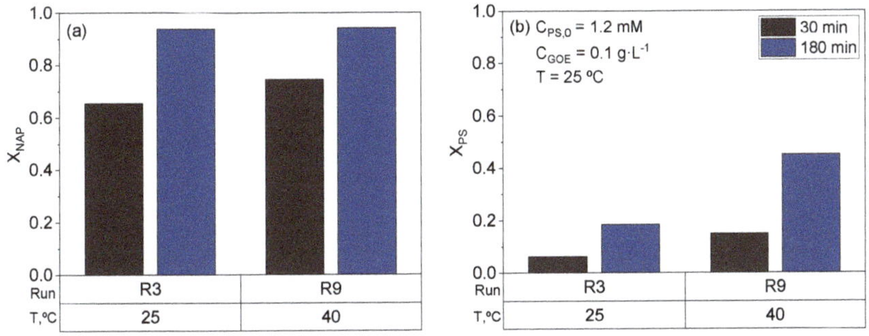

Figure 3. Effect of temperature in (**a**) conversion of NAP; (**b**) conversion of PS at 30 min and 180 min. $C_{NAP,0} = 0.1$ mM, $C_{PS,0} = 1.2$ mM, $I = 0.18$ W·cm^{-2} and $C_{GOE} = 0.1$ g·L^{-1}.

At 40 °C, PS activation by temperature can be performed, generating more sulfate radicals than those obtained at 25 °C [33]. This fact can be experimentally observed since the higher the temperature, the higher the conversion of PS and NAP [34,35].

2.4. *Effect of PS Concentration*

The oxidant concentration was tested within 0–4.8 mM using the same GOE concentration and irradiance values. The obtained conversions of NAP and PS are summarised in Figure 4. As shown in Figure 4, negligible NAP oxidation occurred without PS (B2). In runs R6 and R3, using 0.6 mM and 1.2 mM PS concentrations, a similar conversion of NAP was obtained after 30 min of reaction. On the other hand, an initial concentration of PS higher than 2.4 mM resulted in slightly lower values of NAP conversion at 30 min. The observed effect of the higher PS concentration and the lower NAP conversion can be explained by the reactions in Equations (7) and (8). This phenomenon was in agreement with Zeng [13] et al. and Kusic et al. [36]. They concluded that the production of $SO_4^{-}\bullet$ increases with the increase in PS. The radical sulfate in high concentrations reacts with the PS (Equation (7)), resulting in the unproductive reaction of the oxidant. In addition, $SO_4^{-}\bullet$ in high concentrations can be self-consumed in a self-inhibitory reaction (Equation (8)); therefore, the $SO_4^{-}\bullet$ available to abate the NAP decreased [13,36].

$$S_2O_8^{2-} + SO_4^{-}\bullet \rightarrow S_2O_8^{-}\bullet + SO_4^{2-} \tag{7}$$

$$2\,SO_4^{-}\bullet \rightarrow S_2O_8^{-2} \tag{8}$$

The NAP conversion achieved at 180 min was similar at all PS concentrations, close to unity.

2.5. *Ions and Organic Matter Effects*

The effects of different ions (Cl^{-} and HCO_3^{-}) and organic matter (humic acid (HuAc)) found in natural waters on the NAP conversion have been studied, comparing the NAP degradation in the presence and absence of these compounds in the aqueous phase. The experimental values of NAP and PS conversions in runs S1, S2, and S3 are summarised in Figure 5 and compared with the results obtained in R3.

All reactions in Figure 5 were carried out under the same initial PS and NAP concentrations and irradiance conditions. As shown in Figure 5a, the degradation of NAP was slightly reduced in the presence of chloride ions, indicating that this specie does not exert a remarkable effect on NAP conversion. However, a significant decrease in NAP conversion was found when bicarbonate or humic acid were added to the reaction media.

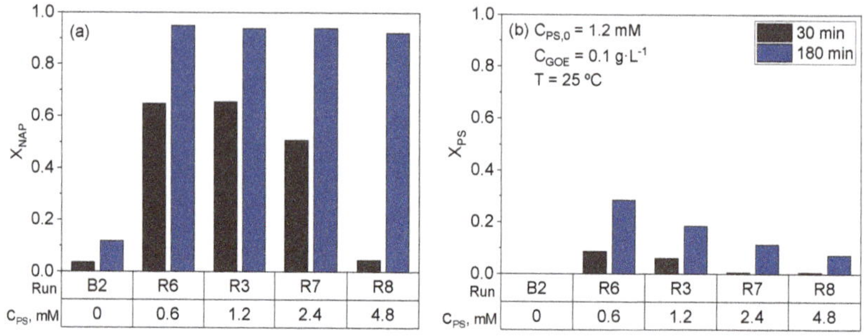

Figure 4. Effect of PS concentration in (**a**) conversion of the NAP; (**b**) conversion of PS at 30 min and 180 min. $C_{NAP,0} = 0.1$ mM, $I = 0.18$ W·cm^{-2}, $C_{GOE} = 0.1$ g·L^{-1}·cm^2 and 25 °C.

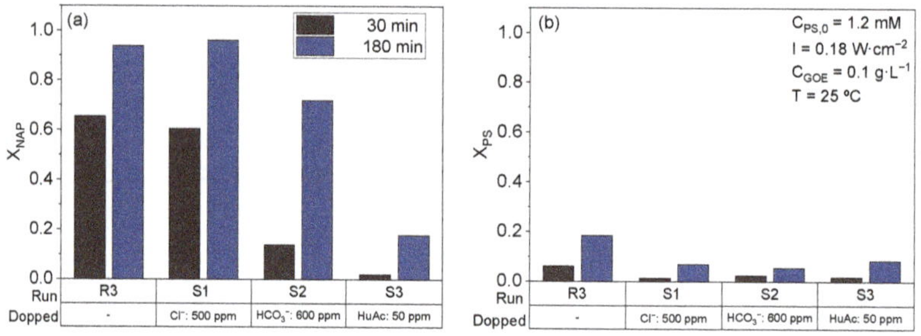

Figure 5. Effect of ions and organic matter in (**a**) Conversion of the NAP; (**b**) Conversion PS at 30 min and 180 min. $C_{NAP,0} = 0.1$ mM, $I = 0.18$ W·cm^{-2}, $C_{GOE} = 0.1$ g·L^{-1}·cm^2, $C_{PS,0} = 1.2$ mM, $pH_0 = 7$ and 25 °C.

In the presence of a high bicarbonate concentration (run S2), NAP conversion at 30 min was one-third of the value obtained without bicarbonate (R3). The scavenger effect of sulfate radicals by bicarbonate ions has been already reported in the literature [24] and results in the production of $HCO_3\bullet$ (Equation (9)) or $CO_3^-\bullet$ (Equation (10)) radicals.

$$CO_3H^- + SO_4^-\cdot \rightarrow SO_4^{2-} + HCO_3\cdot \quad (9)$$

$$CO_3^{2-} + SO_4^-\cdot \rightarrow SO_4^{2-} + CO_3^-\cdot \quad (10)$$

The redox potential of carbonate and bicarbonate radicals (1.63 V) is much lower than the redox potential of sulfate radicals (2.70 V) [37], explaining the inhibitory effect of bicarbonate on NAP conversion.

The effect of the presence of natural organic matter in the reaction medium was studied using 50 mg·L^{-1} of humic acid (HuAC) (Run S3). As shown in Figure 5a, a remarkable decrease in NAP conversion was obtained in the presence of HuAc at 30 min. At 180 min, NAP conversion was reduced by about 50% to that obtained in the absence of HuAc (Run R3). The inhibitory effect of HuAc on NAP conversion was even more remarkable than that noticed with bicarbonate. These findings can be explained by turbidity and the colour of the solution obtained with the solubilisation of HuAC. The appearance of the reaction medium in the runs carried out in the presence (run S3) and absence of HuAc (run R3) are shown in Figure S4 of the Supplementary Materials. As shown in Figure S4a, before the lamp is switched on, the reaction mixture of run S3 showed an intense colouration compared to the appearance of the reaction medium without HuAC Figure S4c (run R3). The intense colouration associated with HuAC in the reaction medium can modify the light capable of activating the goethite, thus decreasing the NAP conversion. Despite the intense colouration observed in the presence of HuAC, a remarkable conversion of NAP was obtained at 180 min (50%). The colouration of natural water is scarcely as intense as that obtained in run S3, and the inhibition expected in these natural waters will be lower than that found in S3.

2.6. Toxicity Analysis

Toxicity tests were performed using a standard process (Microtox®) to monitor the bacterium Photobacterium phosphoreum. The toxicity of the reaction samples after 180 min was measured at 15 min and 15 °C, expressed as toxicity units (*TU*). *TU* was used as a direct measurement of the acute toxicity and it was calculated from the IC_{50} values at 15 min and 15 °C, applying Equation (11). The *TU* obtained from the untreated aqueous solution of 0.1 mM in NAP was also measured giving a value of *TU* = 1.76. This value is considered the reference value, TU_0. For each reaction, the *TU* of the aqueous sample at the final time (t = 180 min) was measured and the ratio *TU* to TU_0 has been calculated. The corresponding results are summarised in Table 3.

$$TU = \frac{100}{IC_{50}} \tag{11}$$

TU/TU_0 ratios lower than unity indicate that the toxicity of the reaction mixture was reduced after the oxidation treatment. As seen in Table 3, the higher the lamp irradiance (runs R2, R3, R4), the lower the toxicity of the reaction media. Only the lower irradiance used (run R2) increases the initial toxicity. On the contrary, *TU*s obtained in runs R3 and R4 are lower than the initial value. The GOE concentration does influence the toxicity obtained at 180 min, but the final toxicity is always lower than the initial one.

The PS concentration has a remarkable effect on the final toxicity of the reaction media, as seen in Table 3. The higher the PS concentration, the higher the reduction of the initial toxicity. At the lowest PS concentration (0.6 mM), the toxicity of the reaction media at 180 min is higher than the initial one. On the contrary, at a PS concentration above 1.2 mM, the final toxicity at 180 min is lower than the initial value. At the highest PS concentration (4.8 mM), a reduction of 85% of the initial toxicity was found at 180 min.

The differences in *TU*s obtained at 180 min can be explained by the byproducts produced and reacted during NAP oxidation. In most of the runs seen in Table 3, the NAP conversion at 180 min was close to unity. Unfortunately, no organic intermediates were significantly identified by ionic chromatography, GC/MS chromatography, or HPLC. Only traces of DEHP (Di (2-ethylhexyl) phthlate) were detected in some runs, agreeing with what was reported in the literature. However, the toxicity decrease found at 180 min ensures that non-toxic intermediates remain in the reaction media. The toxicity is remarkably reduced using a sufficient PS concentration, confirming the treatment benefits.

Lastly, the effect of temperature on *TU*s was studied. As can be seen in Table 3, an increase in the temperature does not produce a reduction in the initial toxicity. This can be explained as a higher unproductive consumption of PS occurs when the temperature increases. PS is not reacting with the byproducts but with an unproductive consumption of PS.

Table 3. Ratio of *TU* of the reaction samples after 180 min to initial Toxicity (*TUo* = 1.76).

Experiment	T (°C)	C_{PS} (mM)	C_{GOE} ($g{\cdot}L^{-1}$)	I ($W{\cdot}cm^{-2}$)	TU/TU_o 15 min
NAP					1.00
Effect of irradiance					
R2	25	1.2	0.1	0.10	1.54
R3	25	1.2	0.1	0.18	0.79
R4	25	1.2	0.1	0.24	0.62
Effect of goethite concentration					
R3	25	1.2	0.1	0.18	0.79
R5	25	1.2	0.2	0.18	0.79
Effect of PS concentration					
R6	25	0.6	0.1	0.18	1.36
R3	25	1.2	0.1	0.18	0.79
R7	25	2.4	0.1	0.18	0.36
R8	25	4.8	0.1	0.18	0.14
Effect of temperature					
R3	25	1.2	0.1	0.18	0.79
R9	40	1.2	0.1	0.18	1.06

3. Material and Methods

3.1. Chemicals

Naphthalene (NAP, $C_{10}H_8$) of analytical quality was purchased from Sigma-Aldrich (Darmstadt, Germany). This compound was dissolved in methanol (Fisher-Chemical, Hampton, CV, USA) to prepare the standards used in the calibration curve.

Sodium persulfate (PS, $Na_2S_2O_8$), used as oxidant, and Goethite (GOE), used as catalyst, were supplied by Sigma-Aldrich (Darmstadt, Germany). GOE is composed of 57.3 wt. % Fe (III) with a specific surface area (S_{BET}) of 10.24 $m^2{\cdot}g^{-1}$ [32]. Potassium iodide (KI, Fisher-Chemical, Hampton, CV, USA) and sodium hydrogen carbonate ($NaHCO_3$, Panreac, Darmstadt, Germany) were used for PS quantification. The latter was also used for the ions studied along with sodium chloride (NaCl), supplied by Sigma Aldrich (Darmstadt, Germany). Humic acid (H.Ac) 45–70% (Acros Organics, Hampton, CV, USA) was used to analyse the influence of organic matter in the aqueous media. All the reagents employed were of analytical grade. Solutions were prepared with ultra-pure water produced by a deionising system (Millipore Direct-Q).

3.2. Experimental Setup

The experiments were carried out in a cylindrical batch reactor made of borosilicate glass with an operating volume of 100 mL. The reaction mixture was agitated using a magnetic plate (IKA C-MG HS 7, Staufen, Germany) to remove the mass transfer limitations. For temperature control of the reactor, an outer jacket coupled to a pumped water bath was used. The temperature was controlled using a PID controller. The reaction medium was illuminated with a monochromatic LED lamp (470 nm) located at the top of the reactor. The distance between the lamp and the reaction surface was 11 cm. The light emitter was located at the focal plane of a collimating lens with a clear aperture of 1 cm (pre-adjusted in the factory). The collimator light sources produce an optical power output of up to 4.17 W. The power can be adjusted using a Mightex BLS-13000-1E LED controller. The output current (0–13 A) can be controlled manually. The light LED source emits over a surface reaction of 11 cm^2, coincident with the reactor window. Figure S1 shows the experimental setup.

3.3. Experimental Procedure

Three sets of experiments were carried out to study the removal of NAP. The first one corresponds to blank runs carried out in the absence of oxidant, catalyst, or light (B1–B3). The second set studies the influence of variables such as temperature, PS concentration, and irradiance (R1–R9). The third set studies the effect of commons ions and natural organic matter in natural water on NAP conversion (S1–S3). The experimental conditions of the runs carried out are summarised in Table 1.

The experiments were carried out by dissolving NAP in water at 25 °C to achieve a concentration of 0.1 mM. This solution of 0.1 mM was used to charge the reactor with the corresponding mass of the catalyst to obtain GOE concentrations ranging from 0.1 $g \cdot L^{-1}$ to 0.2 $g \cdot L^{-1}$. A volume of a concentrated PS solution (400 mM) was added to obtain a final PS concentration in the reaction media ranging from 0.6 mM to 4.8 mM. Once the goethite presented a good suspension (keeping an agitation of 500 r.p.m), the LED light was turned on (zero time). All the experiments were carried out at natural pH.

Reaction samples were analysed at different times. An amount of 2 mL of the reaction media was withdrawn and the goethite was separated from the aqueous phase using a nylon filter (0.22 μm). NAP concentration was analysed in the filtrate by HPLC, and the remaining PS concentration was determined by iodometric titration. At the end of the reaction (180 min), the toxicity of the samples was measured by the Microtox® bioassay. Runs were carried out in triplicate, with the standard deviation lower than 5% in all cases.

3.4. Analytical Methods

The aqueous NAP concentration was analysed using high-performance liquid chromatography (Agilent 1100 Series) equipped with a UV detector. A Poroshell 120 SB-C18 2.7 μm (4.6 × 100 mm) column was used. The mobile phase used was methanol–water (80:20, *v/v*) at a flow rate of 0.5 $mL \cdot min^{-1}$ and the effluent was monitored at 254 nm. This method is similar to that in the literature for NAP quantification [13,15].

The PS concentration changes with the reaction times were determined by iodometric titration using a potentiometric titration analyser (Metrohm, Tiamo 2.3, Gallen, Switzerland).

The LED lamp irradiance and intensity were measured with a Flame UV-VIS spectrometer in the 200–850 nm wavelength range coupled to a monacolin UV-Visible optical fibre of 100 μm core diameter using a cosine corrector (Ocean Insight, Duiven, The Netherlands). The spectrometer and fibre were radiometrically calibrated using the cosine corrector (CC-3) with an HL-3P-CAL lamp (Ocean Insight, Duiven, The Netherlands).

The acute toxicity of the liquid samples at the final reaction time was determined by a standard bioassay following the Microtox test procedure (ISO 11348-3, 2009) using a Microtox® M500 analyser (Azur Environmental). This test is based on bioluminescence inhibition of the bacteria *Vibrio Fischeri* when exposed to a toxicant. The method was performed at 15 °C and 15 min of exposure. The samples were fixed at pH 6–7 and persulfate neutralised with sodium thiosulfate to carry out the experiments. Phenol was used as a reference toxicant, and a test was conducted in each set of experiments to ensure that comparisons between the two sets of experiments could be made. The pH was analysed in all experiments with a Metrohm 914 pH/conductometer.

4. Conclusions

From the results obtained, it can be inferred that activation of persulfate by goethite intensified by VIS-LED light is effective on the oxidation of naphthalene in an aqueous solution. This system allowed working at an initial neutral pH at room temperature and with the iron source in a solid phase, overcoming several of the barriers of the traditional Fenton and photo-Fenton processes.

This method found that the conversion of naphthalene reached was 0.94 in only 180 min of reaction time at room temperature when 1.2 mM of oxidant is employed independently of the irradiance and catalyst concentration.

Regarding the effect of ions and organic matter, a strong influence on the oxidation of NAP when there are HCO_3^- ions and humic acid in the aqueous media was observed. This influence is because part of the PS is lost in attacking these species or generating radicals with lower oxidising power. In addition, an increase in the turbidity and colour was noted with HuAc, which may hinder the photon flux.

Additionally, the toxicity units of the aqueous phase after treatment were determined using the Microtox® technique to determine if the NAP oxidation byproducts are more toxic than naphthalene. The results showed that the final toxicity was notably reduced when the PS concentration increased. More slight or null effects of lamp irradiation and goethite concentration, respectively, were noticed on the final TUs, and the negative effect of temperature on the TUs was noticed. Only at the lowest lamp irradiance or the lowest PS concentration employed was an increase in the initial toxicity measured. A total NAP conversion and a reduction of 85% of the initial toxicity were achieved at 4.8 mM of PS, indicating the effectivity of the treatment applied.

The technique proposed in this manuscript has been demonstrated to be effective for the removal of NAP, achieving in most cases a conversion of 0.92 at 180 min. Future research should be directed towards the use of real water polluted with naphthalene based on the knowledge obtained from the study of the variables.

Supplementary Materials: The following are available online at https://www.mdpi.com/article/10.3390/catal12070732/s1, Figure S1. Schematic representation of the experimental setup; Table S1. Initial and end pH of the different experiments summarized in Table 1; Figure S2 Temporal evolution of NAP and PS conversion, (R1). Determination of reaction time and study. Maintaining $C_{NAP,0} = 0.1$ mM, $C_{PS,0} = 1.2$ mM, $I = 0.18$ W·cm^{-2}, $C_{GOE} = 0.1$ g·L^{-1} and 25 °C; Figure S3. Discretized spectral irradiance measured at the reactor window for different lamp nominal powers (Pn); Figure S4 Images of the physical appearance of the reactor medium corresponding to run S3 $C_{NAP,0} = 0.1$ mM, $C_{PS,0} = 1.2$ mM, $I = 0.18$ W·cm^{-2}, $C_{H.Ac} = 50\ ppm$ and $C_{GOE} = 0.1$ g·L^{-1}. (a) Medium without illumination (b) Medium illuminated by LED lamp. Images of the physical appearance of the reactor medium corresponding to run R3 $C_{NAP,0} = 0.1$ mM, $C_{PS,0} = 1.2$ mM, $I = 0.18$ W·cm^{-2} and $C_{GOE} = 0.1$ g·L^{-1}. (c) Medium without illumination (d) Medium illuminated by LED lamp.

Author Contributions: Conceptualization, D.L. and A.S.-Y.; methodology, D.L.; software, P.S.; validation, D.L. and A.S.; formal analysis, A.S.; investigation, P.S. and A.S.-Y.; resources, A.S. and A.R.; data curation, P.S. and A.S.-Y.; writing—original draft preparation, P.S. and A.S.-Y.; writing—review and editing, D.L.; supervision, A.S. and D.L.; project administration, A.S. and A.R.; funding acquisition, A.S. and A.R. All authors have read and agreed to the published version of the manuscript.

Funding: This work was supported by the Regional Government of Madrid, project CARESOIL (S2018/EMT-4317), and the Spanish Ministry of Economy, Industry and Competitiveness, project PID2019-105934RB-I00. Andrés Sánchez-Yépes acknowledges the FPI grant from the Spanish Ministry of Science and Innovation (Ref. PRE2020-09319).

Conflicts of Interest: The authors declare no conflict of interest.

References

1. Mojiri, A.; Zhou, J.L.; Ohashi, A.; Ozaki, N.; Kindaichi, T. Comprehensive review of polycyclic aromatic hydrocarbons in water sources, their effects and treatments. *Sci. Total Environ.* **2019**, *696*, 133971. [CrossRef] [PubMed]
2. Patel, A.B.; Shaikh, S.; Jain, K.R.; Desai, C.; Madamwar, D. Polycyclic Aromatic Hydrocarbons: Sources, Toxicity, and Remediation Approaches. *Front Microbiol.* **2020**, *11*, 562813. [CrossRef] [PubMed]
3. Wise, S.A.; Sander, L.C.; Schantz, M.M. Analytical Methods for Determination of Polycyclic Aromatic Hydrocarbons (PAHs)—A Historical Perspective on the 16 U.S. EPA Priority Pollutant PAHs. *Polycycl. Aromat. Compd.* **2015**, *35*, 187–247. [CrossRef]
4. Alshabib, M. Removal of naphthalene from wastewaters by adsorption: A review of recent studies. *Int. J. Environ. Sci. Technol.* **2021**, *19*, 4555–4586. [CrossRef]
5. Sarria-Villa, R.; Ocampo-Duque, W.; Paez, M.; Schuhmacher, M. Presence of PAHs in water and sediments of the Colombian Cauca River during heavy rain episodes, and implications for risk assessment. *Sci. Total Environ.* **2016**, *540*, 455–465. [CrossRef]

6. Froehner, S.; Rizzi, J.; Vieira, L.M.; Sanez, J. PAHs in Water, Sediment and Biota in an Area with Port Activities. *Arch Environ. Contam Toxicol.* **2018**, *75*, 236–246. [CrossRef]
7. Solano, R.A.; De León, L.D.; De Ávila, G.; Herrera, A.P. Polycyclic aromatic hydrocarbons (PAHs) adsorption from aqueous solution using chitosan beads modified with thiourea, TiO2 and Fe3O4 nanoparticles. *Environ. Technol. Innov.* **2021**, *21*, 101378. [CrossRef]
8. Gaurav, G.K.; Mehmood, T.; Kumar, M.; Cheng, L.; Sathishkumar, K.; Kumar, A.; Yadav, D. Review on polycyclic aromatic hydrocarbons (PAHs) migration from wastewater. *J. Contam. Hydrol.* **2021**, *236*, 103715. [CrossRef]
9. Yang, R.; Zeng, G.; Xu, Z.; Zhou, Z.; Huang, J.; Fu, R.; Lyu, S. Comparison of naphthalene removal performance using H_2O_2, sodium percarbonate and calcium peroxide oxidants activated by ferrous ions and degradation mechanism. *Chemosphere* **2021**, *283*, 131209. [CrossRef]
10. Yang, R.; Zeng, G.; Xu, Z.; Zhou, Z.; Zhou, Z.; Ali, M.; Sun, Y.; Sun, X.; Huang, J.; Lyu, S. Insights into the role of nanoscale zero-valent iron in Fenton oxidation and its application in naphthalene degradation from water and slurry systems. *Water Environ. Res.* **2022**, *94*, e10710. [CrossRef]
11. Zhang, J.W.; Fan, S.S.; Lu, B.; Cai, Q.H.; Zhao, J.X.; Zang, S.Y. Photodegradation of naphthalene over Fe_3O_4 under visible light irradiation. *R. Soc. Open Sci.* **2019**, *6*, 181779. [CrossRef]
12. McQuillan, R.V.; Stevens, G.W.; Mumford, K.A. Assessment of the electro-Fenton pathway for the removal of naphthalene from contaminated waters in remote regions. *Sci. Total Environ.* **2021**, *762*, 143155. [CrossRef]
13. Zeng, G.; Yang, R.; Fu, X.; Zhou, Z.; Xu, Z.; Zhou, Z.; Qiu, Z.; Sui, Q.; Lyu, S. Naphthalene degradation in aqueous solution by Fe(II) activated persulfate coupled with citric acid. *Sep. Purif. Technol.* **2021**, *264*, 118441. [CrossRef]
14. Li, Y.H.; Zhao, L.; Chen, F.L.; Jin, K.S.; Fallgren, P.H.; Chen, L. Oxidation of nine petroleum hydrocarbon compounds by combined hydrogen peroxide/sodium persulfate catalyzed by siderite. *Environ. Sci. Pollut. Res.* **2020**, *27*, 25655–25663. [CrossRef]
15. Liang, C.; Guo, Y.-Y. Mass Transfer and Chemical Oxidation of Naphthalene Particles with Zerovalent Iron Activated Persulfate. *Environ. Sci. Technol.* **2010**, *44*, 8203–8208. [CrossRef]
16. Beninca, C.; Boni, E.C.; Goncalves, F.F.; Primel, E.G.; Freire, F.B.; Zanoelo, E.F. Photo-fenton and UV photo degradation of naphthalene with zero- and two-valent iron in the presence of persulfate. *Chem. Eng. Commun.* **2019**, *206*, 1–11. [CrossRef]
17. Sreeja, P.H.; Sosamony, K.J. A Comparative Study of Homogeneous and Heterogeneous Photo-fenton Process for Textile Wastewater Treatment. *Procedia Technol.* **2016**, *24*, 217–223. [CrossRef]
18. Shokri, A.; Fard, M.S. A critical review in Fenton-like approach for the removal of pollutants in the aqueous environment. *Environ. Chall.* **2022**, *7*, 100534. [CrossRef]
19. Lorenzo, D.; Santos, A.; Sánchez-Yepes, A.; Conte, L.Ó.; Domínguez, C.M. Abatement of 1,2,4-trichlorobencene by wet peroxide oxidation catalysed by goethite and enhanced by visible led light at neutral ph. *Catalysts* **2021**, *11*, 139. [CrossRef]
20. Vicente, F.; Rosas, J.M.; Santos, A.; Romero, A. Improvement soil remediation by using stabilizers and chelating agents in a Fenton-like process. *Chem. Eng. J.* **2011**, *172*, 689–697. [CrossRef]
21. Baciocchi, R. Principles, Developments and Design Criteria of In Situ Chemical Oxidation. *Water Air Soil Poll.* **2013**, *224*, 1717. [CrossRef]
22. Baciocchi, R.; D'Aprile, L.; Innocenti, I.; Massetti, F.; Verginelli, J. Development of technical guidelines for the application of in-situ chemical oxidation to groundwater remediation. *J. Clean. Prod.* **2014**, *77*, 47–55. [CrossRef]
23. Siegrist, R.L.; Crimi, M.; Simpkin, T.J. *In Situ Chemical Oxidation for Groundwater Remediation*; Springer: New York, NY, USA, 2011. [CrossRef]
24. Waclawek, S.; Lutze, H.V.; Grubel, K.; Padil, V.V.T.; Cernik, M.; Dionysiou, D.D. Chemistry of persulfates in water and wastewater treatment: A review. *Chem. Eng. J.* **2017**, *330*, 44–62. [CrossRef]
25. Tsitonaki, A.; Petri, B.; Crimi, M.; Mosbaek, H.; Siegrist, R.L.; Bjerg, P.L. In Situ Chemical Oxidation of Contaminated Soil and Groundwater Using Persulfate: A Review. *Crit. Rev. Environ. Sci. Technol.* **2010**, *40*, 55–91. [CrossRef]
26. Sra, K.S.; Thomson, N.R.; Barker, J.F. Stability of Activated Persulfate in the Presence of Aquifer Solids. *Soil Sediment. Contam.* **2014**, *23*, 820–837. [CrossRef]
27. Dahmani, M.A.; Huang, K.; Hoag, G.E. Sodium Persulfate Oxidation for the Remediation of Chlorinated Solvents (USEPA Superfund Innovative Technology Evaluation Program). *Water Air Soil Pollut. Focus* **2006**, *6*, 127–141. [CrossRef]
28. Rodriguez, S.; Santos, A.; Romero, A. Oxidation of priority and emerging pollutants with persulfate activated by iron: Effect of iron valence and particle size. *Chem. Eng. J.* **2017**, *318*, 197–205. [CrossRef]
29. Lominchar, M.A.; Santos, A.; de Miguel, E.; Romero, A. Remediation of aged diesel contaminated soil by alkaline activated persulfate. *Sci. Total Environ.* **2018**, *622*, 41–48. [CrossRef]
30. Liu, J.L.; Liu, Z.H.; Zhang, F.J.; Su, X.S.; Lyu, C. Thermally activated persulfate oxidation of NAPL chlorinated organic compounds: Effect of soil composition on oxidant demand in different soil-persulfate systems. *Water Sci. Technol.* **2017**, *75*, 1794–1803. [CrossRef]
31. Abdelhaleem, A.; Chu, W. Prediction of carbofuran degradation based on the hydroxyl radical's generation using the FeIII impregnated N doped-TiO_2/H_2O_2/visible LED photo-Fenton-like process. *Chem. Eng. J.* **2020**, *382*, 122930. [CrossRef]
32. Lorenzo, D.; Dominguez, C.M.; Romero, A.; Santos, A. Wet Peroxide Oxidation of Chlorobenzenes Catalyzed by Goethite and Promoted by Hydroxylamine. *Catalysts* **2019**, *19*, 553. [CrossRef]

33. Dominguez, C.M.; Romero, A.; Lorenzo, D.; Santos, A. Thermally activated persulfate for the chemical oxidation of chlorinated organic compounds in groundwater. *J. Environ. Manag.* **2020**, *261*, 110240. [CrossRef] [PubMed]
34. Huling, S.G.; Ko, S.; Park, S.; Kan, E. Persulfate oxidation of MTBE- and chloroform-spent granular activated carbon. *J. Hazard. Mater.* **2011**, *192*, 1484–1490. [CrossRef] [PubMed]
35. Hutson, A.; Ko, S.; Huling, S.G. Persulfate oxidation regeneration of granular activated carbon: Reversible impacts on sorption behavior. *Chemosphere* **2012**, *89*, 1218–1223. [CrossRef]
36. Kusic, H.; Peternel, I.; Ukic, S.; Koprivanac, N.; Bolanca, T.; Papic, S.; Bozic, A.L. Modeling of iron activated persulfate oxidation treating reactive azo dye in water matrix. *Chem. Eng. J.* **2011**, *172*, 109–121. [CrossRef]
37. Liang, C.; Wang, Z.S.; Mohanty, N. Influences of carbonate and chloride ions on persulfate oxidation of trichloroethylene at 20 degrees C. *Sci. Total Environ.* **2006**, *370*, 271–277. [CrossRef]

catalysts

MDPI

Article

An Evaluation of the Kinetic Properties Controlling the Combined Chemical and Biological Treatment of Toxic Recalcitrant Organic Compounds from Aqueous Solution

Seshibe Makgato [1,*] and Evans Nkhalambayausi-Chirwa [2]

1 Department of Chemical Engineering, College of Science, Engineering and Technology, University of South Africa (UNISA), c/o Christiaan de Wet & Pioneer Avenue, Johannesburg 1710, South Africa
2 Water Utilisation and Environmental Engineering Division, Department of Chemical Engineering, University of Pretoria, Pretoria 0002, South Africa
* Correspondence: makgato2001@yahoo.com; Tel.: +27-11-670-9276

Abstract: Due to their high toxicity, propensity for cancer, teratogenicity, mutagenicity, and genotoxicity, hazardous water-soluble phenolic compounds must be controlled immediately. In this study, a model was created to simulate the degradation of harmful recalcitrant organic compounds in a combined chemical and biological treatment system. The parameter estimations with inhibition coefficient (Haldane model) and without inhibition coefficient (Michaelis-Menten model) were assessed over a wide range of initial concentrations using the Monod-like model. The kinetic parameters were optimized using AQUASIM 2.0 software. At a 50 mg·L^{-1} feed concentration of 4-chlorophenol, removal efficiencies of more than 98% were attained under these circumstances. The primary kinetic parameters were identified and their values models were validated using the fitted parameter values that reached a good degree of agreement (R^2 = 0.998). We may better comprehend and make use of the complex phenolic compounds' biodegradation processes, such as progress optimization and scale-up, by understanding the mechanisms of substrate interaction and the new kinetic models that have been provided in this work.

Keywords: kinetics; chemical reactor; biodegradation; toxic recalcitrant; aqueous solution

1. Introduction

The manufacturing of plastics, resin and dyes, insecticides, pharmaceuticals, paper and other petroleum products all heavily rely on phenols or phenolic chemicals [1]. As a result, numerous contaminants are frequently detected in the effluent water from these enterprises, the most prevalent of which is chlorophenol [2], which is reportedly exceedingly harmful to both humans and aquatic life [3]. The levels of 4-chlorophenol (4-CP) in the effluents from the manufacturers of these products range in concentration from 10 mg·L^{-1} to 200 mg·L^{-1} [4]. Numerous derivatives of chlorophenol are refractory, which makes them resistant to biological deterioration. These compounds' recalcitrance is caused by the extremely difficult-to-cleave carbon-hydrogen bond [2]. Additionally, the toxicity of chlorophenols may rise with an increase in their chlorine atom, and the toxicity is also influenced by the difficulty of its microbial degradation [5]. The US Environmental Protection Agency (EPA) and the European Union (EU) have designated chlorophenols as "priority pollutants", meaning that their presence in the aquatic environment needs to be continuously monitored [6]. Because of their high toxicity [3], carcinogenic potential [1], teratogenic potential [7], mutagenic potential [8], genotoxicity [8], mutagenicity, and genotoxicity, hazardous water-soluble phenolic compounds need to be addressed urgently. Over the past few decades, research into various emerging treatment approaches for wastes and waste streams that are chlorophenol-rich has received much attention due to the awareness that chlorophenols are poisonous and harmful.

A few researchers [9–16] examined ways to improve the physical, chemical and biological processes of degrading chlorophenols from wastewater in order to address these problems. However, the first two techniques can be pricey [10], and in certain situations, they may not completely remove the contaminants, causing issues with the generation of dangerous by-products of even more toxic species [17] that call for additional treatments for further mineralization or disposal [18]. Advanced oxidation processes (AOPs), on the other hand, have restricted applications, incur high costs due to the need for energy and chemicals, are slow at higher concentrations or largely useless at low contamination levels, and cannot destroy recalcitrant organic matter [19,20]. All of these significant flaws render these technologies uncompetitive and restrict their ability to be used in real-world scenarios.

More effort has been put into figuring out how to use cultures of microorganisms to treat chlorophenol-contaminated sediments and water. However, the most environmentally sound, economically viable and ecologically sound method of mineralizing biodegradable micropollutants is through biological processes [16]. However, as these treatments are designed to decompose biodegradable organic pollutants, direct biodegradation treatment is not appropriate due to the abundance of non-biodegradable and recalcitrant organic contaminants [17]. As a result, the use of AOPs as pre-treatment technologies could be used to increase biodegradability and decrease the charge of these molecules in any given wastewater. By integrating chemical and biological processes, uncommon advancements have been made in alternative techniques for efficiently removing pollutants in wastewater [16]. Sarria et al. [18] stated that AOPs have been generally shown to be effective when paired with other cutting-edge technologies. In a combined chemical and biological process, AOPs pre-treatment aimed to alter the structure of pollutants by converting them into less toxic and quickly biodegradable intermediates, enabling the subsequent biological degradation to be completed in less time and at a lower cost [18].

The kinetic characteristics of the organic contaminants involved and the kinetics of the processes must be understood when designing combined chemical and biological treatments [21]. According to Malato et al. [22], scaling up of combined chemical and biological treatments is nevertheless hampered by the absence of kinetic parameter da-ta to compare integrated chemical and biological processes with other technologies. Therefore, it is essential to have a kinetics equation that, for the design of a combined chemical and biological treatment system and the optimization of operating settings, specifies the dependence of the reaction rate on some other relevant quantity [21]. Even while Wade et al. [23] have shown that deterministic modelling of reactor systems, often using a set of coupled ordinary differential equations, gives a vital knowledge of these frequently complicated processes, it has been proved to be successful, particularly in identifying adjustments to the dynamic systems given perturbations in the inputs. Currently, there is no open literature on the kinetic characteristics evaluation of a combined chemical and biological treatment system of harmful refractory organic compounds in a two-stage continuous stirred tank reactors system (CSTRs) system, despite the mathematical complexity and simplicity of both batch and continuous systems. However, there hasn't been a lot of research done on the analysis of kinetic data and optimization of process parameters for the degradation of 4-CP pollutants at higher concentrations (i.e., 2000 mg·L^{-1}). The objective of the study is to create a model of the controlling the combined chemical and biological treatment system simulated harmful recalcitrant organic compound degradation process. The Monod-like model was used to evaluate the parameter estimations with inhibition coefficient (Haldane model) and without inhibition coefficient (Michaelis-Menten model).

2. Results and Discussion

2.1. Chemical Reactor Rate Coefficients Estimation

The 4-chlorophenol is utilized for cell upkeep, metabolic products, and cell biomass. Equations (1) and (2) are used in order to create a model for the photochemical reactor:

$$r_1 = kC_1C_{oh} \tag{1}$$

where k = the first-order maximum reaction rate coefficient (T^{-1}), C_1 = chemical reactor-calculated effluent concentration ($M{\cdot}L^{-3}$), and C_{oh} = calculated concentration of ($OH^{\bullet}$) radicals ($M{\cdot}L^{-3}$).

$$C_{oh} = C_{ohmax} + \frac{a}{\left(1 + e^{\frac{-(t-t_0)}{b}}\right)^c} \tag{2}$$

where C_{ohmax} = Initial concentration of ($OH^{\bullet}$) radicals ($M{\cdot}L^{-3}$), t = Time (T), t_0 = Initial boundary value for accumulation of ($OH^{\bullet}$) radicals (T), a = Kinetic parameter constant for accumulation of ($OH^{\bullet}$) radicals (T), b = Kinetic parameter constant for accumulation of ($OH^{\bullet}$) radicals (T), and c = Kinetic parameter constant for accumulation of ($OH^{\bullet}$) radicals (T).

Now, substituting Equation (1) on Equation (2) gives:

$$r_1 = kC_1\left(C_{ohmax} + \frac{a}{\left(1 + e^{\frac{-(t-t_0)}{b}}\right)^c}\right) \tag{3}$$

Substituting for Equation (3) on Equation (4) gives Equation (5):

$$-r_S = \left(\frac{Q}{V}\right)(C_{in} - C) \tag{4}$$

$$kC_1\left(C_{ohmax} + \frac{a}{\left(1 + e^{\frac{-(t-t_0)}{b}}\right)^c}\right) = \left(\frac{Q}{V}\right)(C_{in} - C) \tag{5}$$

A straightforward model, Equation (5), illustrates the wide range of behaviors of chemical reactors in a combined chemical and biological treatment system. As a result, utilizing Equation (5) and optimization, a satisfactory solution for the kinetic parameters of a chemical reactor was discovered based on the goodness of fit (R^2 = 0.98) over a range of influent concentrations. Table 1 lists the computed average kinetic parameters as follows: a = 648.78 $mg{\cdot}L^{-1}$, b = 14.27 h^{-1}, c = 249.39, t_0 = 10.13 h, k = 1.97 $\times 10^{-5}$ $L^{-3}{\cdot}mg^{-1}{\cdot}h^{-1}$, C_{ohmax} = 534.82 $mg{\cdot}L^{-1}$, and χ^2 = 1046. 55.

Table 1. The operating conditions for the chemical reactor.

Conc ($mg{\cdot}L^{-1}$)	C_{in} ($mg{\cdot}L^{-1}$)	C_{out} ($mg{\cdot}L^{-1}$)	η	a ($mg{\cdot}L^{-1}$)	b (h)	c	t_0 (h)	k_{ms} (h^{-1})	C_{ohmax} ($mg{\cdot}L^{-1}$)
50	55	6.17	88.78	377.8	710.73	877.07	974.86	0.046	648.78
100	107	36.02	66.34	243.95	302.80	414.54	17.85	0.057	414.27
50 + H_2O_2	55	21.19	61.47	100.21	508.46	690.72	137.85	0.064	249.39
100 +H_2O_2	106	21.32	79.89	9.66	6.04	104.81	8.07	0.082	10.13
200 + H_2O_2	191	10.11	94.71	1.55×10^{-5}	2.19×10^{-5}	1.25×10^{-5}	1.29×10^{-5}	0.089	2.30

The k_{ms} value is 0.046 at a dose of 50 $mg{\cdot}L^{-1}$ and it rises to 0.057 at a concentration of 100 $mg{\cdot}L^{-1}$. However, k_{ms} rose from 0.064 to 0.089 when hydrogen peroxide (H_2O_2) was added as a liquid catalyst. The k_{ms} values for the chemical reactor system were found to generally rise with rising chlorophenol concentration and further rise with rising H_2O_2 dosage. Constant a, b and c values, on the other hand, were found to follow the opposite pattern from the k_{ms} values. Hydrogen peroxide (H_2O_2) was added as a liquid catalyst and caused the a, b, and c values in the instances of those values to decrease. The chemical reactor's kinetic parameter values matched those in the literature in a reliable manner. According to Yang et al. [24], k_{ms} values for 50, 100, and 200 $mg{\cdot}L^{-1}$ with additional H_2O_2 were reported to be 0.0997, 0.0480, and 0.0307, respectively. These values compared well with the k_{ms} values in the current investigation. Due to variations in the experimental procedures used by Albarrán and Mendoza [25], there is a slight discrepancy between the values found in the literature and the values acquired in this work. Additionally, Ghaly et al. [26] found that the pseudo-first-order rate constants (k_{ms}) of chlorophenol with ($OH^\bullet$) radicals were 0.15 in another scenario. Even while the result of Ghaly et al. [26] appears high in comparison with our results, this is because Ghaly et al. [26] employed light intensity measurements of 60 W, which is seven times greater than the 9 W light used in the current investigation. In terms of engineering, Equation (5) is more intriguing since it offers the chance to push the system to the edge of its operational viability without sacrificing its capabilities.

2.2. Steady-State Estimation of Biological Parameters

A model was created for a biological reactor that took into account both the Haldane and Monod growth models, similar to how one would for a photochemical reaction chamber. Since the majority of the samples were taken during the logarithmic phase of bacterial development, as was the case with Patel et al. [4], we made the assumption that no by-products of 4-CP's biodegradation would be produced. Additionally, we believed that cells' maintenance coefficient would be negligible [27,28]. In order to obtain the following Equations (6) and (7), two reaction rate models were replaced into Equation (6):

$$C_{in} - C = \left(\frac{V}{Q}\right)\left[-\frac{k_{ms}C\left(X_0 e^{\left(\frac{Yk_{ms}}{K_S} - k_d\right)}\right)t}{K_S + C}\right] \quad (6)$$

where C_{in} = influent substrate concertation ($M{\cdot}L^{-3}$), C = effluent substrate concentration ($M{\cdot}L^{-3}$), r_s = reaction rate ($M{\cdot}L^{-3}{\cdot}T^{-1}$), Q = flow rate across the reactor ($L^3{\cdot}T^{-1}$), V = operating volume of the reactor (L^3). k_{ms} = maximum specific reaction rate coefficient (T^{-1}), K_s = half velocity concentration ($M{\cdot}L^{-3}$), X_0 = the viable cell concentration in the reactor ($M{\cdot}L^{-3}$), Y = The cell growth yield, k_d = Endogenous decay coefficient (T^{-1}) and t = Time (T).

And Equation (7)

$$C_{in} - C = \left(\frac{V}{Q}\right)\left[-\frac{k_{ms}C\left(X_0 e^{\left(\frac{Yk_{ms}}{K_S} - k_d\right)}\right)t}{K_S + C + \frac{C^2}{K_I}}\right] \quad (7)$$

where K_I = coefficient of inhibition ($M{\cdot}L^{-3}$).

The Haldane model, which takes into account the inhibitory impact of the substrate exerts during biodegradation, is utilized to characterize the biodegradation kinetics in Equation (7). The vast spectrum of biological reactor behavior when a chemical and a biological treatment system are combined is therefore represented by quasi-second-order simple models in Equations (6) and (7). The relative toxicity of 4-CP concentration can be evaluated quantitatively under various experimental settings thanks to the inclusion of the Haldane equation in the kinetic model and the application of the inhibition coefficient K_i. The values of the kinetic parameters K_s, k_{ms}, and K_I in the continuous process were calculated from the experimental data in a quasi-steady state and are provided as indicated in Table 2.

Table 2. Kinetic parameter estimation for the biological reactor at different initial concentrations using measured data (Haldane model).

Conc ($mg \cdot L^{-1}$)	C_{in} ($mg \cdot L^{-1}$)	C_{out} ($mg \cdot L^{-1}$)	Removal Efficiency	K_I ($mg \cdot L^{-1}$)	k_{ms} (h^{-1})	K_s ($mg \cdot L^{-1}$)	χ^2 ($mg \cdot L^{-1}$)	R^2
50	6.17	1.14	81.52	25	0.184	5	405	0.976
100	36.02	19.01	47.22	144	0.228	7	430	0.997
50 + H_2O_2	21.19	20.36	3.92	150	0.226	8	411	0.995
100 +H_2O_2	21.32	16.10	24.48	259	0.328	23	423	0.991
200 + H_2O_2	10.11	6.00	40.65	274	0.316	30	468	0.994

It was discovered that the bacteria's specific growth rate increased as the concentration of 4-CP increased and that 4-CP acts as an inhibitor based on the fact that k_{ms} decreased as the original 4-CP concentration increased. The first-order maximum reaction rate coefficient, k_{ms}, in the chemical reactor was higher than the maximum specific reaction rate coefficient, kms, in the biological reactor. This shows that the chemical reactor destroyed 4-CP more quickly than the biological reactor. According to the results of the current investigation, removal rates were decreased at high 4-CP concentrations, and the Haldane substrate-inhibition model could be used to predict removal rates at the concentrations examined.

The half velocity constant (K_s) and maximal specific removal rate fell within the ranges described in the literature [29]. In order to find the model that best captures the intended process, the Michaelis–Menten and Haldane models were also evaluated.

When the initial 4-CP concentration was increased from its lower level of 50 $mg \cdot L^{-1}$ to its higher level of 200 $mg \cdot L^{-1}$ with H_2O_2, K_I values in the range of 25–274 $mg \cdot L^{-1}$ were reported. Higher K_I values signify a stronger inhibitory effect due to the C^2/K_I term's contribution to the Haldane equation, which is sensitive to the substrate's inhibitory characteristics. The substrate-inhibition constant, on the other hand, was substantially greater than Assadi et al. [29] had previously observed (K_I = 609.7 $mg \cdot L^{-1}$). As a result, only a short period of time was needed to achieve the maximum removal rate from mixed-acclimated sludge, which suggests that it has a high tolerance for the harmful chemical 4-CP. This finding is consistent with that of Sahinkaya and Dilek [30] who have noted significant inhibition constants for phenol breakdown in mixed cultures. In contrast to the Haldane model, the Michaelis–Menten model was shown to have an extremely low substrate constant (K_s) as shown in Table 3. The Haldane model's high K_I suggests that the inhibition was quite strong, which indicates that the bacteria were likely thriving primarily on 4-CP intermediates rather than the genuine 4-CP. Additionally, the K_I values were fairly comparable with those of other researchers [31–34]. The findings demonstrate that the Haldane model is more accurate since it takes into account the inhibition coefficient, which in this study's circumstance could not have been disregarded.

Table 3. Kinetic parameter estimation for the biological reactor at different initial concentrations using measured data (Michaelis–Menten model).

Conc ($mg \cdot L^{-1}$)	C_{in} ($mg \cdot L^{-1}$)	C_{out} ($mg \cdot L^{-1}$)	Removal Efficiency	k_{ms} (h^{-1})	K_s ($mg \cdot L^{-1}$)	χ^2 ($mg \cdot L^{-1}$)	R^2
50	6.17	1.14	81.52	0.046	0.94	405	0.993
100	36.02	19.01	47.22	0.057	2.93	430	0.994
50 + H_2O_2	21.19	20.36	3.92	0.064	3.98	411	0.995
100 +H_2O_2	21.32	16.10	24.48	0.082	3.71	423	0.999
200 + H_2O_2	10.11	6.00	40.65	0.089	3.84	468	0.989

Overall, the degradation rates of 4-CP in inoculum samples were 88.78%, 66.3%, 61.70%, 79.89%, and 94.71% for 50 $mg \cdot L^{-1}$, 100 $mg \cdot L^{-1}$, 50 $mg \cdot L^{-1}$ + H_2O_2, 100 $mg \cdot L^{-1}$ + H_2O_2, and

200 mg·L^{-1} + H_2O_2, respectively, in 200 h. Therefore, 200 mg·L^{-1} with 96% additional H_2O_2 provided the maximum removal effectiveness for 4-CP degradation.

2.3. Developed Model Validations

True steady state conditions were not possible because the reactor was run at 2–3 HRTs every cycle. However, the reaction-rate kinetics were assessed under a transient condition using AQUASIM 2.0 software in order to create a more realistic modelling effort. The findings of comparing experimental and modelled concentration values for the various initial 4-CP concentrations utilized in the current study are shown in Figures 1–5. These findings show that with the exception of Figure 3, where the model was unable to accurately monitor the 4-CP concentration profile, the model's simulation of the experimental data and its fitting to those data are both satisfactory. A perfect stoichiometric chloride release was recorded, and prior research on the bio-degradation of chlorophenols revealed that the accumulation of intermediates was low [35]. This is supported by the experimental data's reasonably strong fit ($R^2 > 0.97$) given that the Haldane model was said to be insufficient in the presence of those metabolic intermediates [33].

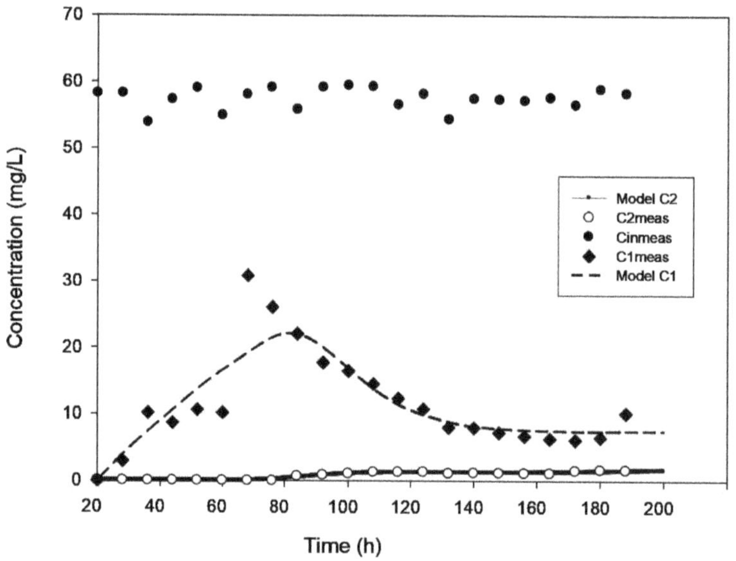

Figure 1. Model validation for the concentration of 50 mg·L^{-1}.

In order to compare these data with the experimental results, an effort was made to match the researched kinetic parameter values on an optimized model. Both the Haldane model's and the Michaelis-Menten model's kinetic parameter values for the 4-CP degradation were examined using AQUASIM 2.0, and the results are listed in Tables 4 and 5. As the original 4-CP concentration increased, we discovered that the values of k_{ms}, K_s and K_I did exhibit a continuous trend of increasing. The results show that the Haldane model is more accurate since it considers the inhibition coefficient, which in the context of this investigation could not have been ignored. This outcome is fairly consistent with Oh et al's experimental findings [33]. Overall, the degradations of 4-CP in inoculum samples were about 98%, 82%, 63%, 85%, and 97% for 50 mg·L^{-1}, 100 mg·L^{-1}, 50 mg·L^{-1} + H_2O_2, 100 mg·L^{-1} + H_2O_2, and 200 mg·L^{-1} + H_2O_2, respectively, in 200 h. Consequently, a 98% removal efficiency for 4-CP degradation was attained at 50 mg·L^{-1}. However, the biodegradation rate was further enhanced when hydrogen peroxide (H_2O_2) was used as a liquid catalyst, increasing the rate to 200 mg·L^{-1} with a 97% overall efficiency. However, with an influent concentration of 4-chlorophenol of 300 mg·L^{-1} and higher, the system was unable to reach a quasi-steady-state condition and was instead characterized by a large loss of viable biomass because it was unable to scavenge OH radicals under the conditions of high H_2O_2 concentrations. According to Fu et al. [8], when there are too many unreacted ($OH^{\bullet}$) radicals present and there is not enough organic feed, the radicals recombine to produce water with no beneficial interaction with the organics. This remark is in line with the findings of the study.

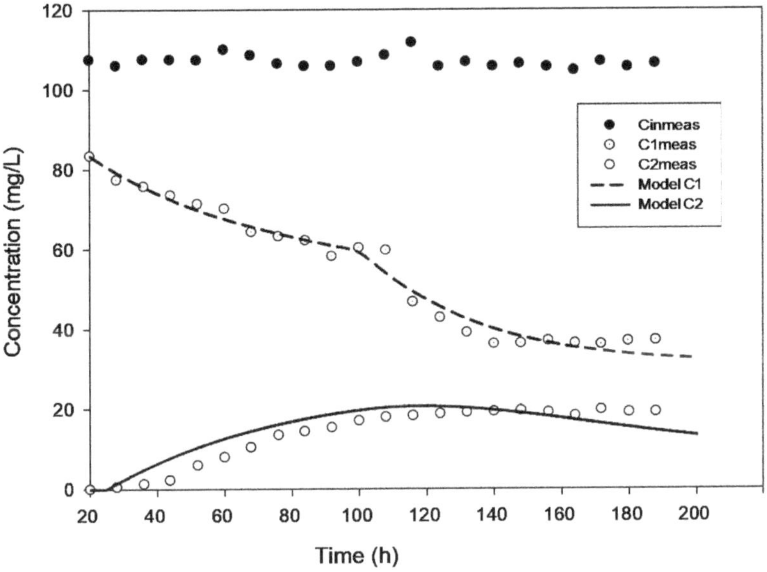

Figure 2. Model validation for the concentration of 100 mg·L^{-1}.

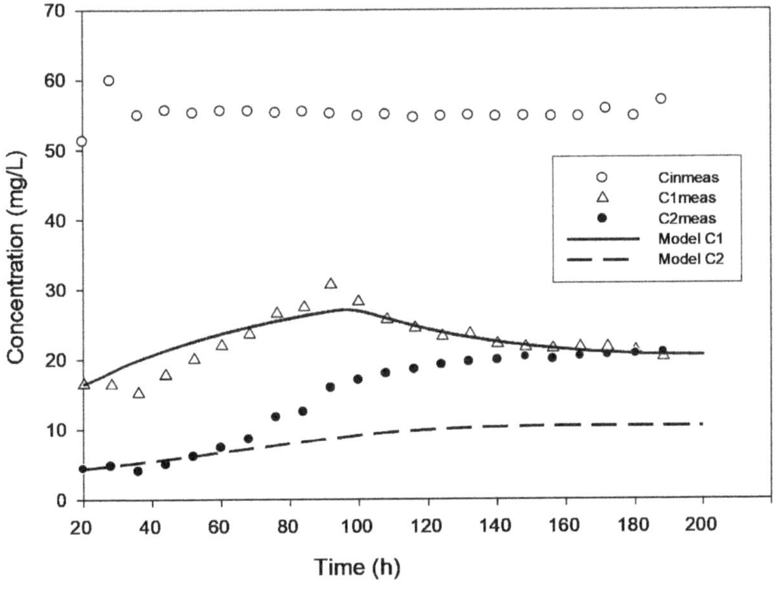

Figure 3. Model validation for the concentration of 50 mg·L^{-1} with added H_2O_2.

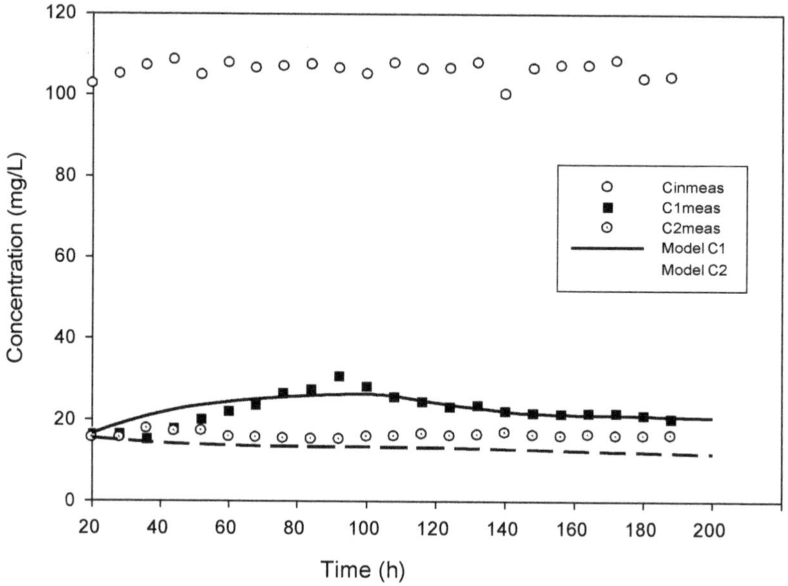

Figure 4. Model validation for the concentration of 100 mg·L^{-1} with added H_2O_2.

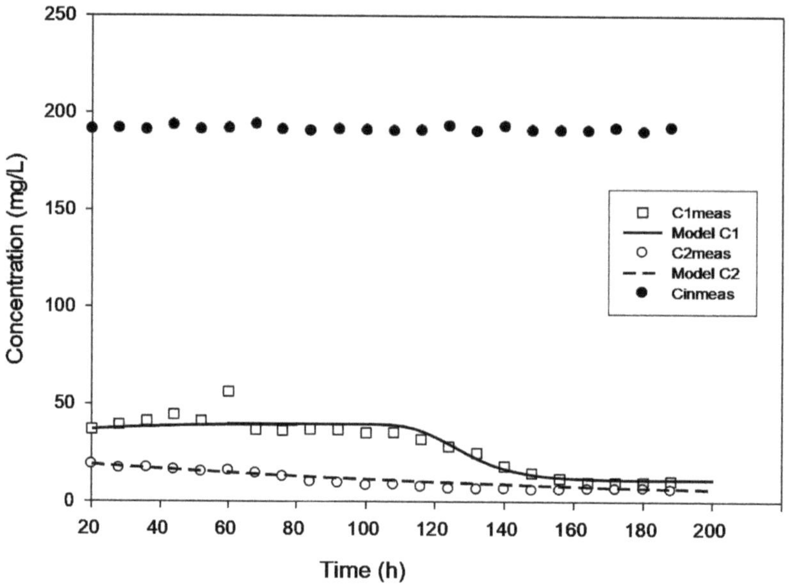

Figure 5. Model validation for the concentration of 200 mg·L^{-1} with added H_2O_2.

Table 4. Kinetic parameter estimation for the biological reactor at different initial concentrations using AQUASIM (Haldane model).

Conc ($mg{\cdot}L^{-1}$)	C_{in} ($mg{\cdot}L^{-1}$)	C_{out} ($mg{\cdot}L^{-1}$)	Overall Efficiency	K_I ($mg{\cdot}L^{-1}$)	k_{ms} (h^{-1})	K_s ($mg{\cdot}L^{-1}$)	χ^2 ($mg{\cdot}L^{-1}$)	R^2
50	55	1.14	98%	165	0.181	8	405	0.921
100	107	19.01	82%	284	0.205	10	430	0.995
50 + H_2O_2	55	20.36	63%	290	0.221	11	411	0.991
100 +H_2O_2	106	16.10	85%	399	0.260	26	423	0.989
200 + H_2O_2	191	6.00	97%	424	0.320	33	468	0.992

Table 5. Kinetic parameter estimation for the biological reactor at different initial concentrations using AQUASIM (Michaelis-Menten model).

Conc ($mg{\cdot}L^{-1}$)	C_{in} ($mg{\cdot}L^{-1}$)	C_{out} ($mg{\cdot}L^{-1}$)	Overall Efficiency	k_{ms} (h^{-1})	K_s ($mg{\cdot}L^{-1}$)	χ^2 ($mg{\cdot}L^{-1}$)	R^2
50	55	1.14	98%	0.081	0.96	405	0.93
100	107	19.01	82%	0.092	1.20	430	0.94
50 + H_2O_2	55	20.36	63%	0.099	1.33	411	0.95
100 +H_2O_2	106	16.10	85%	0.117	3.13	423	0.99
200 + H_2O_2	191	6.00	97%	0.144	3.98	468	0.98

3. Materials and Methods

3.1. Experimental Set-Up

The combined chemical and biological treatment system employed in this study included a 22.8 L photoreaction chamber, followed by a 32.7 L biological reactor, both made of Perspex glass, a 9-Watt (W) UV lamp, a feed pump, and a UV-recycle pump. The biological reactor's dimensions were 447 × 348 × 246 mm (L × B × H), whereas the photoreactor's dimensions were 300 × 295 × 295 × 295 mm (L × B × H). The two reactors were run as CSTRs that were connected in series. It was anticipated that Reactor 2 will use phenolic compound breakdown products as carbon sources. A continuous influent feed was introduced at a flowrate of 0.200 Lh^{-1} to allow enough retention for degradation of 4-CP in only one pass through the reactor. Since the model had to account for these radicals, a degradation reaction was created. The functions of effluent C_1 from reactor 1, a chemical reactor, and C_2 from reactor 2, a biological reactor, are used to represent the reaction rates in Figure 6. Equations (1) and (8) are the formulas for the generalized rate equation for the degradation of 4-CP.

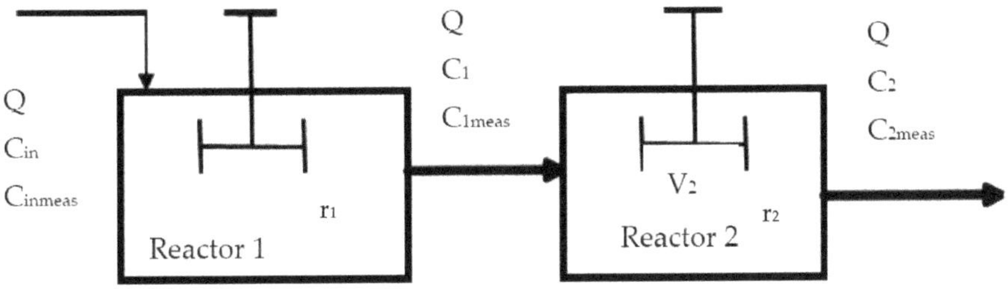

Figure 6. Schematic representation of the combined system.

$$r_2 = \frac{(k_{ms}C_2X_0)}{\left(K_S + C_2 + \frac{C_2^2}{K_I}\right)} \tag{8}$$

where: k_{ms} = Maximum specific reaction rate coefficient (T^{-1}), C_2 = Biological reactor calculated effluent concentration ($M{\cdot}L^{-3}$), X_0 = Attached viable cells concentration in the reactor ($M{\cdot}L^{-3}$), K_S = Half velocity concentration ($M{\cdot}L^{-3}$), K_I = Coefficient of inhibition ($M{\cdot}L^{-3}$).

And

$$X = X_0 e^{\left(\frac{Y k_{ms} C_2}{K_s}\right)t} \quad (9)$$

where Y = The cell growth yield and t = Time (T).

3.2. Operating Conditions for Chemical and Biological Reactors

One experimental run was operated under 4-CP feed concentration range from 50 to 1000 mg·L^{-1}. The amount of H_2O_2 necessary to achieve the highest treatment efficiency is another crucial factor to take into account in AOPs and biological treatment procedures that are coupled. Using an H_2O_2 dose of 0.1 mg·L^{-1}, the impact of oxidant H_2O_2 addition was studied over the loading range of 50 to 1000 mg·L^{-1}. The breakdown of hydrogen peroxide into water and oxygen is a quick process that takes less than 5 min to complete. The authors did not anticipate finding considerable levels of H_2O_2 in the influent of the bioreactor because the retention time of the chemical reactor was significantly longer (about 6 h). The additional peroxide's 0.1 mg·L^{-1} concentration was intended to be adequate for a speedy reaction and the elimination of all residuals in the effluent.

3.3. Chemicals, Media, and Microorganisms

A mixed culture of activated sludge bacteria was initially gathered from sand-drying beds at a neighboring wastewater treatment plant (WWTP) in Brits (Northwest Province). In a chemical reactor, the first stage of AOP dehalogenation was carried out (1st stage). Because of this, highly aromatic particular microorganisms are not necessarily required in the biological reactor's second stage. This study did not include the selection and acclimatization step that would have produced a highly specialized culture.

The culture was grown in basal mineral medium (BMM), which was made by dissolving the following ingredients in one liter of distilled water: 0.54 g of NH_4Cl, 10.74 g of NaH_2PO_4, 2.72 g of K_2HPO_4, 0.05 g of $MgSO_4$, 7.95 g of $FeSO_4$, 0.01 g of $ZnCl_2$, 0.03 g of $CuCl_2$, 0.01 g of NaBr, and 0.01 g of Na_2MoO_4. BMM is a media for the growth of bacteria. These flasks served as the biological reactor's inoculation source once the bacteria entered their exponential growth phase. To begin the experiment with a healthy culture, 2 g·L^{-1} of D-glucose was added to the feed solution at starting. The reagents used in the experiments, which were of analytical purity and used as received, were supplied by MERCK Chemicals (Johannesburg, South Africa). Ultrapure water was used in all the experiments.

3.4. Chlorophenol Concentration Determination

Using HPLC (high-performance liquid chromatography, Waters Alliance 2695), the concentration of 4-CP was determined (Meadows Instrumentation Inc., Bristol, WI, USA). The Waters 626 LC System, 717 Plus Autosampler, and 996 Photodiode Array Detector made up the HPLC system. The column was a Waters PAH C18 Symmetry Column (4.6 mm 250 mm, 5 m). First, a 5 mL syringe filter with a 0.45 m pore size was used to filter the samples. An amount of 10 L was used for the injection, and a 254 nm wavelength was used for the detection. One percent acetic acid in water and one percent acetic acid in acetonitrile made up the mobile phases A and B, respectively. The Millennium 2010 Chromatography Manager translated the data. Filtered samples from the collection process were quantified using HPLC Waters 9625 after being passed through a 0.2 m mixed cellulose filter.

3.5. Simulation Analysis Using AQUASIM 2.0 Software

Data simulation with AQUASIM 2.0 uses the DASSL algorithm, which is based on the implicit (backward differencing), variable-step, variable-order Gear integration technique [36]. This method uses a system of partial and ordinary differential equations that are numerically integrated in time while also resolving the algebraic equations. Using conservative finite difference techniques, partial differential equations are spatially discretized [36]. The equation below was created using AQUASIM 2.0 and was obtained using differential conservation law. Additionally, the DASSL algorithm implementation permits the use of the full or branded Jacobian matrix Equation (10) in the solution of the nonlinear system of algebraic equations [36]:

$$\frac{\partial \widehat{p}}{dt} = \frac{\partial \widehat{j}}{\partial z} + \widehat{r} \tag{10}$$

Equation (10) is discretized as Equation (11):

$$\frac{d}{dt}\widehat{p}(x_i, t) = \frac{\widehat{j}_{num}(x_{i+0.5}, t) - \widehat{j}_{num}(x_{i-0.5}, t)}{x_{i+0.5} - x_{i-0.5}} + \widehat{r}(x_i, t) \tag{11}$$

$$J_{-} = \frac{\partial F}{\partial y} \tag{12}$$

$$J_{-} = \begin{bmatrix} \frac{\partial f_1}{\partial x_1} \cdots \frac{\partial f_1}{dy} \\ \frac{\partial f_2}{\partial x_2} \cdots \frac{df_2}{dy} \end{bmatrix} \tag{13}$$

3.6. Parameter Estimation

Using the fourth-order Runge–Kutta method, AQUASIM 2.0's mass balance was quantitatively assessed (RK-4). AQUASIM 2.0's simplex method was used to determine the parameters by minimizing the Chi-square (χ^2) values between the model data and the actual data. By minimizing Equation (13) with the constraints ($\ell_{min,i} \le \ell_i \le \ell_{max,i}$), Aquasim's estimate model parameters are represented by constant variables [36]. The weighted deviations between the measurements and the predicted model results are expressed as the sum of the squares in this equation [36]. A simplex method or secant algorithm is used to minimize Equation (14) [36].

$$x^2(\ell) = \sum_{i=1}^{n} \left[\frac{U_{meas,i} - U_i(\ell)}{\alpha_{meas,i}} \right]^2 \tag{14}$$

where $U_{meas,i}$ = i-th measurement, $\alpha_{meas,i}$ = standard deviation, $U_i(\ell)$ = calculated value of the model variable corresponding to the i-th measurement and evaluated at the time and location of this measurement, $\ell = (\ell_i, \ldots, \ell_m)$ = model parameters; $\ell_{min,i}$ and $\ell_{max,i}$ = minimum and maximum constant variable representing ℓ_i; n = the number of points; and x^2 = the sum of the deviation for all the fit targets.

3.7. Sensitivity Analysis of the Estimated Parameters

Combining identifiability and uncertainty analysis allows AQUASIM 2.0 to address sensitivity analysis problems [36]. In order to determine if the model parameters can be estimated exclusively using the available data, identifiability analysis with AQUASIM 2.0 is used. This study also aims to calculate the uncertainty of the parameter estimations. This is accomplished by estimating the parameters' correlation coefficients and standard errors during parameter estimation [36]. AQUASIM 2.0 distinguishes Equations (15)–(18):

$$\delta_{y,\ell}^{a,a} = \frac{\partial y}{\partial \ell} \tag{15}$$

$$\delta_{y,\ell}^{r,a} = \frac{1}{y}\frac{\partial y}{\partial \ell} \tag{16}$$

$$\delta_{y,\ell}^{a,r} = p\frac{\partial y}{\partial \ell} \tag{17}$$

$$\delta_{y,\ell}^{r,r} = \frac{p}{y}\frac{\partial y}{\partial \ell} \tag{18}$$

where y = arbitrary variable calculated by AQUASIM and ℓ = model parameter by a constant.

$$\delta_y = \sqrt{\sum_{i=1}^{m}\left(\frac{\partial y}{\partial \ell_i}\right)^2 \delta_{\ell_i}^2} \tag{19}$$

where ℓ_i= uncertainty model parameter, δ_i= standard deviations, $y(\ell_i, \ldots, \ell_m)$ = solution of the model equations for a given variable at a given location and time, δy = approximate standard deviation of the model result. The error contribution of each parameter is given as:

$$\delta_{y\ell}^{err} = \frac{\partial y}{\partial \ell}\delta_\ell \tag{20}$$

AUASIM 2.0 calculates Equations (15)–(18) and (20) using the derivatives as follows:

$$\frac{\partial y}{\partial \ell_i} \approx \frac{y(\ell_i + \Delta\ell_i) - y(\ell_i)}{\Delta\ell_i} \tag{21}$$

where $\Delta\ell_i$ = 1% of the standard deviation $\delta\ell_i$, of the parameter $\Delta\ell_i$.

3.8. General Mass Balance of 4-Chlorophenol in CSTR

The mass balances of 4-CP were suggested in order to model the reaction and acquire the theoretical evolution of the concentrations of the dangerous reactive organic molecules. According to the findings of various authors [27,28], the following reasonable assumptions were made: (1) differential conversion at each pass in the photoreaction chamber, (2) properly mixed reaction medium, (3) reaction at the solid–liquid interface, (4) negligible photolysis, and (5) mass transfer constraints can be disregarded. Fogler [37] developed the generalized mass balance surrounding a fully mixed photoreactor, and Equation (22) can be used to express it for 4-CP:

$$V\frac{dC}{dt} = Q(C_{in} - C) + r_s V \tag{22}$$

where C_{in} = influent substrate concentration ($M{\cdot}L^{-3}$), C = effluent substrate concentration ($M{\cdot}L^{-3}$), r_s = reaction rate ($M{\cdot}L^{-3}{\cdot}T^{-1}$), Q = flow rate across the reactor ($L^3{\cdot}T^{-1}$), and V = operating volume of the reactor (L^3). Therefore, during steady-state operation, the left-hand derivative approaches zero i.e., $V{\cdot}\frac{dC}{dt} \Rightarrow 0$. As a result, Equation (22) simplifies to Equation (23):

$$-r_s = \left(\frac{Q}{V}\right)(C_{in} - C) \tag{23}$$

Now, considering the single-reactant kinetics in the photoreaction chamber, we assume first order rate kinetics, $-r_s = k_s.C$, such that, the rate constant can be estimated during steady-state operation from the influent and effluent concentration as shown in Equation (24):

$$\frac{C}{C_{in}} = \frac{1}{1 + k_s\tau} \tag{24}$$

where τ = the hydraulic retention time (V/Q) (T) and k_s = first-order maximum reaction rate coefficient (T^{-1}).

Two reaction rate models, however, for the biological reactor, were assessed. The first was the reaction without inhibition as given by the Michaelis-Menten substrate utilization rate kinetics.

$-r_s = \frac{k_{ms}CX}{K_s+C}$ and the reaction with substrate level inhibition, which was widely used to evaluate the growth kinetics when the cells grew on a single substrate with the Briggs and Haldane model [38]: $-r_s = \frac{k_{ms}CX}{K_s+C+\frac{C^2}{K_I}}$, where k_{ms} = maximum specific reaction rate coefficient (T^{-1}), K_s = half velocity concentration ($M{\cdot}L^{-3}$), K_I = coefficient of inhibition

($M{\cdot}L^{-3}$), and X = the viable cell concentration in the reactor ($M{\cdot}L^{-3}$). The reaction rate is substituted into the reactor mass balance in Equation (22):

$$V\frac{dC}{dt} = Q(C_{in} - C) + \left(-\frac{k_{ms}CX}{K_s + C}\right)V \quad (25)$$

which resolves to:

$$\frac{X\tau}{C_{in} - C} = \left(\frac{K_s}{k_{ms}}\right)\frac{1}{C} + \frac{1}{k_{ms}} \quad (26)$$

The cell production in the reactor is proportional to the substrate used by the bacteria there, as shown by Equation (26) derived from the cell mass balance across the reactor, i.e., X = mass of cells per mass of substrate utilized. Only if hydraulic retention time (HRT) is brief enough to make cell death inside the reactor insignificant does Equation (26)'s expression hold true. However, as can be shown in Equation (27), the mass balance equation that resulted from the replacement of the inhibitory kinetics was more complicated:

$$V\frac{dC}{dt} = Q(C_{in} - C) + \left(-\frac{k_{ms}CX}{K_s + C + \frac{C^2}{K_I}}\right)V \quad (27)$$

which resolves to:

$$C_{in} - C = \left(\frac{V}{Q}\right)\left(-\frac{k_{ms}C.X}{K_s + C + \frac{C^2}{K_I}}\right)V \quad (28)$$

Equation (28) can be solved numerically.

4. Conclusions

The modelling of the system behavior was successfully carried out in a combined chemical and biological treatment system under 4-CP concentrations loading range of 50–1000 $mg{\cdot}L^{-1}$. Optimized operating parameters were achieved using AQUASIM 2.0 software. Moreover, it led to an overall degradation efficiency of 98% and 97% at a 50 $mg{\cdot}L^{-1}$ and 200 $mg{\cdot}L^{-1}$ with added H_2O_2 feed concentrations respectively, which is significantly higher as compared to the photocatalytic performance of chlorophenols degradation reported in previous studies. The results indicated that the Monod model with inhibition coefficient (Haldane model) well described the 4-CP degradation. The findings demonstrate that the Haldane model is more accurate since it takes into account the inhibition coefficient, which in this study's circumstance could not have been disregarded. Moreover, the degradation kinetics values and the experimental results matched well with a good degree of agreement (R^2 = 0.998). The results of the current study are encouraging enough to support additional investigation utilizing continuous or semi-continuous systems. They may also be useful for wastewater compositions containing 4-CP and other chlorinated monophenols and can be utilized as a starting point for design calculations that could eventually lead to the scaling up of continuous combined chemical and biological treatment system treating 4-CP in industrial waste streams.

Author Contributions: S.M.; writing–original draft preparation, conceptualization, methodology, AQUASIM 2.0 software, validation, and formal analysis, E.N.-C.; writing–review and editing, funding acquisition. All authors have read and agreed to the published version of the manuscript.

Funding: This research received no external funding.

Data Availability Statement: Not Applicable.

Conflicts of Interest: The authors declare no conflict of interest.

Nomenclature

Parameter	Unit	Parameter description
a	(T)	Kinetic parameter constant for accumulation of ($OH^{\bullet}$) radicals
b	(T)	Kinetic parameter constant for accumulation of ($OH^{\bullet}$) radicals
c	(T)	Kinetic parameter constant for accumulation of ($OH^{\bullet}$) radicals
C_{in}	($M \cdot L^{-3}$)	Influent targeted substrate concentration
C_1	($M \cdot L^{-3}$)	Chemical reactor calculated effluent concentration
C_2	($M \cdot L^{-3}$)	Biological reactor calculated effluent concentration
C	($M \cdot L^{-3}$)	Concentration at steady-state
C_{ohmax}	($M \cdot L^{-3}$)	Initial concentration of ($OH^{\bullet}$) radicals
C_{oh}	($M \cdot L^{-3}$)	Calculated concentration of ($OH^{\bullet}$) radicals
V	(L^3)	Reactor volume
Q	($L^3 \cdot T^{-1}$)	Flow rate across the reactor
k_{ms}	(T^{-1})	Maximum specific reaction rate coefficient
K_s	($M \cdot L^{-3}$)	Half velocity concentration
K_I	($M \cdot L^{-3}$)	Coefficient of inhibition
X_0	($M \cdot L^{-3}$)	Attached viable cells concentration in the reactor
Y		The cell growth yield
t_0	(T)	Initial boundary value for accumulation of ($OH^{\bullet}$) radicals
X	(g)	Mass of cells per mass of substrate utilized.
k_d	(T^{-1})	Endogenous decay coefficient

Symbols

$U_{meas,i}$	i-th measurement
$\alpha_{meas,i}$	standard deviation
$U_i(\ell)$	calculated value of the model variable corresponding to the i-th measurement and evaluated at the time and location of this measurement
$\ell = (\ell_i, \ldots, \ell_m)$	model parameters
$\ell_{min,i}, \ell_{max,i}$	minimum and maximum constant variable representing
ℓ_i n	number of points
x^2	the sum of the deviation for all the fit targets
ℓ_i	uncertainty model parameter
δ_i	standard deviations
$y(\ell_i, \ldots, \ell_m)$	solution of the model equations for a given variable at a given location and time
δ_y	approximate standard deviation of the model result
y	arbitrary variable calculated by AQUASIM
ℓ	model parameter by a constant
4-CP	4-chlorophenol

References

1. Wang, J.; Ma, X.; Liu, S.; Sun, P.; Fan, P.; Xia, C. Biodegradation of phenol and 4-chlorophenol by Candida tropicalis W1. *Procedia Environ. Sci.* **2012**, *16*, 299–303. [CrossRef]
2. Mohammadi, M.; Sabbaghi, S. Photo-catalytic degradation of 2,4-DCP wastewater using MWCNT/TiO_2 nano-composite activated by UV and solar light. *Environ. Nanotechnol. Monit. Manag* **2014**, *1*, 24–29. [CrossRef]
3. Sgherza, D.; Pentassuglia, S.; Altieri, V.; Mascolo, G.; De Sanctis, M.; Di Iaconi, C. Integrating biodegradation and ozone-catalysed oxidation for treatment and reuse of biomass gasification wastewater. *J. Water Process Eng.* **2021**, *43*, 102297. [CrossRef]
4. Patel, N.; Shahane, S.; Bhunia, B.; Mishra, U.; Chaudhary, V.K.; Srivastav, A.L. Biodegradation of 4-chlorophenol in batch and continuous packed bed reactor by isolated Bacillus subtilis. *J. Environ. Manag.* **2021**, *301*, 113851. [CrossRef]
5. Zhang, J.; Liu, X.; Xu, Z.; Chen, H.; Yang, Y. Degradation of chlorophenols catalyzed by laccase. *Int. Biodeterior. Biodegradation* **2008**, *61*, 351–356. [CrossRef]
6. Hamdaoui, O.; Naffrechoux, E. Sonochemical and photosonochemical degradation of 4-chlorophenol in aqueous media. *Ultrason. Sonochemistry* **2008**, *15*, 981–987. [CrossRef]
7. Ahmadi, M.; Samarbaf, S.; Golshan, M.; Jorfi, S.; Ramavandi, B. Data on photo-catalytic degradation of 4- chlorophenol from aqueous solution using UV/ZnO/persulfate. *Data Brief* **2018**, *20*, 582–586. [CrossRef]
8. Fu, T.; Gong, X.; Guo, J.; Yang, Z.; Liu, Y. Zn-CNTs-Cu catalytic in-situ generation of H_2O_2 for efficient catalytic wet peroxide oxidation of high-concentration 4-chlorophenol. *J. Hazard. Mater.* **2020**, *401*, 123392. [CrossRef]
9. Gomez, M.; Murcia, M.D.; Christofi, N.; Gómez, E.G.; Gómez, J. Photodegradation of 4-chlorophenol using XeBr, KrCl and Cl_2 barrier-discharge excilamps: A comparative study. *Chem. Eng. J.* **2010**, *158*, 120–128. [CrossRef]

10. Rodriguez-Narvaez, O.M.; Peralta-Hernandez, J.M.; Goonetilleke, A.; Bandala, E.R. Treatment technologies for emerging contaminants in water: A review. *Chem. Eng. J.* **2017**, *323*, 361–380. [CrossRef]
11. Wang, J.; Bai, Z. Fe-based catalysts for heterogeneous catalytic ozonation of emerging contaminants in water and wastewater. *Chem. Eng. J.* **2017**, *312*, 79–98. [CrossRef]
12. Deng, J.; Wu, G.; Yuan, S.; Zhan, X.; Wang, W.; Hu, Z. Ciprofloxacin degradation in UV/chlorine advanced oxidation process: Influencing factors, mechanisms and degradation pathways. *J. Photochem. Photobiol. A* **2019**, *371*, 151–158. [CrossRef]
13. Wang, J.; Wang, S. Preparation, modification and environmental application of biochar: A review. *J. Clean. Prod.* **2019**, *227*, 1002–1022. [CrossRef]
14. Long, M.; Zeng, C.; Wang, Z.; Xia, S.; Zhou, C. Complete dechlorination and mineralization of para-chlorophenol (4-CP) in a hydrogen-based membrane biofilm reactor (MBfR). *J. Clean. Prod.* **2020**, *276*, 123257. [CrossRef]
15. Zhang, K.; Yang, W.; Liu, Y.; Zhang, K.; Chen, Y.; Yin, X. Laccase immobilized on chitosan-coated Fe3O4 nanoparticles as reusable biocatalyst for degradation of chlorophenol. *J. Mol. Struct.* **2020**, *1220*, 128769. [CrossRef]
16. Patidar, R.; Srivastava, V.C. Mechanistic and kinetic insights of synergistic mineralization of ofloxacin using a sono-photo hybrid process. *Chem. Eng. J.* **2020**, *403*, 125736. [CrossRef]
17. Munoz, M.; de Pedro, Z.M.; Casas, J.A.; Rodriguez, J.J. Improved wet peroxide oxidation strategies for the treatment of chlorophenols. *Chem. Eng. J.* **2013**, *228*, 646–654. [CrossRef]
18. Sarria, V.; Parra, S.; Adler, N.; Péringer, P.; Benitez, N.; Pulgarin, C. Recent developments in the coupling of photoassisted and aerobic biological processes for the treatment of biorecalcitrant compounds. *Catal. Today* **2002**, *76*, 301–315. [CrossRef]
19. González, L.F.; Sarria, V.; Sanchez, O.F. Degradation of chlorophenols by sequential biological-advanced oxidative process using Trametes pubescens and TiO_2/UV. *Bioresour. Technol.* **2010**, *101*, 3493–3499. [CrossRef] [PubMed]
20. Gaya, U.I.; Abdullah, A.H.; Zainal, Z.; Hussein, M.Z. Photocatalytic treatment of 4-chlorophenol in aqueous ZnO suspensions: Intermediates, influence of dosage and inorganic anions. *J. Hazard. Mater.* **2009**, *168*, 57–63. [CrossRef]
21. Camera-Roda, G.; Santarelli, F.; Martin, C.A. Design of photocatalytic reactors made easy by considering the photons as immaterial reactants. *Sol. Energ.* **2005**, *79*, 343–352. [CrossRef]
22. Malato, S.; Fernandez-Ibañez, P.; Maldonado, M.I.; Blanco, J.; Gernjak, W. Decontamination and disinfection of water by solar photocatalysis: Recent overview and trends. *Catal. Today* **2009**, *147*, 1–59. [CrossRef]
23. Wade, M.J.; Pattinson, R.W.; Parker, N.G.; Dolfing, J. Emergent behavior in a chlorophenol-mineralising three-tiered microbial 'food web'. *J. Theor. Biol.* **2016**, *389*, 171–186. [CrossRef]
24. Yang, K.; Zhao, Y.; Ji, M.; Li, Z.; Zhai, S.; Zhou, X.; Wang, Q.; Wang, C.; Liang, B. Challenges and opportunities for the biodegradation of chlorophenols: Aerobic, anaerobic and bioelectrochemical processes. *Water Res.* **2021**, *193*, 116862. [CrossRef] [PubMed]
25. Albarrán, G.; Mendoza, E. Radiolysis induced degradation of 1,3-dichlorobenzene and 4-chlorophenol in aqueous solution. *Radiat. Phys. Chem.* **2020**, 109318. [CrossRef]
26. Ghaly, M.Y.; Hartel, G.; Mayer, R.; Haseneder, R. Photochemical oxidation of p-chlorophenol by UV/H_2O_2 and photo-Fenton process. A comparative study. *Waste Manag.* **2001**, *21*, 41–47.
27. Tolosana-Moranchel, A.; Manassero, A.; Satuf, M.L.; Alfano, A.M.; Casas, J.A.; Bahamonde, A. TiO_2-rGO photocatalytic degradation of an emerging pollutant: Kinetic modelling and determination of intrinsic kinetic parameters. *J. Environ. Chem. Eng.* **2019**, *7*, 103406. [CrossRef]
28. Satuf, M.L.; Brandi, R.J.; Cassano, A.E.; Alfano, O.M. Photocatalytic degradation of 4-chlorophenol: A kinetic study. *Appl. Catal. B Environ.* **2008**, *82*, 37–49. [CrossRef]
29. Assadi, A.; Alimoradzadeh, R.; Movahedyan, H.; Amin, M.M. Intensified 4-chlorophenol biodegradation in an aerobic sequencing batch reactor: Microbial and kinetic properties evaluation. *Environ. Technol. Innov.* **2020**, *21*, 101243. [CrossRef]
30. Sahinkaya, E.; Dilek, F.B. Biodegradation of 4-CP and 2.4-DCP mixture in a rotating biological reactor (RBC). *Biochem. Eng. J.* **2006**, *31*, 141–147. [CrossRef]
31. Sharma, S.; Mukhopadhyay, M.; Murthy, Z.V.P. Rate parameter estimation for 4-chlorophenol degradation by UV and organic oxidants. *J. Ind. Eng. Chem.* **2012**, *18*, 249–254. [CrossRef]
32. Czaplicka, M. Photo-degradation of chlorophenols in the aqueous solution. *J. Hazard. Mater.* **2006**, *134*, 45–59. [CrossRef]
33. Oh, W.D.; Lim, P.E.; Seng, S.E.; Sujari, A.N.A. Kinetic modeling of bioregeneration of chlorophenol-loaded granular activated carbon in simultaneous adsorption and biodegradation processes. *Bioresour. Technol.* **2012**, *114*, 179–187. [CrossRef] [PubMed]
34. Konya, I.; Eker, S.; Kargi, F. Mathematical modelling of 4-chlorophenol inhibition on COD and 4-chlorophenol removals in an activated sludge unit. *J. Hazard. Mater.* **2007**, *143*, 233–239. [CrossRef]
35. Kayan, I.; Oz, N.A.; Kantar, C. Comparison of treatability of four different chlorophenol-containing wastewater by pyrite-Fenton process combined with aerobic biodegradation: Role of sludge acclimation. *J. Environ. Manag.* **2020**, *279*, 111781. [CrossRef] [PubMed]
36. Rechert, P. *Computer Program for the Identification and Simulation of Aquatic Systems, AQUASIM 2.0: User Manual*; Swiss Federal Institute for Environmental Science and Technology (EAWAG): Dübendorf, Switzerland, 1998.
37. Fogler, H.S. *Elements of Chemical Reaction Engineering*, 3rd ed.; Pearson Education Limited: Hongkong, China, 1999; ISBN 9780135317167.
38. Briggs, G.E.; Haldane, J.B.S. A Note on the Kinetics of Enzyme Action. *Biochem. J.* **1925**, *19*, 338–339. [CrossRef]

catalysts

MDPI

Communication

Study on Optimum Preparation Conditions of $ZnIn_2S_4$ to Effectively Reduce Cr(VI) under Visible Light Radiation

Yali Song [1,*], Huan Zhang [1], Ziyan Li [1], Long Huang [2], Junhao Xie [1], Long Han [1], Ruoying Zheng [1], Yuqing Zhang [1] and Hongzhong Zhang [1,*]

[1] Department of Material and Chemical Engineering, Henan Collaborative Innovation Centre of Environmental Pollution Control and Ecological Restoration, Zhengzhou University of Light Industry, Zhengzhou 450001, China

[2] School of Ecology and Environment, Zhengzhou University, Zhengzhou 450001, China

* Correspondence: songyl@zzuli.edu.cn (Y.S.); zhz@zzuli.edu.cn (H.Z.)

Abstract: Previous studies have displayed various conclusions about the effect of preparation factors on the photoreduction property of $ZnIn_2S_4$. Therefore, it is not easy to figure out the optimal preparation conditions of $ZnIn_2S_4$ for Cr(VI) photoreduction. To ensure Cr(VI) reduction efficiency, various $ZnIn_2S_4$ photocatalysts were prepared in different solvents (i.e., water and ethylene glycol) and temperatures (i.e., 120 °C, 150 °C and 180°C). Different characterization methods were used to explain the difference in optical performance and photocatalytic property among the obtained samples. The results show that all the samples exhibit a similar band gap. The reaction solvent and temperature have a great influence on the surface morphology and optical property, leading to the different photocatalytic properties. $ZnIn_2S_4$ synthesized at 120 °C in the solvothermal condition shows the optimal efficiency on Cr(VI) photoreduction due to the effective utilization of photo-induced carriers. The reasonable analysis and effective conclusion presented may provide the optimal synthesis method of $ZnIn_2S_4$ to effectively remove Cr(VI) from water environment.

Keywords: $ZnIn_2S_4$; Cr(VI); solvothermal; hydrothermal; photoreduction

1. Introduction

With the rapid development of society, compounds containing hexavalent chromium (Cr(VI)) have been used in industries, including electroplating, mining, pigment manufacturing, and smelting [1–3]. Cr(VI) can enter water environment through sewage effluent, rainfall, and runoff and possesses the characteristic of easy migration in environment and high solubility in water [4]. Cr(VI) can be untaken by animals, plants, and even humans via the food chain, causing potential and persistent threat for humans and ecosystem [5]. Therefore, various strategies have been developed to remove Cr(VI) from water environments, such as physical adsorption, chemical precipitation, chemical reduction, and membrane filtration [5]. However, the traditional approaches have the disadvantages of high cost and power consumption, complicated operation, and membrane pollution [6]. It is urgent to find sustainable, renewable, and effective ways to remove Cr(VI). In the situation of environmental pollution and energy crisis, photocatalytic reduction is an extremely efficient method to reduce Cr(VI) by photogenerated electrons due to high efficiency, low cost, and environmental friendliness [2].

In recent years, layered $ZnIn_2S_4$, especially hexagonal $ZnIn_2S_4$, has exhibited excellent photocatalytic activity in hydrogen generation, Cr(VI) reduction and carbon dioxide reduction under visible light radiation [7,8]. $ZnIn_2S_4$ nanopowder synthesized via hydrothermal method as a visible light catalyst was applied to reduce Cr(VI) in aqueous solution for the first time [9]. After that, various methods for the reduction of aqueous Cr(VI) were developed. Among the variety of methods for synthesizing micro/nano-structure $ZnIn_2S_4$, both hydrothermal and solvothermal approaches are prevalently used in reported studies. Gou et al. [10] adjusted reaction solvent (i.e., pyridine, alcohol, acetonitrile) and temperatures (120–180 °C) to control the shape of $ZnIn_2S_4$. However, the photocatalytic activity of these $ZnIn_2S_4$ photocatalysts has not been investigated, which is significant on its application. Tian et al. [11] used the hydrothermal method to synthesize $ZnIn_2S_4$ by changing reaction temperature from 120 °C to 200 °C for hydrogen generation, and $ZnIn_2S_4$ synthesized at 160 °C showed the best activity. Based on the above researches, it can be obtained that solvent and temperature have a great influence on the shape and photoreduction property of $ZnIn_2S_4$. In the previous research, the influence of preparation conditions on the morphology and property of $ZnIn_2S_4$ has been studied. However, the suitable conditions for synthesizing bulk $ZnIn_2S_4$ with efficient reduction activities have not systematically been concluded and elaborated. Therefore, it is difficult to figure out the optimum synthesis conditions of $ZnIn_2S_4$ for Cr(VI) reduction from water environment. However, this is convenient and essential to evaluate material properties.

In the work, various $ZnIn_2S_4$ photocatalysts were prepared in different synthesis solvent and temperatures, in order to investigate the optimum condition for the application of Cr(VI) reduction. The photocatalytic properties of different $ZnIn_2S_4$ photocatalysts were evaluated by Cr(VI) reduction under visible light radiation and the difference in reduction efficiency has been analyzed and discussed.

2. Results and Discussion

2.1. Character of Different $ZnIn_2S_4$ Photocatalysts

XRD was performed to study the phase structures of the $ZnIn_2S_4$ photocatalysts. Figure 1 displays that all the samples show the same characteristic peaks at 2θ of 22.2°, 27.7°, 32.5°, 41.3°, 47.3°, 54.5, 57.1°, and 77.5°, which correspond to (006), (102), (104), (108), (110), (116), (022), and (123) crystal faces of $ZnIn_2S_4$, respectively (JCPDS, No. 65-2023) [12]. Besides, no other impurity peaks can be found in the XRD patterns. This result indicates that $ZnIn_2S_4$ can be successfully prepared via both the hydrothermal process after reacting at 120 °C for 2 h (DI-120) and solvothermal process after reacting at 120 °C (EG-120), 150 °C (EG-150) and 180 °C (EG-180) for 2 h. From the diffraction peak intensity, it should be noted that diffraction peaks (i.e., the crystal faces of (006), (104), (116), (022)) of DI-120 are narrower and sharper in comparison with EG-120, EG-150, and EG-180. This means that the hydrothermal system is beneficial to the crystallization of $ZnIn_2S_4$ as compared with the solvothermal system.

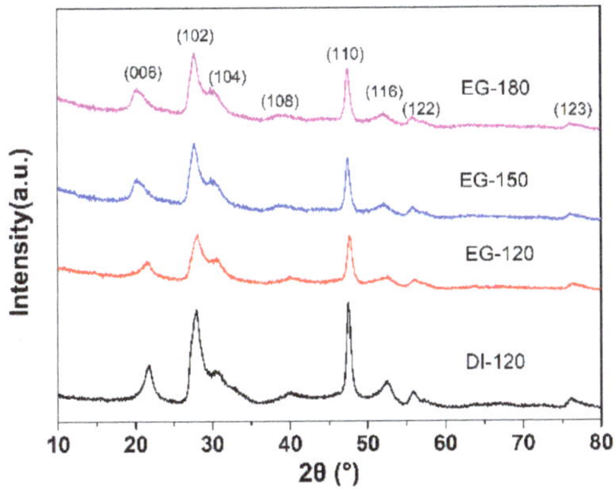

Figure 1. XRD patterns of different $ZnIn_2S_4$ photocatalysts.

The morphologies of different $ZnIn_2S_4$ photocatalysts were characterized by SEM (Figure 2a–d). It can be seen that all the samples display spherical structures consisting of numerous sheets due to the Oswald ripening [8]. DI-120 displays marigold-like spherical structure, which is similar with that prepared at 180 °C by the hydrothermal method [13].

However, EG-120, EG-150 and EG-180 show rose-like spherical structure. The petal gap of EG-120 further increases when compared with DI-120, which can provide more active sites to enhance photocatalytic activity. However, the previous study reported that the reaction temperature should be increased to 200 °C in the hydrothermal process to achieve wider petal gap [11]. The thickness of petal in synthesized $ZnIn_2S_4$ is increased when raising temperature from 120 °C to 180 °C, which is similar with the result of TEM (Figure S1a,b). This means that the higher temperature may lead to the increased thickness in the $ZnIn_2S_4$ synthesis process. Besides, the lattice fringes and good crystallinity can be clearly observed in the corresponding TEM image of EG-120 (Figure S2). The lattice spacing of 0.333 nm and 0.410 nm corresponds to the (102) and (006) planes of the tetragonal $ZnIn_2S_4$, respectively [14].

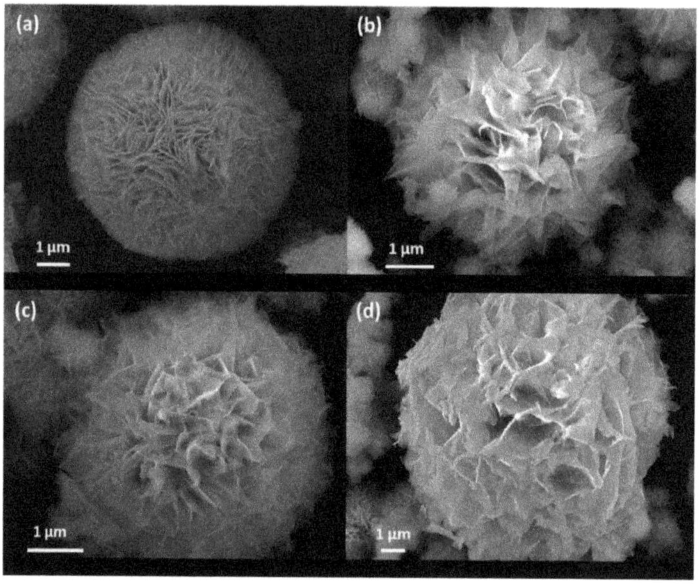

Figure 2. SEM of (**a**) DI-120, (**b**) EG-120, (**c**) EG-150 and (**d**) EG-180.

2.2. Comparison on Photocatalytic Activities of Different $ZnIn_2S_4$ Photocatalysts

To evaluate the photocatalytic performance of different $ZnIn_2S_4$ photocatalysts, the removal efficiencies of Cr(VI) under visible light irradiation were investigated. First, adsorption removal in the dark by different $ZnIn_2S_4$ photocatalysts was studied. Figure S3 shows that only less than 5% of Cr(VI) can be removed in the adsorption processes. When exposed to visible light for 30 min, the removal efficiencies of Cr(VI) by DI-120, EG-120, EG-150, and EG-180 are 53.3%, 94.3%, 57.1%, and 78.4% (Figure 3a). The result indicates that Cr(VI) is mainly removed by the photoreduction action of different $ZnIn_2S_4$ due to the highest valence state of chromium and the limited adsorption effect of $ZnIn_2S_4$. Photocatalytic processes by DI-120, EG-120, EG-150, and EG-180 all obey the first order dynamic models and the kinetic constant in EG-120 process is 4, 3.43, and 1.85 times than that in DI-120, EG-150, and EG-180 processes, respectively (Figure 3b). The result shows that $ZnIn_2S_4$ prepared via solvothermal process in 120 °C exhibits the excellent photoreduction activity, which is inconsistent with the previous study [15]. This may be due to the different raw materials, resulting in different effects of solvents on the photoreduction performance of $ZnIn_2S_4$.

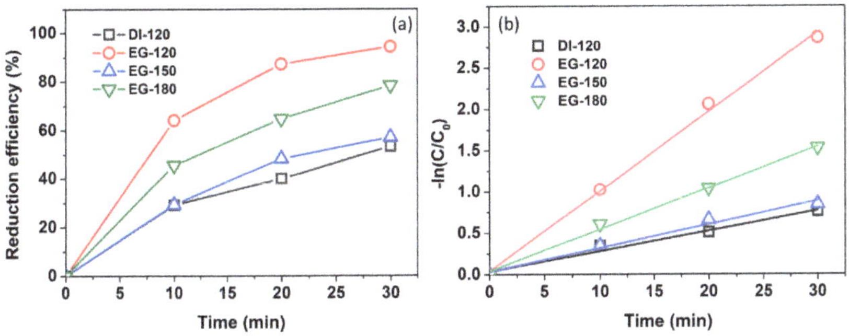

Figure 3. (**a**) Photoreduction efficiency of Cr(VI) and (**b**) kinetic models by different $ZnIn_2S_4$ photocatalysts under visible light irradiation.

Besides, different quenching agents were added in the photoreduction process to investigate the mechanism of Cr(VI) removal by EG-120. Tertiary butanol (t-BuOH, 1 mM), benzoquinone (BQ, 1 mM), ethylenediamine tetraacetate (EDTA-2Na, 1 mM) and $AgNO_3$ (1 mM) were employed as scavengers for OH·, $\cdot O_2^-$, holes, and photo-induced electrons, respectively [5,16]. Figure S4 shows that photo-induced electrons play a major role during the removal process of Cr(VI) by EG-120. $\cdot O_2^-$ shows a minor effect on Cr(VI) removal and the other two active substances almost never work. The quenching experimental result indicates that Cr(VI) is mainly removed by the reducing action of photo-induced electrons in EG-120 photocatalytic system.

2.3. Comparison on Optical Property of Different $ZnIn_2S_4$ Photocatalysts

When $ZnIn_2S_4$ and Cr(VI) solution are simultaneously irradiated by visible light, Cr(VI) is reduced by photo-generated electrons due to the separation of photo-generated electrons and holes. The generation, separation, and combination of electron-hole pairs in $ZnIn_2S_4$ influence the Cr(VI) reduction efficiency. Therefore, the optical properties of different $ZnIn_2S_4$ were investigated to expound the utilization of electron-hole pairs and confirm the difference in Cr(VI) reduction.

First, UV–vis DRS was used to investigate the visible light absorption of $ZnIn_2S_4$. Commonly, the wide visible light region is beneficial to the increase of photocatalytic activity. As shown in Figure 4a, all the $ZnIn_2S_4$ photocatalysts exhibit strong absorption intensity in the UV and partially visible region, and the absorption edges with a steep shape locate at 500~554 nm. The corresponding band gaps (Figure 4b) of $ZnIn_2S_4$ photocatalysts were estimated via the Kubelka–Munk method [17]. As shown, the Eg values are calculated as 2.74 eV for EG-180, 2.71 eV for EG-150, 2.58 eV for EG-120, and 2.47 eV for DI-120. Relatively, DI-120 is superior to absorb visible light but demonstrates little difference with EG-120.

Besides, electrochemical impedance spectroscopy (EIS) and photoluminescence (PL) spectra were conducted to figure out the combination of electron-hole pairs. Normally, the small radius in an EIS Nyquist indicates the inhibited recombination of photoinduced carriers, suggesting the effective separation of electron-hole pairs [18,19]. Figure 5a presents EIS of different photocatalysts irradiated with visible light. As shown, the radius of EG-120 is significantly smaller than the other three $ZnIn_2S_4$ photocatalysts, suggesting that EG-120 can exhibit the excellent electroconductivity. EG-120 is expected to possess the effective transfer and separation efficiency of photoinduced electron-hole pairs when irradiated with visible light, which is beneficial for the enhancement of photocatalytic activity. Meanwhile, the PL spectra of different $ZnIn_2S_4$ photocatalysts at an excitation wavelength of 315 nm have been measured and compared under visible light irradiation. As shown in Figure 5b, the emission intensity of DI-120 is similar with that of EG-180 and exhibits high PL intensity, indicating the fast recombination rate of photogenerated pairs. EG-150 and EG-120 also show the similar PL intensity. Compared with DI-120 and EG-180, a decreased intensity can be obviously obtained, suggesting the more effective separation of photoinduced carriers. Furthermore, transient photocurrent responses were performed to investigate the transportation behaviors of electron-hole pairs. As shown in Figure 5c, all the samples exhibit a quick response to the light on, but the generated photocurrent responses are different. In comparison with that in the dark, the photocurrent density of EG-120, DI-120, EG-150, and EG-180 can be increased by 0.17, 0.10, 0.08, and 0.05 $\mu A \cdot cm^{-2}$ at 0 V vs. RHE when irradiated with visible light, respectively. The response value of EG-120 is 1.7, 2.1, and 3.4 times higher than that of DI-120, EG-150, and EG-180 at the same experimental condition. In addition, the photocurrent intensity does not change significantly after four light on-off cycles, which indicates the excellent stability of the four $ZnIn_2S_4$ photocatalysts [20]. However, among the methods for improving $ZnIn_2S_4$ activity, Au doping and $Sn^{2+}/^{4+}$ co-doping $ZnIn_2S_4$ shows the decreased photocurrent intensity with increasing cycle number [21,22]. The above synthesis regarding optical property shows that the photogenerated electrons and holes of EG-120 are separated effectively and the carrier life is extended, which is be beneficial to utilize photo-induced electrons for Cr(VI) reduction. This is consistent with the results of photoreduction activity of $ZnIn_2S_4$ (Figure 3).

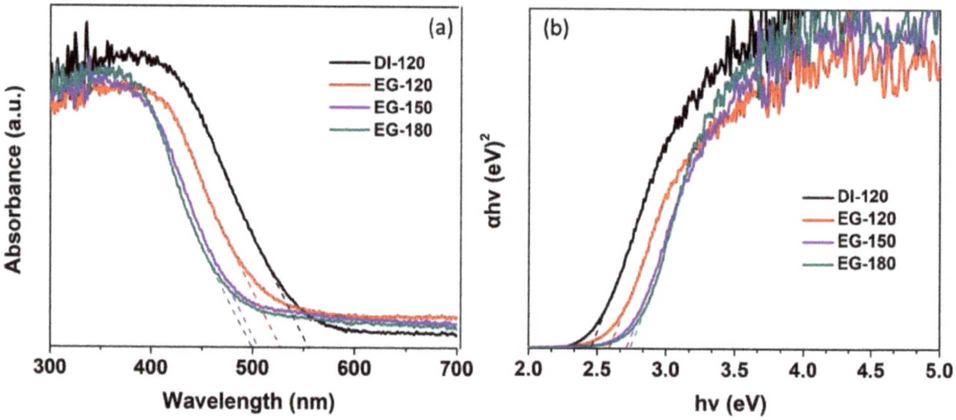

Figure 4. (**a**) UV–vis DRS and (**b**) Plot of $(\alpha h\nu)^2$ vs. hν for the band gap energy of different $ZnIn_2S_4$ photocatalysts.

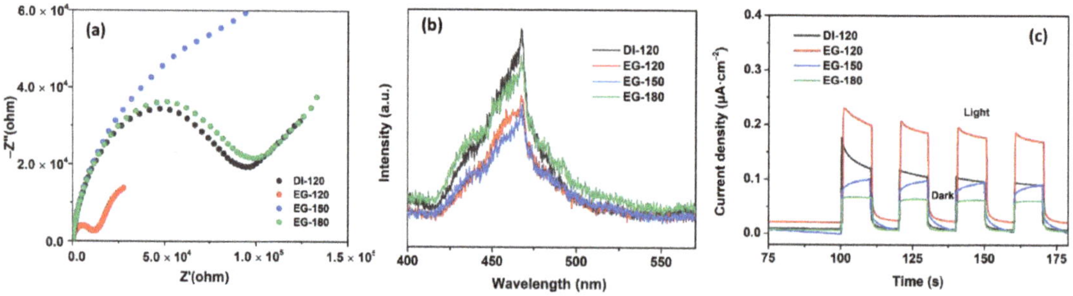

Figure 5. (**a**) electrochemical impedance spectroscopy, (**b**) photoluminescence spectra, and (**c**) transient photocurrent response of different $ZnIn_2S_4$ photocatalysts irradiated with visible light.

3. Materials and Methods

3.1. Reagents

Thioacetamide, indium chloride, and zinc chloride were purchased from Aladdin Chemical Reagent. Ethylene glycol, benzoquinone, ethylenediamine tetraacetate, silver nitrate, and tertiary butanol were purchased from Sinopharm Chemical Reagent. Potassium bichromate was purchased from Tianjin Fengchuan Chemical Reagent Technology Co. Ltd. (Tianjin, China). All the reagents were in analytically grade and used without any further processing.

3.2. Preparation of Different $ZnIn_2S_4$

Briefly, 0.442 g of indium chloride and 0.1363 of zinc chloride were mixed in 60 mL ethylene glycol (EG) and treated by ultrasonic wave for 30 min. Then 0.3 g of thioacetamide was added into the above solution. After stirring for 1 h, the obtained solution was transferred to a 100 mL Teflon-lined autoclave and heated at 120 °C for 2 h. While cooling down to room temperature, $ZnIn_2S_4$ was obtained by centrifugation, washed with deionized water/ethanol, dried, and marked as EG-120. When $ZnIn_2S_4$ was prepared in ethylene glycol at 150 °C and 180 °C, the as-prepared $ZnIn_2S_4$ was marked as EG-150 and EG-180, respectively. While $ZnIn_2S_4$ was prepared in deionized water instead of ethylene glycol at 120 °C, the obtained $ZnIn_2S_4$ was marked as DI-120.

3.3. Characterization Analysis of $ZnIn_2S_4$

Crystal phases and structures of the samples were investigated by X-ray diffraction analysis (XRD, Bruker D8 ADVANCE, Bruker AXS GmbH., Karlsruhe, Germany) with Cu-Kα radiation, λ = 0.15406 nm. Morphology was analyzed by transmission electron microscopy (TEM, JEM 2100, JEOL, Tokyo, Japan) and scanning electron microscopy (SEM, JSM-7001F, JEOL, Tokyo, Japan). Light adsorption was characterized by UV–vis diffuse reflection spectra (DRS, U-3900H, Hitachi, Tokyo, Japan). The transient photocurrent responses and electrochemical impedance spectroscopy (EIS) were performed by using an electrochemical workstation (CHI660E, Shanghai CH Instrument Company, Shanghai, China) with three electrodes. In addition, a 300 W xenon lamp (PLS-SXE300, Beijing, China) equipped with a 420 nm cut-off filter was adopted as the optical source. The photoluminescence (PL) spectra were conducted by a fluorescence spectrophotometer (FLS980, Bain Square, UK) at an excitation wavelength of 315 nm.

3.4. Photoreduction Cr(VI) by $ZnIn_2S_4$ under Visible Light Irradiation

The photoreduction experiments were performed in a 150 mL beaker at room temperature. Hence, 10 mg of $ZnIn_2S_4$ was dispersed in 100 mL of 10 mg L^{-1} Cr(VI) solution by ultrasonic wave for 5 min and stirred in the dark for 30 min to achieve adsorption equilibrium between Cr(VI) and the photocatalyst. Afterwards, the obtained suspension was exposed to visible light irradiation and the illumination radiation intensity was approximately 100 mW cm^{-2}. After irradiation for some time, 2 mL of reaction solution was taken out, filtered, and analyzed by ultraviolet spectrophotometry.

3.5. Determination of Cr(VI) Concentration

Briefly, 1 mL of sample solution, 0.5 mL of H_2SO_4, 0.5 mL H_3PO_4, 1 mL 1,5-diphenylcarbazide/acetone/water ternary solution, and 47 mL deionized water were uniformly mixed. The mixed solution stood for 10 min and measured at 540 nm by ultraviolet spectrophotometry and the measured absorbance at 540 nm was denoted as A_i. For comparison, 0.5 mL of H_2SO_4, 0.5 mL H_3PO_4, 1 mL 1,5-diphenylcarbazide/acetone/water ternary solution, and 47 mL deionized water were uniformly mixed as well to eliminate the interference of deionized water. After standing for 10 min, the mixed solution was measured at 540 nm and the absorbance was denoted as B_0. The removal efficiency of Cr(VI) was calculated as Equation (1).

$$\eta Cr_{(VI)}\ (\%) = \frac{(A_0 - B_0) - (A_i - B_0)}{A_0 - B_0} \times \% \quad (1)$$

where A_0 represents the absorbance of initial concentration, B_0 represents the absorbance of deionized water, and A_i represents the absorbance of reaction solution after being irradiated for a certain time.

4. Conclusions

To identify the optimum preparation conditions, a series of $ZnIn_2S_4$ photocatalysts were synthesized at 120 °C in water and at different temperatures (120 °C, 150 °C and 180 °C) in ethylene glycol. All the synthesized $ZnIn_2S_4$ photocatalysts display spherical structures. Among them, DI-120 shows a marigold-like spherical structure. EG-120, EG-150, and EG-180 exhibit a rose-like spherical structure. Therein, EG-120 shows the widest petal gap, providing more active sites to improve photocatalytic activity. UV-vis DRS indicates there is no obvious difference among the absorption edges of all the samples, which are located at 500~554 nm. EIS and PL spectra show that EG-120 exhibits the effective transfer and low combination efficiency of photo-induced electron pairs. Besides, transient photocurrent response also reveals that EG-120 shows the effective separation of photo-induced carriers, the photocurrent value of which is 1.7, 2.1, and 3.4 times higher than that of DI-120, EG-150, and EG-180, respectively. As expected, EG-120 exhibits the optimum reduction efficiency, due to the surface morphology and the excellent optical property.

Supplementary Materials: The following supporting information can be downloaded at: https://www.mdpi.com/article/10.3390/catal12111429/s1, Figure S1: TEM of EG-120 (a) and EG-180 (b); Figure S2: TEM (a) and HRTEM (b) of EG-120; Figure S3: Adsorption removal efficiencies of Cr(VI) by different $ZnIn_2S_4$ in the dark; Figure S4: Photocatalytic removal efficiencies of Cr(VI) by EG-120 in the presence of different quenchers.

Author Contributions: Methodology, Y.S. and H.Z. (Hongzhong Zhang); investigation, Z.L. and L.H. (Long Han); data curation, H.Z. (Huan Zhang), J.X., R.Z. and Y.Z.; writing—original draft preparation, Y.S.; writing—review and editing, L.H. (Long Huang); project administration, H.Z. (Hongzhong Zhang). All authors have read and agreed to the published version of the manuscript.

Funding: National Natural Science Foundation of China [Nos. 22006139, 52000162], the Open Project of State Key Laboratory of Urban Water Resource and Environment, Harbin Institute of Technology (No. QA202016).

Data Availability Statement: Not applicable.

Conflicts of Interest: The authors declare no conflict of interest.

References

1. Wang, D.F.; Zhang, G.L.; Dai, Z.Y.; Zhou, L.L.; Bian, P.; Zheng, K.; Wu, Z.Y.; Cai, D.Q. Sandwich-like nanosystem for simultaneous removal of Cr(VI) and Cd(II) from water and soil. *ACS Appl. Mater. Interfaces* **2018**, *10*, 18316–18326. [CrossRef] [PubMed]
2. Li, K.X.; Huang, Z.A.; Zhu, S.Y.; Luo, S.L.; Yan, L.S.; Dai, Y.H.; Guo, Y.H.; Yang, Y.X. Removal of Cr(VI) from water by a biochar-coupled g-C_3N_4 nanosheets composite and performance of a recycled photocatalyst in single and combined pollution systems. *Appl. Catal. B Environ.* **2019**, *243*, 386–396. [CrossRef]
3. Kholisa, B.; Matsena, M.; Chirwa, E.M.N. Evaluation of Cr(VI) reduction using indigenous bacterial consortium isolated from a municipal wastewater sludge: Batch and kinetic studies. *Catalysts* **2021**, *11*, 1100. [CrossRef]
4. Nafiey, A.A.; Addad, A.; Sieber, B.; Chastanet, G.; Barras, A.; Szunerits, S.; Boukherroub, R. Reduced graphene oxide decorated with Co_3O_4 nanoparticles (rGO-Co_3O_4) nanocomposite: A reusable catalyst for highly efficient reduction of 4-nitrophenol, and Cr(VI) and dye removal from aqueous solutions. *Chem. Eng. J.* **2017**, *322*, 375–384. [CrossRef]
5. Zhang, G.P.; Chen, D.Y.; Li, N.J.; Xu, Q.F.; Li, H.; He, J.H.; Lu, J.M. Preparation of $ZnIn_2S_4$ nanosheet-coated CdS nanorod heterostructures for efficient photocatalytic reduction of Cr(VI). *Appl. Catal. B Environ.* **2018**, *232*, 164–174. [CrossRef]
6. Liu, B.B.; Liu, X.J.; Li, L.; Li, J.W.; Li, C.; Gong, Y.Y.; Niu, L.Y.; Zhao, X.S.; Sun, C.Q. $ZnIn_2S_4$ flowerlike microspheres embedded with carbon quantum dots for efficient photocatalytic reduction of Cr(VI). *Chin. J. Catal.* **2018**, *39*, 1901–1909. [CrossRef]

7. Janani, R.; Preethi, R.; Singh, S.; Rani, A.; Chang, C.T. Hierarchical ternary sulfides as effective photocatalyst for hydrogen generation through water splitting: A review on the performance of $ZnIn_2S_4$. *Catalysts* **2021**, *11*, 277. [CrossRef]
8. Pan, Y.; Yuan, X.Z.; Jiang, L.B.; Yu, H.B.; Zhang, J.; Wang, H.; Guan, R.P.; Zeng, G.M. Recent advances in synthesis, modification and photocatalytic applications of micro/nano-structured zinc indium sulfide. *Chem. Eng. J.* **2018**, *354*, 407–431. [CrossRef]
9. Wang, T.X.; Xu, S.H.; Yang, F.X. $ZnIn_2S_4$ nanopowder as an efficient visible light-driven photocatalyst in the reduction of aqueous Cr(VI). *Mater. Lett.* **2012**, *83*, 46–48. [CrossRef]
10. Gou, X.L.; Cheng, F.Y.; Shi, Y.H.; Zhang, L.; Peng, S.J.; Chen, J.; Shen, P.W. Shape-controlled synthesis of ternary chalcogenide $ZnIn_2S_4$ and $CuIn(S,Se)_2$ nano-/microstructures via facile solution route. *J. Am. Chem. Soc.* **2006**, *128*, 7222–7229. [CrossRef]
11. Tian, F.; Zhu, R.S.; Song, K.L.; Niu, M.L.; Ouyang, F.; Cao, G. The effects of hydrothermal temperature on the photocatalytic performance of $ZnIn_2S_4$ for hydrogen generation under visible light irradiation. *Mater. Res. Bull.* **2015**, *70*, 645–650. [CrossRef]
12. Gao, Y.H.; Lin, H.X.; Zhang, S.Y.; Li, Z.J. $Co(dmgH)_2pyCl$ as a noble-metal-free co-catalyst for highly efficient photocatalytic hydrogen evolution over hexagonal $ZnIn_2S_4$. *RSC Adv.* **2016**, *6*, 6072–6076. [CrossRef]
13. Yang, L.Q.; Zhao, J.; Wang, Z.; Wang, L.L.; Zhao, Z.J.; Li, S.Y.; Li, G.H.; Cai, Z.Y. Facile construction of $g\text{-}C_3N_4/ZnIn_2S_4$ nanocomposites for enhance Cr(VI) photocatalytic reduction. *Spectrochim. Acta Part A* **2022**, *276*, 121184. [CrossRef]
14. Zhang, J.Z.; Yuan, G.Q.; Wang, H.H.; Wu, J.B.; Yang, G.D.; Jia, Q.L.; Zhang, S.W.; Li, F.L.; Zhang, H.J. Preparation of core/shell-structured $ZnFe_2O_4@ZnIn_2S_4$ catalysts and its ultrafast microwave catalytic reduction performance for aqueous Cr(VI). *Chem. Eng. J.* **2023**, *451*, 138182. [CrossRef]
15. Shen, S.H.; Zhao, L.; Guo, L.J. Morphology, structure and photocatalytic performance of $ZnIn_2S_4$ synthesized via a solvothermal/hydrothermal route in different solvents. *J. Phys. Chem. Solids* **2008**, *69*, 2426–2432. [CrossRef]
16. Yang, Z.F.; Li, S.J.; Xia, X.N.; Liu, Y.T. Hexagonal MgAl-LDH simultaneously facilitated active facet exposure and holes storage over $ZnIn_2S_4$/MgAl-LDH heterojunction for boostin photocatalytic activities and anti-photocorrosion. *Sep. Purif. Technol.* **2022**, *300*, 121819. [CrossRef]
17. Yang, H.C.; Cao, R.Y.; Sun, P.X.; Yin, J.M.; Zhang, S.W.; Xu, X.J. Constructing electrostatic self-assembled 2D/2D ultra-thin $ZnIn_2S_4$/protonated $g\text{-}C_3N_4$ heterojunctions for excellent photocatalytic performance under visible light. *Appl. Catal. B Environ.* **2019**, *256*, 117862. [CrossRef]
18. Guo, F.; Cai, Y.; Guan, W.S.; Shi, W.D. Ag_3PO_4 nanoparticles decorated on sheet-on-sheet structured $g\text{-}C_3N_4/ZnIn_2S_4$ for enhanced photocatalytic activity. *Mater. Lett.* **2017**, *201*, 62–65. [CrossRef]
19. Wang, Q.; Wang, W.; Zhong, L.L.; Liu, D.M.; Cao, X.Z.; Cui, F.Y. Oxygen vacancy-rich 2D/2D $BiOCl\text{-}g\text{-}C_3N_4$ ultrathin heterostructure nanosheets for enhanced visible-light-driven photocatalytic activity in environmental remediation. *Appl. Catal. B Environ.* **2018**, *220*, 290–302. [CrossRef]
20. Xing, C.W.; Yu, G.Y.; Chen, T.; Liu, S.S.; Sun, Q.Q.; Liu, Q.; Hu, Y.J.; Liu, H.Y.; Li, X.Y. Perylenetetracarboxylic diimide covalently bonded with mesoporous $g\text{-}C_3N_4$ to construct direct Z-scheme heterojunctions for efficient photocatalytic oxidative coupling of amines. *Appl. Catal. B Environ.* **2021**, *298*, 120534. [CrossRef]
21. Zhu, H.B.; Gu, S.; Zhang, C.M.; Gao, K.Y.; Xia, J.J.; Cheng, C.L.; Wang, X.F. Regulating Electron-transfer over $ZnIn_2S_4$ by Sn(II)/Sn(IV) Co-Doping for Efficient Photocatalytic Hydrogen Production. *Catal. Lett.* 2022, in press. [CrossRef]
22. Li, P.; Jia, X.R.; Zhang, J.P.; Li, J.Q.; Zhang, J.Q.; Wang, L.J.; Wang, J.M.; Zhou, Q.F.; Wei, W.; Zhao, X.L.; et al. The roles of gold and silver nanoparticles on $ZnIn_2S_4$/silver (gold)/tetra(4-carboxyphenyl)porphyrin iron(III) chloride hybrids in carbon dioxide photoreduction. *J. Colloid Interface Sci.* **2022**, *628*, 831–839. [CrossRef] [PubMed]

catalysts

MDPI

Article

Pristine TiO_2 and Sr-Doped TiO_2 Nanostructures for Enhanced Photocatalytic and Electrocatalytic Water Splitting Applications

Mohd Fazil and Tokeer Ahmad *

Nanochemistry Laboratory, Department of Chemistry, Jamia Millia Islamia, New Delhi 110025, India; fazilzeus@gmail.com

* Correspondence: tahmad3@jmi.ac.in; Tel.: +91-11-26981717 (ext. 3261)

Abstract: Pristine TiO_2 and Sr-doped TiO_2 (1%, 2.5% and 5%) nanoparticles were synthesized at low temperatures via an eco-friendly hydrothermal route for water-splitting applications. XRD, EDAX and Raman analysis were performed to analyze the crystallinity, purity and structure of the as-synthesized materials. TEM, SEM, BET and UV-DRS studies were carried out to elucidate the size, morphology, surface area and optoelectronic properties of the nanoparticles. High surface areas of 169, 182, 178 and 141.16 $m^2\ g^{-1}$ for pristine TiO_2 (12 $\pm$ 0.6 nm) and 1% (11.1 $\pm$ 0.6 nm), 2.5% (12.1 $\pm$ 0.6 nm) and 5% (13 $\pm$ 0.7 nm) Sr-doped TiO_2 nanoparticles were obtained, respectively. One-percent Sr-doped TiO_2 nanoparticles were found to be active photocatalysts, as they showed higher hydrogen production (26.30 $mmolg_{cat}^{-1}$). Furthermore, electrocatalysis was investigated for HER and OER in 0.5 N H_2SO_4 and 0.1 N KOH electrolytic solutions using calomel as a reference electrode, revealing that 1% and 5% Sr-doped TiO_2 showed maximum current density for both HER ($\approx$10 mA/cm^2) and OER ($\approx$2.49 mA/cm^2), with an onset potential of 0.96 V for HER and 1.55 V for OER, and Tafel slopes of 84.09 and 91.60 mV/dec, respectively.

Keywords: Sr-doped TiO_2 nanoparticles; hydrothermal synthesis; photocatalysis; electrocatalysis; water splitting; hydrogen energy

1. Introduction

Hydrogen production through photocatalytic water splitting is considered to be the most renewable and sustainable energy production technology, with almost negligible impact on the environment [1]. Hydrogen is considered to be the most promising and clean source of energy for the future, as the only combustion product is water. Furthermore, with the process of development, the demand for energy is also increasing. As a result, present energy resources cannot last long and fulfill our future requirements. Advanced functional nanomaterials have shown potential applications in nanocatalysis for H_2 generation to meet the global energy demand [2–4]. The utilization of non-renewable energy resources has become one of the major causes of environmental pollution due to the release of CO_2, which causes the greenhouse effect and global warming [5–8]. Hence, the researchers are trying to develop sustainable and environment-friendly energy resources with high gravimetric energy density that are easily accessible. There are several environment-friendly techniques for green fuel/energy on which researchers are working; among them, photo- and electrocatalytic techniques have attracted significant attention -due to their relatively decreased dangers to the environment [2,4,5,9–11]. In today's era, hydrogen energy has become the center of attraction, and water-splitting via electro-catalysis or photo-catalysis is considered to be the best method for hydrogen generation, presenting a clean and renewable source of energy for fuel cells and batteries [12]. To store renewable electricity, electrochemical water-splitting is one of the most attractive ways of achieving hydrogen energy [13]. Nanosized metal oxide nanoparticles are widely used in the semiconductor industry as catalysts for hydrogen generation due to their low toxicity, high stability and low cost [14,15].

Transition metal-based oxides have wide band gap semiconducting properties, which are large enough for water splitting and charge transfer. This meets the thermodynamic requirement, showing excellent hydrogen evolution performance as photo- and electrocatalysts because of their exceptional size dependency [16,17]. These materials have extraordinary optoelectronic and chemical properties, along with advanced chemical and thermal stabilities with these processes encompassing charge carrier (e^-/h^+) transformation [18–21]. Recently, several metals and non-metal oxide-based catalysts have been explored as photo- and electrocatalysts for significant hydrogen evolution activity [22–29]. The use of metal oxides as catalysts was uncovered by the use of TiO_2 as a catalyst in photocatalytic and electrocatalytic water splitting. Honda and Fujishima (in 1972) used TiO_2 with a Pt electrode for the electrocatalysis of water [30]. However, it has very limited application as a photocatalyst, though due to it low cost, non-toxicity and excellent physical and chemical properties, TiO_2 was found to be very useful in the paint industry, cosmetics industry, gas sensors, optoelectronic devices, protective coating, white pigment, catalysis and electronics, etc [16,17,31–35].

The photocatalysis mechanism is based on e^- and h^+ pair formation with the irradiation of a light source [36–39]. Due to wide bandgap (3.2 eV) energy, fast charge carrier recombination rate, and low utilization of lights, the photocatalytic activity of TiO_2 towards hydrogen evolution is limited [36,38]. To overcome these issues, researchers have tried to develop TiO_2 materials to reduce the band gap and charge separation enhancement in many ways, such as by coupling it with other semiconductors and doping with foreign elements [40]. Metal doping to TiO_2 host lattice causes surface plasmon resonance (SPR), which is free electrons on the metal-doped nano-catalyst. Owing to SPR, the charge transfer efficiency of metal-doped TiO_2 is enhanced through electron trapping via efficient absorption of solar radiation [41,42]. In addition, metal doping introduces interstitial site defects and oxygen vacancies by the formation of Ti^{3+} in the TiO_2 lattice. Furthermore, metal doping and heterostructured composites create an empty impurity band above the valence band (V.B.) and bridge most of the gaps of valence band maxima and conduction band (C.B.) minima in pristine TiO_2, with less charge (e^-/h^+) separation and charge transfer. This leads to the tailoring of the bandgap and hence enhances the photo- and electrocatalytic performance of the nanocatalysts [43–46]. For instance, iron, cobalt, molybdenum, strontium, magnesium or noble metals (Ag, Pt, Au) and non-metal elements (C, N, S) have been exploited as dopants to improve the photocatalytic and electrocatalytic performance of pristine TiO_2 [47–50]. Strontium is a significant alkaline earth metal extensively used in many research fields, electronics, the military industry, the chemical industry, optics, metallurgy, etc. In recently reported literature related to Sr and TiO_2 systems, most researchers reported Sr-doped composite materials or nanostructures. Film Sr-doped TiO_2 nanotube shows a 65% incident photon conversion efficiency. Sr-doping to TiO_2 enhances the visible light response and generates active site defects in the pristine TiO_2 host lattice; it thereby enhances the activity by a decrease in the band gap and suppresses the charge recombination rate [49,51]. The precious nature of metals (Ru, Ir, W, etc.) and oxides, which are the most efficient electro-active materials for OER, is becoming a bottleneck for commercial applications [52–55]. Since OER follows a complex multistep reaction mechanism, it involves four electrons in either an acidic or alkaline medium (OER, $2H_2O \rightarrow O_2 + 4H^+ + 4e^-$ in acidic solution, $4OH^- \rightarrow O_2 + 2H_2O + 4e^-$ in alkaline solution); hence, OER reactions have high overpotential and sluggish kinetics even after using active electrocatalysts such as RuO_2 and IrO_2, which is hinders practical application of water electrocatalysis [56–58]. For OER, RuO_2 and IrO_2 are considered the best electrocatalysts owing to their high stability, excellent electrolytic activity and low overpotential [59,60].

The major concern of the current research is to develop highly efficient, cheap and earth-abundant electrocatalysts and photocatalysts for use in sustainable energy. Herein, we report a simple, environmentally friendly, low-temperature hydrothermal approach to the synthesis of pristine and Sr-doped TiO_2 nanocatalysts. As-synthesized material was further investigated with XRD, Raman, TEM, SEM, EDX, BET and UV-vis DRS. In addition, detailed photo and electrocatalytic water-splitting studies were carried out for the application of as-prepared Sr-doped TiO_2 nanoparticles for hydrogen production.

2. Experimental Section

2.1. Materials

The following chemicals and reagents were used without further purification: glacial acetic acid (Fisher Scientific, Heysham, Lancashire, UK), titanium isopropoxide (Alfa Aesar, 97%, Heysham, Lancashire, UK), sodium hydroxide (97%, Merck Life Science Pvt. Ltd., Vikhroli, Mumbai, India), strontium acetate (Aldrich, Sigma-Aldrich Chemie GmbH, Steinheim, Germany), ethanol (Merck, 99.9%, Darmstadt, Germany), sodium sulfide (Na_2S, 60%, SRL Pvt. Ltd., Taloja, Maharashtra, India), sodium sulfite (Na_2SO_3, Alfa Aesar, 98%, Heysham, England), and double distilled water.

2.2. Synthesis of Pristine and Sr-Doped TiO_2 Nanoparticles (NPs)

Pristine TiO_2 NPs were synthesized through a hydrothermal route using a laboratory autoclave. 7.4 mL glacial acetic acid was taken in a round-bottom flask to prevent the direct oxidation of titanium isopropoxide to TiO_2, which helps to carry out the reaction in a controlled manner by forming stable hydroxyl titanium acylate [61,62]; then, 0.74 mL of titanium isopropoxide was added after making an inert atmosphere with continuous stirring. After, 16.86 mL of distilled water was added such that the total volume became 25 mL; the resulting solution is referred to as 0.1 M Ti^{4+} [61,62]. This solution was stirred on a magnetic stirrer for about 30 min with slow addition of 25 mL of 0.1 M NaOH. The solution was then kept in an autoclave and placed in a vacuum oven at 150 °C for 12 h. The solution was cooled to room temperature and then centrifuged, washed with distilled water and ethanol several times, and dried overnight at 60 °C. Then, the dried white precipitate was ground and white powders were obtained.

For the preparation of three different compositions, 25 mL of strontium acetate solution of 0.001 M was prepared by doping 1.0% of strontium in 0.1 M of TiO_2. In addition, a solution of titanium metal ions was prepared using a similar procedure to that stated above. These two solutions were mixed and stirred on a magnetic stirrer with slow addition of 45 mL of 0.1 M NaOH solution for about 30 min [29], followed by a normal synthesis procedure to obtain washed and dried precipitate, which was ground to obtain a white-yellowish powder. Similarly, the procedure for the doping of 2.5% and 5.0% of Sr in TiO_2 nanoparticles was performed separately with 0.0025 M strontium acetate solution for 2.5% doping and 0.005 M for 5.0% doping of Sr in TiO_2 nanoparticles. The reaction scheme was depicted schematically in Figure 1 below.

2.3. Characterizations

The powder X-ray diffraction (XRD) technique was employed to analyze the crystal structure and phase purity of the as-prepared samples by employing a Rigaku Japan D/max 2500 diffractometer operating at a scan rate of 5° min^{-1} with Cu Kα radiation (1.5406 Å) with a step size of 0.05 °/s to carry out the XRD studies in the 2θ range of 20–80°. A RENISHAW inVia Raman microscope (Leica Microsystems CMS GmbH, Wetzlar, Germany) with a He–Ne laser and an output of 15 mW was used to analyze the Raman spectral studies using a 532 nm excitation wavelength. TEM analyses were carried out on a TALOS instrument operating at 200 kV to investigate the size and crystal structure of undoped pristine and Sr-doped TiO_2 nanoparticles. The samples for TEM analysis were prepared by making creating a suspension in ethanol by sonication and mounting the samples on a carbon-coated copper grid by employing a droplet of suspension. A Nova Nano SEM-450 microscope (FEI Company, Hillsboro, OR, USA) accelerated at 20 kV voltage combined with an energy-dispersive X-ray (EDAX) system was used to perform field emission scanning electron microscopy (FESEM) to evaluate the surface morphological features, elemental analysis and chemical composition. To analyze the FESEM analysis, a dry sample was mounted on an ultrathin layer of gold coated with carbon tape to prevent the surface charging effect. UV-visible DRS (UV-Vis diffused reflectance spectroscopy) studies were performed on PerkinElmer Lambda 365 spectrophotometer. The reflectance spectra of as-synthesized samples were recorded in the 200–800 nm wavelength range to determine the band gap. Furthermore, a Nova 2000e Instrument (Quantachrome Instruments Limited, USA) was used to determine the Brunauer–Emmett–Teller (BET) surface area and pore size distribution of pristine TiO_2 and Sr-doped TiO_2 nanoparticles.

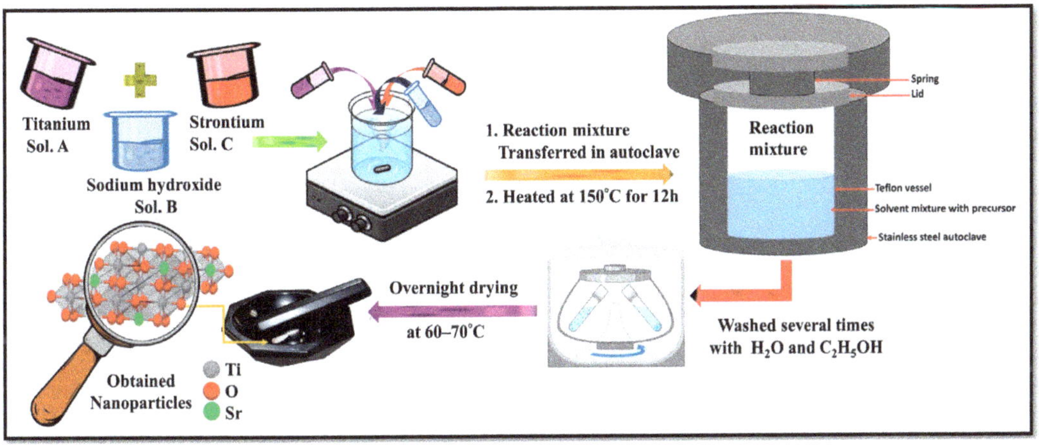

Figure 1. The schematic diagram for the preparation of Sr−doped TiO_2 nanoparticles.

2.4. Photocatalytic Hydrogen Evolution Measurements

The activity of the as-synthesized nanocatalyst towards photocatalysis of water for hydrogen production was estimated with 0.128 M Na_2S (500 mg) and 0.079 M Na_2SO_3 (500 mg) as sacrificial agents, which provide electron donor sites throughout photocatalytic reaction in ambient conditions irradiated under a UV visible light source of 200–2400 nm varying wavelength. In the hydrogen evolution activity, 20 mg of photocatalyst and sacrificial agents were dispersed in 50 mL double distilled water in an airtight photoreactor, which is a specialized cylindrical quartz cell with a bottleneck of 4 cm diameter and a flat base [63]. The prepared solution was stirred with the purging of nitrogen for 30 min in order to create an inert atmosphere and to eliminate gas impurities (such as dissolved oxygen). After stirring, the cylindrical quartz cell containing the solution mixture was fixed on the stirrer and its flat surface was placed 7 cm away from the light source; it was then exposed to the light source (200 W, Hg−Xe arc lamp, Newport, MODEL: 66906–200HXF-R15 Ozone-free), which irradiates at 170 W with constant stirring. Hydrogen production was quantified using a gas chromatography instrument procured from Perkin Elmer, model Clarus 590 GC, using a TCD detector and nitrogen as carrier gas, with sampling done after every 1 h time interval with a specialized airtight glass syringe. For further study, the photocatalysts from the reaction mixture were recovered by centrifugation, dried and then analyzed the recyclability, photostability, reusability and photocatalytic response showing maximum H_2 production by taking consecutive three cycles of 8 h long activity.

2.5. Electrode Preparation and Electrocatalytic Measurements

For the measurement of electrocatalytic activity of pristine TiO_2 and 1, 2.5 and 5% Sr-doped TiO_2 nanostructures, all electrochemical measurements were taken with three electrode systems (i.e., working electrode, reference electrode and counter electrode) in 0.5 N H_2SO_4 (98%, Merck) and 0.1 N KOH (85%, Merck) electrolyte solution for hydrogen evolution reaction (HER) and oxygen evolution reaction (OER) activity, respectively. Measurements were performed using an Autolab PGSTAT204 instrument at room temperature. Here, Pt wire and saturated calomel electrode (SCE) were used as counter and reference electrodes, respectively. The working electrodes of as-synthesized pristine TiO_2 and 1, 2.5 and 5% Sr-doped TiO_2 electrocatalysts for the electrochemical setup were prepared by the coating of catalyst dispersion on pretreated ITO substrates (sheet resistance < 10 ohm/sq, Vritra) of 1 × 1 cm^2 dimension. Before the drop-casting of the dispersion on the substrates, they were cleaned with ethanol (Merck, Emsure), acetone (99.5%, SRL) and isopropanol (99.5%, SRL) for 10 min in each solvent by ultra-sonication. Catalyst dispersions were prepared by making a suspension of 4 mg electrocatalysts, 400 μL isopropanol and 15 μL of Nafion (5 wt %, Alfa Aesar) as a polymer binder. The resulting mixtures were sonicated for about 30 min. 250 μL dispersions were deposited via drop casting on the conducting side (oxide layer side) of ITO substrate and then dried overnight at 60 °C in a vacuum oven, as per the procedure reported elsewhere [64]. For investigation of the electrochemical activity of the as-synthesized nanostructures, cyclic voltammetry (CV) measurements were carried out in the range of 10–100 mV/s, linear sweep voltammetry (LSV) measurements were performed at 100 mV/s and chronoamperometry (CA) measurements were obtained at −0.5 V and 0.5 V for HER and OER, respectively.

3. Results and Discussion

3.1. XRD Analysis

Powder X-ray diffraction was used to identify the phase purity and crystal structure of the as-synthesized nanoparticles. Figure 2 presents the XRD pattern, which was satisfactorily indexed with the anatase phase of TiO_2 with JCPDS card number 71-1167 [65]. It was observed that the synthesized material was pure and highly crystalline. Similarly, the 1%, 2.5% and 5% Sr-doped TiO_2 nanoparticles were indexed with the same anatase phase, as shown in Figure 2. From the XRD, it was observed that there is a slight shift in the peaks after Sr-doping, which shows the incorporation of Sr into the host lattice of TiO_2 nanoparticles. Further, the 2θ values shift to lower values with 1% doping of Sr, which then decrease with 2.5% and 5% doping. In addition, with the increase in Sr concentration in the TiO_2 host lattice, the peak intensity increases; sharpness was also observed, which reflects the increase in overall crystallinity. Since the Sr^{2+} (1.12 Å) ionic radius is much higher than Ti^{4+} (0.605 Å), on substitution, lattice expansions were observed. This also suggests that the solubility limit of Sr^{2+} over the TiO_2 host lattice is not high; hence, Sr-doped TiO_2 nanoparticles were fabricated at small concentrations. Due to the effect of Sr dopant, the crystallinity and crystallite size of Sr-doped TiO_2 were considerably reduced. It also causes the enhancement of point defects such as oxygen vacancy [49]. By using Scherrer's equation, the average crystallite size of all samples was calculated [66] and was found to be 10.21 ± 0.5, 8.92 ± 0.5, 8.99 ± 0.5, and 9.85 ± 0.5 nm for pure TiO_2, 1%, 2.5%, and 5% Sr-doped TiO_2, respectively. The crystal structure of Sr-doped TiO_2 nanoparticles depicting the positions of Sr^{2+} and Ti^{4+} ions is shown in Figure 3.

3.2. TEM Analysis

TEM analysis showed the effect of Sr-doping on TiO_2 nanoparticles. Figure 4a–d presents the micrographs of pristine TiO_2 and Sr-doped TiO_2 nanoparticles. In the TEM micrographs, the appearance of rough cubes with little agglomeration could be seen, along with black spots, showing the appearance of strontium on the TiO_2 host nanoparticles. From the TEM images, particle size histograms were obtained, which revealed the average particle size to be in the range of 5–30 nm and the average particle size to be 12 ± 0.6, 11.1 ± 0.6, 12.1 ± 0.6 and 13 ± 0.7 nm for pristine TiO_2, 1%, 2.5% and 5% Sr-doped TiO_2 nanoparticles, respectively, as shown in Figure 4e–h. It was also observed that with the increase in the Sr concentration, the particle size increases. Furthermore, HRTEM micrographs depict the well-defined lattice fringes separated by the interplaner spacing of 0.35, 0.35, 0.35 and 0.35 nm for the pristine, 1%, 2.5%, and 5% Sr-doped TiO_2 nanoparticles, respectively which correspond to the (101) crystallographic plane of anatase tetragonal TiO_2, as shown in Figure 4i–l.

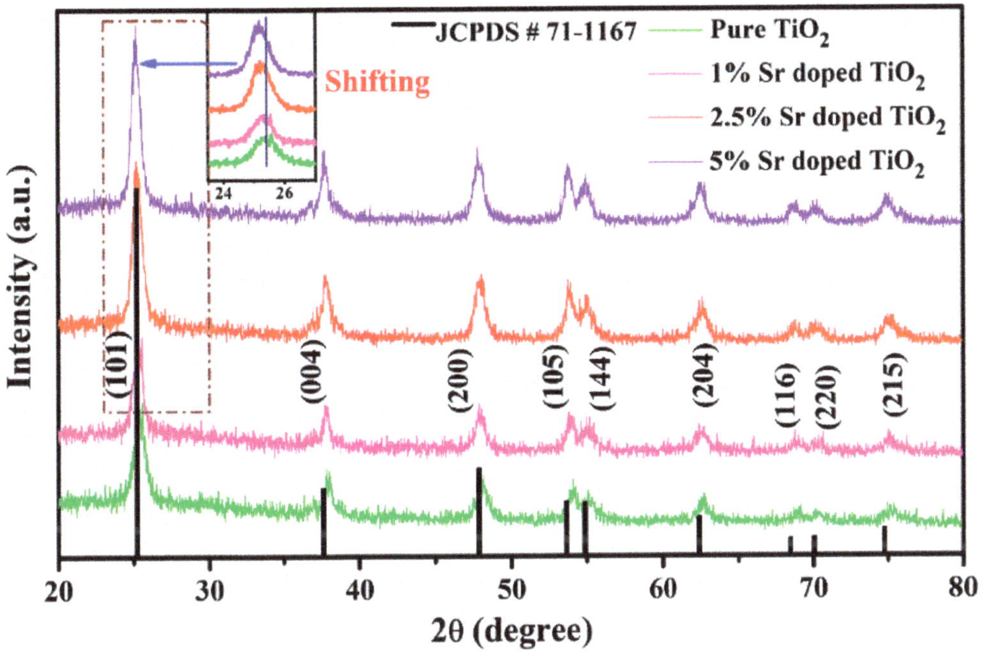

Figure 2. Overlaid XRD patterns of pristine TiO_2, 1%, 2.5% and 5% Sr-doped TiO_2 NPs.

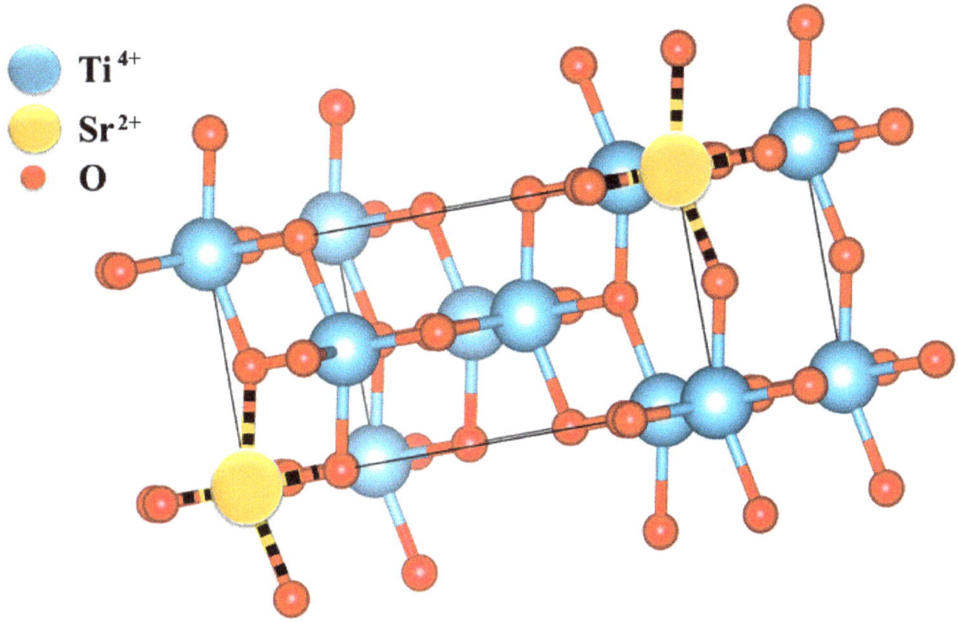

Figure 3. Crystal structure of Sr-doped TiO_2 nanoparticles.

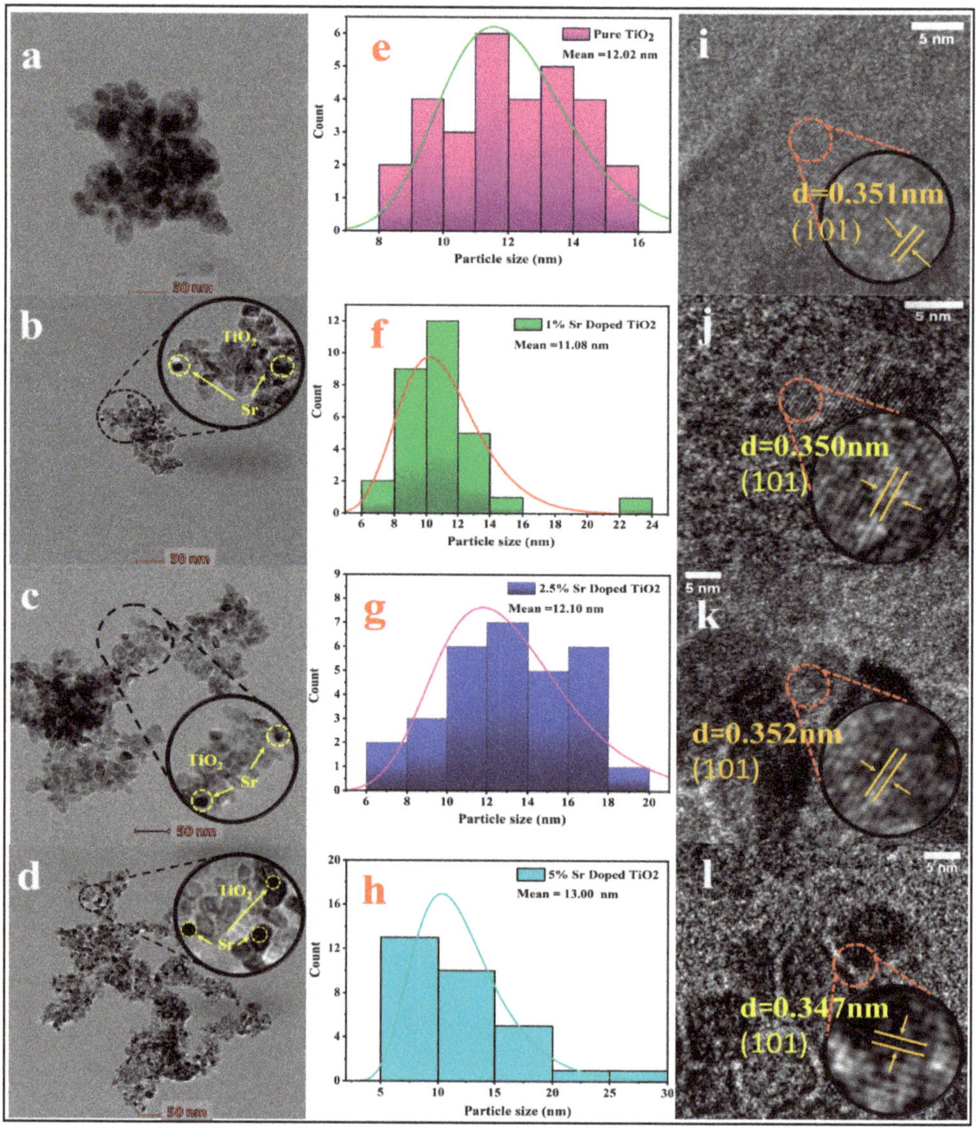

Figure 4. (**a–d**) TEM micrographs, (**e–h**) size distribution plots and (**i–l**) HRTEM micrographs depicting the lattice spacing of the crystallographic planes of pristine TiO_2 and 1%, 2.5% and 5% Sr-doped TiO_2 nanoparticles.

3.3. SEM Analysis

SEM analysis of pristine and Sr-doped TiO_2 nanoparticles was carried out to gather morphological information, as shown in Figure 5a–d. SEM images show the formation of dense nanoparticles with rough and irregular morphology. Careful observation shows that the aggregation and density increase with increasing dopant concentration. The EDAX spectra of pristine TiO_2 and Sr-doped TiO_2 nanoparticles represent the successful doping of Sr into the TiO_2 host lattice, as shown in Figure 5e–h. EDAX spectra show the presence of only the desired elements Sr, Ti and O in the as-synthesized nanoparticles;

theoretically, the derived elemental composition was found to have a close agreement with the experimentally loaded composition. X-ray elemental mapping of pristine TiO_2 and Sr-doped TiO_2 also revealed that Sr was dispersed minutely and Ti and O were present in the synthesized material.

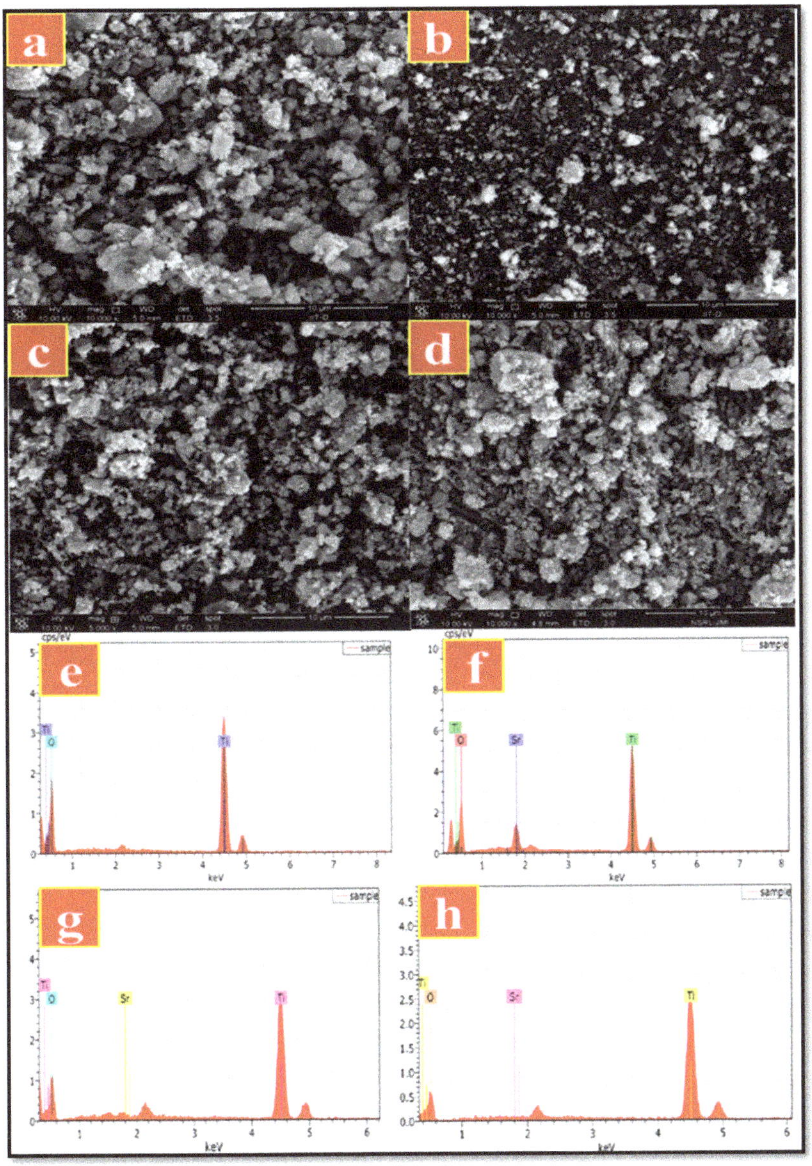

Figure 5. (**a–d**) SEM micrographs and (**e–h**) EDAX spectra of Pristine TiO_2, 1%, 2.5%and 5% Sr-doped TiO_2 nanoparticles.

3.4. UV-Visible DRS Analysis

Pristine TiO_2 and 1, 2.5 and 5% Sr-doped TiO_2 NPs were employed for the analysis of optical properties by UV–visible absorption spectroscopy, as shown in Figure 6a.

Reflectance spectra are presented in Figure 6b. By using the following Kubelka–Munk equation and reflectance data, the optical band gap of the materials was calculated:

$$F(R) = \alpha/s = (1 - R)^2/2R$$

where the terms have their standard and usual meanings. Kubelka–Munk plots of pristine TiO_2 and 1, 2.5 and 5% Sr-doped TiO_2 nanoparticles were drawn using $[F(R\infty)h\nu]^{1/n}$ Vs eV, and with the help of linear extrapolation of $[F(R\infty)h\nu]^{1/n}$, their band gaps were determined [63,64]. The estimated band gap values were found to be 3.19, 3.16, 3.18 and 3.20 eV for pristine TiO_2 and 1%, 2.5% and 5% Sr-doped TiO_2 NPs, respectively. The small changes in the band gap could be due to the incorporation of impurities or vacancy defect levels between V.B. and C.B.; this may result in a change in the activity of the materials in terms of their photo- and electrocatalytic applications [46,67].

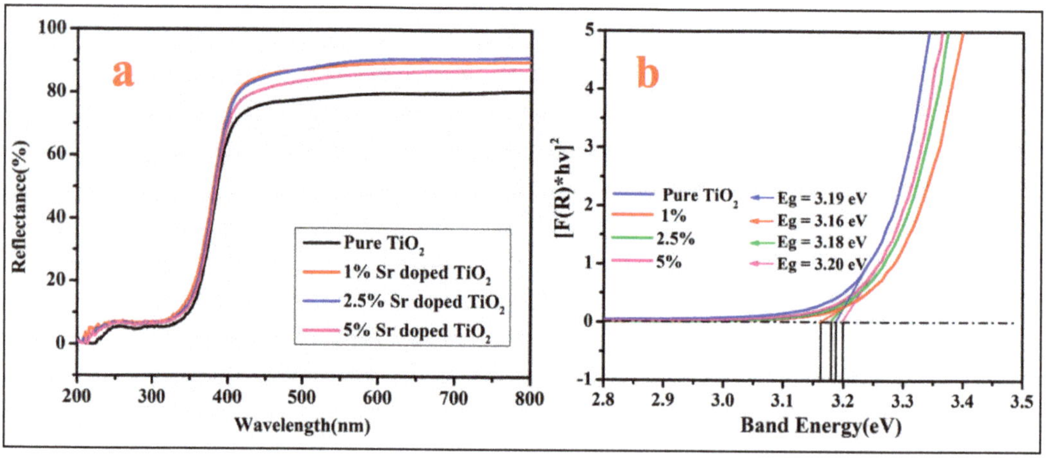

Figure 6. (**a**) UV–vis DRS and (**b**) Kubelka–Munk plots of pristine TiO_2 and 1, 2.5 and 5% Sr-doped TiO_2 NPs.

3.5. Raman Analysis

Raman spectroscopy is a characterization technique used to investigate the structure of complex metal oxides. This technique was used to explore the different vibrational modes of pristine and Sr-doped TiO_2 nanoparticles. For the TiO_2 anatase phase, 15 optical modes were found: $1A_{1g} + 1A_{2u} + 2B_{1g} + 1B_{2u} + 3E_g + 2E_u$ with normal vibrations; six of these are Raman active modes represented by $1A_{1g}$, $2B_{1g}$ and $3E_g$ symmetries. The vibration modes A_{1g}, B_{1g} and E_g arise from antisymmetric, symmetric bending and symmetric stretching of O–Ti–O bending vibrations in the TiO_2 lattice, respectively. As shown in Figure 7, the Raman peaks at 143, 195, 393, 510 and 635 cm^{-1} correspond to the E_g, E_g, B_{1g}, $A_{1g} + B_{1g}$ and Eg modes, respectively [51,68,69]. Comparing the Raman spectrum of pristine and Sr-doped TiO_2, a small blue shift was found due to Sr doping. The principal Raman peak at the E_g (143) mode arises due to external vibration of the anatase structure, which is attributable to the formation of the anatase phase [51,70]. This Raman spectrum was found to be compatible with XRD results.

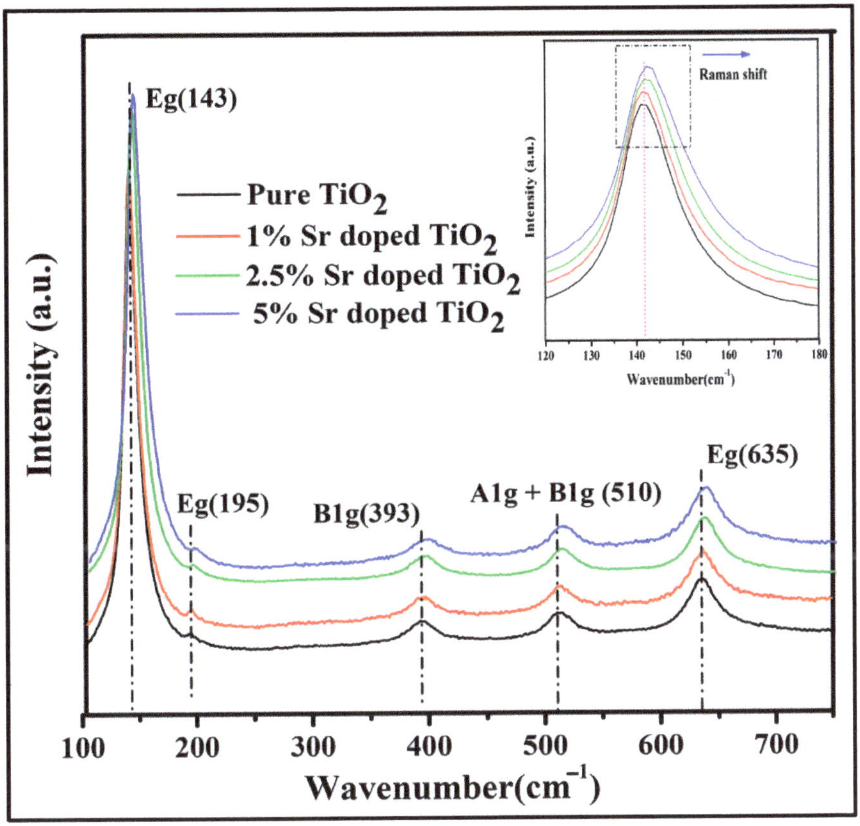

Figure 7. Raman spectra of pristine TiO_2 and Sr-doped TiO_2 nanoparticles.

3.6. BET Surface Area Studies

The catalytic property of materials depends upon the surface area, as the greater surface area of the catalyst is, the greater the number of surface-active sites. Therefore, to examine the catalytic activity of as-prepared nanomaterials, the surface areas of the as-synthesized pristine TiO_2 and 1, 2.5 and 5% Sr-doped TiO_2 nanoparticles were determined. BET results, as shown in Figure 8a for all the as-synthesized samples, have type-IV isotherms and a H3 hysteresis loop. With the help of BET surface area analysis, the specific surface areas of the as-synthesized pristine TiO_2 and 1, 2.5 and 5% Sr-doped TiO_2 nanoparticles were determined and found to be 169, 182, 178 and 141.16 m^2g^{-1}, respectively. The surface area results were associated with the grain size trend, as observed in TEM and SEM studies of Sr-doped TiO_2 NPs. The average pore size distributions calculated by BJH studies of the as-synthesized pristine TiO_2 and 1, 2.5 and 5% Sr-doped TiO_2 nanoparticles were found to be 17.03 $\pm$ 0.85, 16.95 $\pm$ 0.85, 16.97 $\pm$ 0.85 and 16.94 $\pm$ 0.85 Å, respectively, as depicted in Figure 8b, which reveals the mesoporous nature of the samples. The pore radius was calculated by using the Dubinin and Astakov (DA) method and was found to be 7–30 Å; the onset of the plots was observed at 14.0, 14.2, 14.0 and 13.7 Å for the pristine TiO_2 and 1, 2.5 and 5% Sr-doped TiO_2 NPs, respectively, as shown in Figure 8c. The surface area investigation with the band gap of as-synthesized materials is presented in Table 1. Further, from the results of UV-DRS and BET surface area, it was observed that the increase in surface area with the decrease in the particle size causes a decrease in the band gap of the nanoparticles, as reported in the literature [53,63,64].

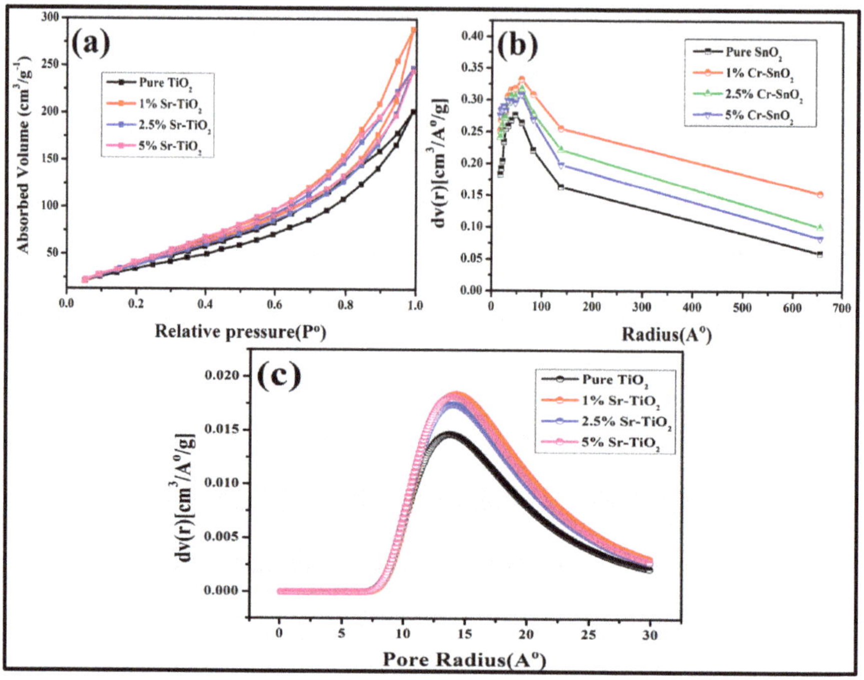

Figure 8. (**a**) Nitrogen adsorption-desorption isotherms, (**b**) BJH pore-size distribution and (**c**) DA average pore-size distribution plots of pristine, 1, 2.5 and 5% Sr-doped TiO_2 NPs.

Table 1. BET surface area (S_{BET}), average pore size, pore volume (V_{total}) and energy band gaps (Eg) of as-synthesized nanocatalysts.

Sample	$S_{BET}/(m^2g^{-1})$	Average Pore Size/Å	$V_{total}/(cm^3g^{-1})$	Eg/eV
Pristine TiO_2	169.9	17.02 ± 0.85	0.359	3.19
1% Sr-doped TiO_2	182.2	16.95± 0.85	0.351	3.16
2.5% Sr-doped TiO_2	178.3	16.97 ± 0.85	0.426	3.18
5% Sr-doped TiO_2	141.2	16.94 ± 0.85	0.294	3.20

3.7. Photocatalytic Water Splitting for Hydrogen Generation

The catalytic activity of as-synthesized nanoparticles was investigated for photocatalytic hydrogen generation by using water-splitting studies. The production of hydrogen is measured as a function of time, as shown in Figure 9a. The total apparent hydrogen evolution after 8 h was found to be 26.3, 16.8, 9.8 and 12.4 $mmolg_{cat}^{-1}$ for 1%, 2.5%, 5% Sr-doped TiO_2 and pristine TiO_2 nanoparticles, respectively, as shown in Figure 9b. The order of photocatalysts on the basis of hydrogen evolution activity is 1% Sr-doped TiO_2 > 2.5% Sr-doped TiO_2 > PristineTiO_2 > 5% Sr-doped TiO_2. Furthermore, 1%, 2.5% and 5% Sr-doped TiO_2 and pristine TiO_2 photocatalysts evolve an average per hour hydrogen of 3.3, 2.1, 1.2 and 1.6 $mmolg_{cat}^{-1}h^{-1}$, respectively, as presented in Figure 9c and comparison of results of the photocatalytic performance of metal-doped TiO_2 synthesized by different methods with present work summarized in Table 2. The photocatalytic activity of 1% Sr-doped TiO_2 was found to be maximum with respect to other compositions and pristine TiO_2 photocatalyst. The increase of the Sr dopant concentration to the host TiO_2 NPs influences the hydrogen evolution rate and decreased the photocatalytic activity after attaining the maximum, as shown in Figure 9b. This study concludes that Sr dopant shows the corrosive effect of increasing concentration towards hydrogen evolution. A similar trend of corrosiveness was previously reported for Ag-doped SnO_2 nanoparticles [25]. The reproducibility experiment was performed under the same experimental conditions to analyze the stability and repeating ability of the as-synthesized photocatalysts for an 8 h reaction time. Here, the quantitative analysis of hydrogen evolution was performed with a gas chromatography analyzer. The rate of hydrogen evolution, reproducibility and stability were examined for three consecutive cycles, as depicted in Figure 9d. The experimental results show that the first and second cycles have approximately no change in hydrogen evolution (26.30 mmol g_{cat}^{-1}) in 8 h, but in the third cycle, a small decrease in hydrogen evolution (25.93 mmol g_{cat}^{-1}) was observed.

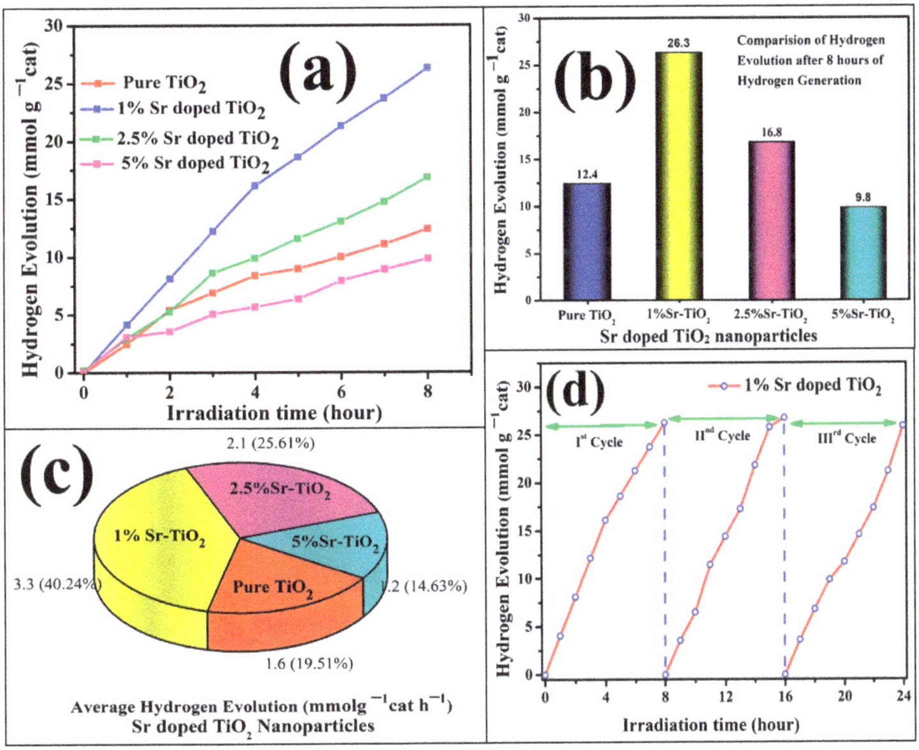

Figure 9. (**a**) Photocatalytic hydrogen evolution curves under UV-visible light irradiation as a function of irradiation time; (**b**) comparison of photocatalytic activity; (**c**) average H_2 evolution per hour of the pristine TiO_2 and 1, 2.5 and 5% Sr-doped TiO_2 NPs; (**d**) stability test of 1% Sr-doped TiO_2 for up to three cycles.

Table 2. Comparison of the photocatalytic performance of metal-doped TiO_2 synthesized by different methods.

S.No.	Dopant	Synthesis Method	Parameters	Hydrogen Production	Ref.
1.	Bismuth, nitrogen	Sol-gel method	Solar light, methanol	1800 μmol/g	[71]
2.	Nitrogen	Solid state/calcination method	UV–Vis light irradiation, Na_2S/Na_2SO_3	18 μmol	[72]
3.	Platinum, nitrogen	Photodeposition method	1150 mL Pyrex vessel, UV light	3200 μmol	[73]
4.	Strontium, silver	Sol–gel method,	500 W Xe arc lamp	49.4 μmol/h	[74]
5.	$Pd/0.2\%K^+$	Hydrothermal method	–	76.6 μmol h−1	[75]
6.	Strontium	Hydrothermal method	200 W, Hg−Xe arc lamp	3.3 $mmolg_{cat}^{-1}h^{-1}$	In this work

3.8. Possible Photocatalytic Reaction Mechanism of Sr-doped TiO_2 Nanoparticles

The hydrogen evolution of pristine TiO_2 and Sr-doped TiO_2 NPs as photocatalysts is illustrated in Figure 10. Here, the photocatalytic efficiency of pristine TiO_2 was observed to be less due to the recombination tendency of photo-induced e^-/h^+. Furthermore, the Sr-doped TiO_2 shows enhanced hydrogen evolution due to a decrease in the recombination tendency by enlarging the band gap and by providing more catalytic active sites, which are generated by enhancing the active surface area [25,63,64]. The photocatalysis efficiency is increased not only by suppressing the e^-/h^+ recombination, but also because of more generation of hydroxyl radicals (•OH). Since Sr-doped TiO_2 NPs involve the replacement of some Sr^{2+} ions by Ti^{4+} ions, which cause charge imbalances, OH^- ions might be adsorbed at the catalyst surface and accept photogenerated holes, generating •OH. This favors the suppuration of photo-induced charge carriers and suppresses the recombination rate. In addition, Sr-doped TiO_2 NPs in solution easily provide electrons to H^+ ions, which facilitates the hydrogen evolution of the system [51,76]. Hence, the holes generated because of electron abstracting by H^+ ions at the valence band of pristine and Sr-doped TiO_2 nanoparticles as photocatalysts were consumed by the sacrificial agents sodium sulphide and sodium sulphite. To summarize, strontium doping causes increased surface area, altering the band gap and inducing defects, which are the reasons for the photocatalytic efficiency of the Sr-doped TiO_2 photocatalysts due to their suppressing the photo-induced e^-/h^+ recombination rate. The possible multi-step reaction for hydrogen evolution based on active sites and adsorbed excited state HS^- species is as follows:

Path A:

$$H_2O + *(\text{active sites}) \rightarrow *OH^- + H^+$$

$$OH^- + h^+ \rightarrow *OH$$

$$2OH \rightarrow H_2O_2 \rightleftharpoons OOH + H^+ \rightleftharpoons O_2^{2-} + 2H^+$$

$$H_2O_2\ (a) + 2h^+ \rightarrow O_2 + 2H^+$$

$$OOH(a) + 2h^+ \rightarrow O_2 + H^+$$

$$O_2^{2-}(a) + 2h^+ \rightarrow O_2$$

$$2H + \rightarrow H_2$$

Path B:

$$HS^-(aq) \rightarrow HS^-(ads) \xrightarrow{hv} [HS^-(ads)]*$$

$$[HS^-(ads)]^* \rightarrow H^* + S^-(ads)$$

$$S^{-}(ads) + [HS^{-}(ads)]* \rightarrow [HS_{2}^{2-}]\dagger$$

$$[HS_{2}^{2-}]\dagger \rightarrow H* + S_{2}^{2-}(ads)$$

$$2H^{*} \rightarrow H_2$$

$$[HS^{-}(ads)]^{*} + H_2O \rightarrow S^{0}(ads) + H_2 + OH^{-}$$

Here, ads = adsorption, * = excited state species and † = species undergoing intramolecular charge transfer [1,77,78].

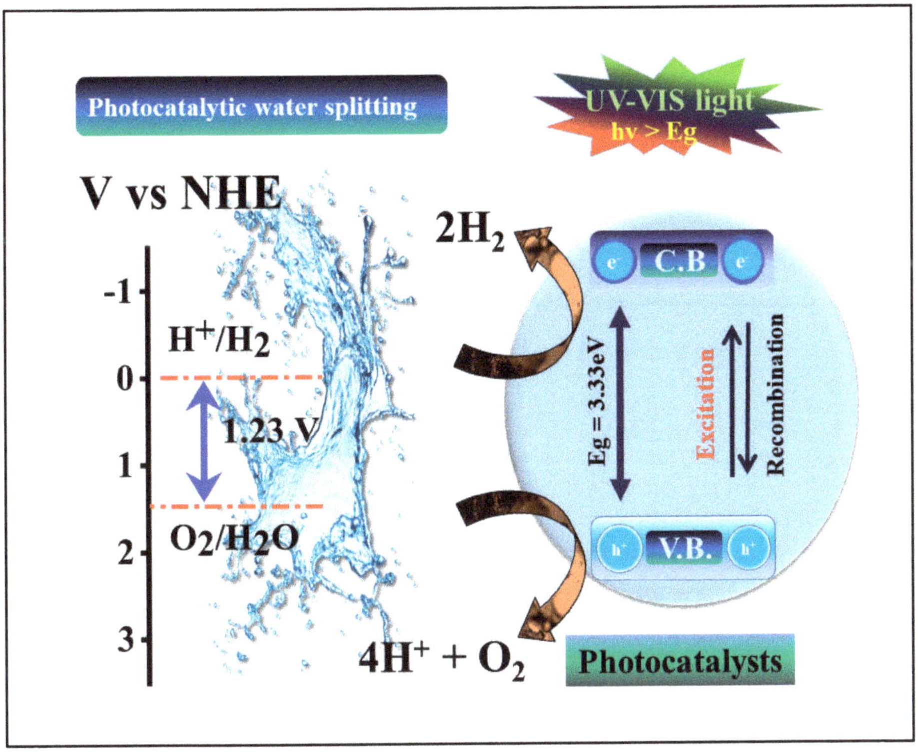

Figure 10. Schematic mechanism of the photocatalytic water splitting for H_2 evolution.

3.9. Electrocatalytic Water Splitting Studies

The electrocatalytic activity of as-synthesized pristine TiO_2 and 1%, 2.5%, 5% Sr-doped TiO_2 NPs as electrocatalysts towards HER (Hydrogen evolution reaction) and OER (Oxygen evolution reaction) performances were determined using a three-electrode configuration in 0.5 N H_2SO_4 and 0.1 N KOH electrolytic solutions under ambient conditions, as demonstrated in Figures 11 and 12. Here, HER responses were investigated by linear sweep voltammetry (LSV) and cyclic voltammetry (CV), as shown in Figure 11a–c. With the increase in the Sr concentration in TiO_2 nanocatalyst, an increase in the overpotential was observed at 10 mA/cm^{-2} cathode current density, which revealed that 1% Sr-doped TiO_2 has high electrocatalytic HER activity. The overpotential values of pristine TiO_2, 1%, 2.5% and 5% Sr-doped TiO_2 electrocatalysts at 10 mA cm^{-2} were 1.00, 0.96, 1.07 and 1.08 V, respectively. The OER performances of pristine TiO_2 and 1%, 2.5% and 5% Sr-doped TiO_2 nanoparticles were evaluated by LSV and CV measurements, as depicted in Figure 13a–c. From the LSV results of OER, superior electrocatalytic activity was shown by 5 % Sr-doped

TiO_2 nanostructures, with an onset potential of 0.8 V as shown in Figure 13a, and an increase in its anodic current density, which produced 2.49 mA/cm^2 at 1.55 V applied potential. Compared with the previously reported current density versus RHE generated by IrO_2, the present works results with 5 % Sr-doped TiO_2 at 1.55 V were ~3.01 and 2.49 mA cm^{-2}, respectively, showing that 5% Sr-doped TiO_2 electrocatalyst has approximately similar potential as IrO_2 electrocatalyst [53,79]. With the help of HER and OER responses, the reaction kinetics and mechanism for electrocatalytic activity were elucidated using Tafel analysis. The reaction kinetics and mechanism are completely correlated with characteristic properties of nanoparticles (such as size, surface area, morphology and optical properties) and orientation of electrocatalysts [4,39,53,80]. Figures 11d and 12d illustrate the Tafel polarization plots of the as-synthesized pristine TiO_2 and 1%, 2.5% and 5% Sr-doped TiO_2 electrocatalysts for the water electrocatalysis. The observed electrodic parameters (HER and OER) are presented in Table 3. The decrease in resistance on the applied potential window were monitored at different scan rates from 10–100 mV. Chronoamperometric (CA) measurements were taken to test the durability and stability of 1% Sr-doped TiO_2 electrocatalyst operating at a voltage of-0.5 V; there was no change found in current density, which was almost stable, as shown in Figure 13. The stability of 5% Sr-doped TiO_2 was also examined through CA at 0.5 V voltage by running for up to 8000 s; it was observed that the current density was stable throughout the measurement (Figure 13). With the help of CV measurement of optimum HER and OER, electrocatalysis of 1% Sr-doped TiO_2 and 5% Sr-doped TiO_2 was carried out, as shown in Figure 11d,e and Figure 12c, respectively.

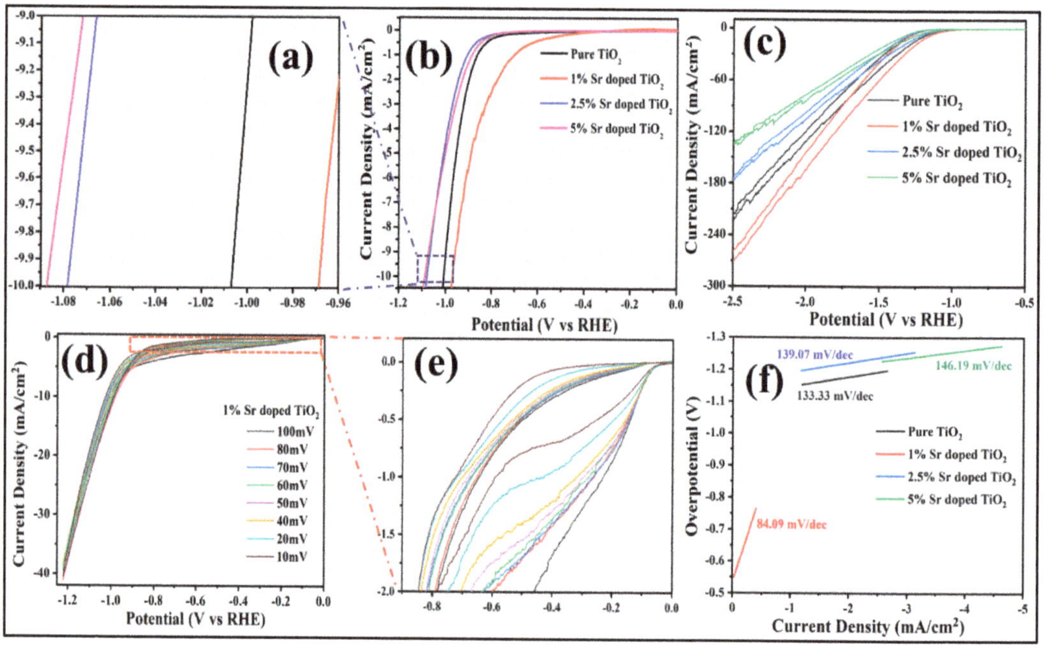

Figure 11. (**a–c**) shows the LSV and CV plots of HER of as-synthesized pristine TiO_2 and 1%, 2.5% and 5% Sr-doped TiO_2; (**d**,**e**) shows the CV plots of 1% Sr-doped TiO_2 toward HER; (**f**) shows HER Tafel plots of pristine TiO_2 and 1%, 2.5% and 5% Sr-doped TiO_2.

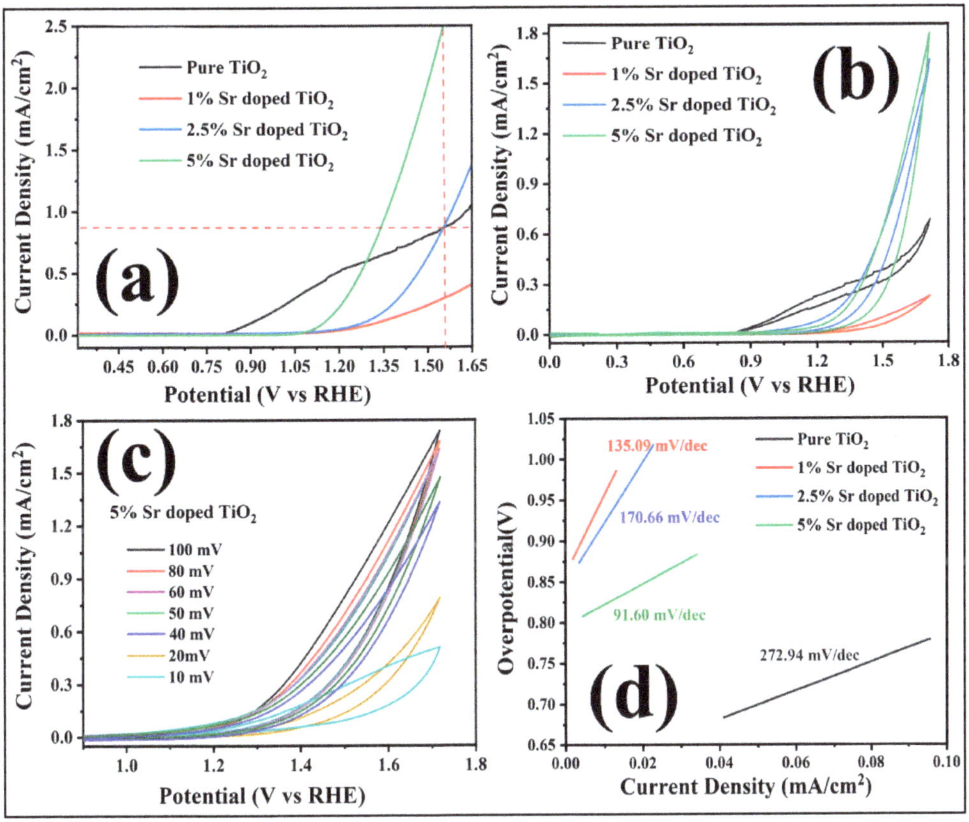

Figure 12. (**a**,**b**) shows the LSV and CV plots of OER of as-synthesized pristine TiO_2 and 1%, 2.5% and 5% Sr-doped TiO_2; (**c**) shows the CV plots of 5% Sr-doped TiO_2 toward OER; (**d**) shows the OER Tafel plots of as-synthesized pristine TiO_2 and 1%, 2.5% and 5% Sr-doped TiO_2.

Table 3. HER and OER electrodic parameters of pristine and Sr-doped TiO_2 electrocatalysts.

S.No	Materials	HER		OER		
		Overpotential (V) to Attain 10 mA/cm^2	Tafel Slope (mV/dec)	Onset Potential (V)	Anodic Current Density (mA/cm^2) at 1.55 V	Tafel Slope (mV/dec)
1.	Pristine TiO_2	1.00	133.33	0.8	0.86	272.94
2.	1% Sr-doped TiO_2	0.96	84.09	1.23	0.30	135.09
3.	2.5% Sr-doped TiO_2	1.07	139.07	1.34	0.87	170.66
4.	5% Sr-doped TiO_2	1.08	146.16	1.1	2.49	91.60

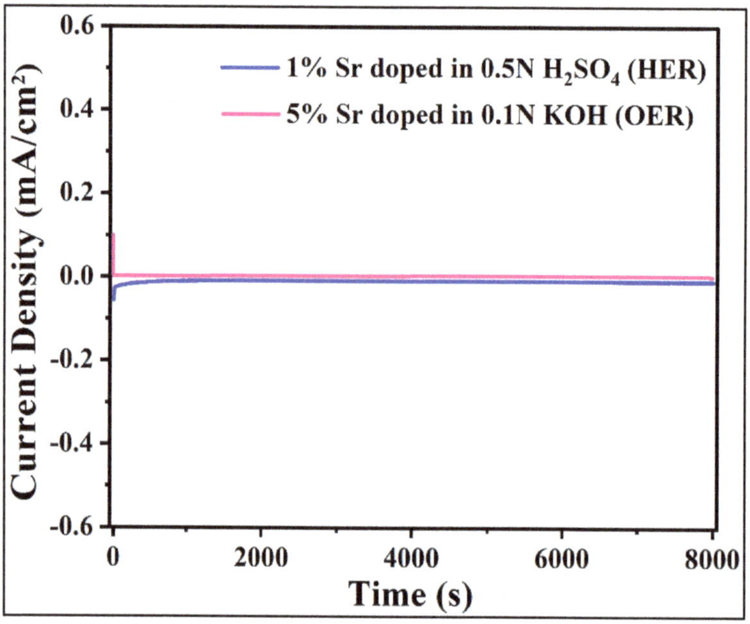

Figure 13. CA analysis at −0.5 V and +0.5 V for 1% and 5% Sr-doped TiO_2 nanoparticles.

4. Conclusions

Pristine anatase TiO_2 and Sr-doped TiO_2 nanocatalysts with highly crystalline and tetragonal structure were successfully synthesized with an eco-friendly low-temperature hydrothermal method. Due to the various reinforcing and synergistic factors of the cubical architectures of the TiO_2 and its dopants, such as increased surface area, the introduction of defects created by doping and the band gap tuning of the material resulted in an enhanced photocatalytic and electrocatalytic performance. One-percent Sr-doped TiO_2 showed the best photocatalytic and electrocatalytic HER activity and 5% Sr-doped TiO_2 displayed optimum electrocatalytic activity towards OER. The maximum H_2 production of 1% Sr-doped TiO_2 was found to be 26.3 mmolg_{cat}^{-1}. Furthermore, 1% and 5% Sr-doped TiO_2 electrocatalysts showed maximum current density for both HER (≈10 mA/cm^2) and OER (≈2.49 mA/cm^2), with an onset potential of 0.96 V for HER and 1.55 V for OER with Tafel slopes of 84.09 and 91.60 mV/dec, respectively. The increase in the photocatalytic and electrocatalytic performances of the photo/electrocatalysts was attributed to the high specific surface areas and the tuning of the band gap introduced by the incorporation of dopants, and resulted in an overall increase in performance. The study thus reveals that the synthesized compounds can act as efficient materials for photo/electrocatalysis for water-splitting hydrogen evolution and current density for HER and OER, which makes them potential candidates for green/renewable energy resources.

Author Contributions: M.F. is responsible experimental investigation, data collection, measurements, analysis and rough draft of the manuscript, whereas T.A. is responsible for analysis, conceptualization, supervision and final draft of the manuscript. All authors have read and agreed to the published version of the manuscript.

Funding: TA thanks UGC, New Delhi, Govt. of India for the Research Grant for In-service Faculty Members. The authors also thank SERB, CSIR and MoE (SPARC/2018-2019/P843/SL) for financial support to Nano Chemistry and Nano Energy Labs.

Institutional Review Board Statement: Not applicable.

Informed Consent Statement: Not applicable.

Data Availability Statement: Not applicable.

Acknowledgments: TA thanks UGC, New Delhi, Govt. of India for the Research Grant for In-service Faculty Members. The authors also thank SERB, CSIR and MoE (SPARC/2018-2019/P843/SL) for financial support to Nano Chemistry and Nano Energy Labs. MF thanks to UGC, New Delhi for the research fellowship. The authors also acknowledge the measurement support provided through the DST PURSE program at CIF, JMI and AIIMS, New Delhi for electron microscopic studies.

Conflicts of Interest: The authors declare no financial competing interest.

References

1. Kumaravel, V.; Mathew, S.; Bartlett, J.; Pillai, S.C. Photocatalytic hydrogen production using metal doped TiO_2: A review of recent advances. *Appl. Catal. B Environ.* **2019**, *244*, 1021–1064. [CrossRef]
2. Mehtab, A.; Banerjee, S.; Mao, Y.; Ahmad, T. Type-II $CuFe_2O_4$/Graphitic Carbon Nitride Heterojunctions for High-Efficiency Photocatalytic and Electrocatalytic Hydrogen Generation. *ACS Appl. Mater. Interfaces.* **2022**, *14*, 44317–44329. [CrossRef] [PubMed]
3. Gautam, A.; Sk, S.; Pal, U. Recent Advances in Solution Assisted Synthesis of Transition Metal Chalcogenides for Photo-electro Catalytic Hydrogen Evolution. *Phys. Chem.* **2022**, *24*, 20638–20673.
4. Mehtab, A.; Alshehri, S.M.; Ahmad, T. Photocatalytic and Photoelectrocatalytic Water Splitting by Porous g-C_3N_4 Nanosheets for Hydrogen Generation. *ACS Appl. Nano Mater.* **2022**, *5*, 12656–12665. [CrossRef]
5. Ali, S.A.; Ahmad, T. Chemical strategies in molybdenum based chalcogenides nanostructures for photocatalysis. *Int. J. Hydrog. Energy* **2022**, *29*, 29255–29283. [CrossRef]
6. Xu, X.; Song, F.; Hu, X. A nickel iron diselenide-derived efficient oxygen-evolution catalyst. *Nat. Commun.* **2016**, *7*, 1–7. [CrossRef] [PubMed]
7. Pandit, N.A.; Ahmad, T. Tin Oxide Based Hybrid Nanostructures for Efficient Gas Sensing. *Molecules* **2022**, *27*, 7038–7082. [CrossRef]
8. Mehtab, A.; Ahmed, J.; Alshehri, S.M.; Mao, Y.; Ahmad, T. Rare earth doped metal oxide nanoparticles for photocatalysis: A perspective. *Nanotechnology* **2022**, *33*, 142001. [CrossRef]
9. Joy, J.; Mathew, J.; George, S.C. Nanomaterials for photoelectrochemical water splitting-review. *Int. J. Hydrog. Energy* **2018**, *43*, 4804–4817. [CrossRef]
10. Naaz, F.; Sharma, A.; Shahazad, M.; Ahmad, T. Hydrothermally Derived Hierarchical CuO Nanoflowers as an Efficient Photocatalyst and Electrocatalyst for Hydrogen Evolution. *ChemistrySelect* **2022**, *7*, e202201800. [CrossRef]
11. Farooq, U.; Ahmed, J.; Alshehri, S.M.; Mao, Y.; Ahmad, T. Self-Assembled Interwoven Nanohierarchitectures of $NaNbO_3$ and $NaNb_{1-x}Ta_xO_3$ ($0.05 \leq x \leq 0.20$): Synthesis, Structural Characterization, Photocatalytic Applications, and Dielectric Properties. *ACS Omega* **2022**, *7*, 16952–16967. [CrossRef] [PubMed]
12. Basavarajappa, P.S.; Patil, S.B.; Ganganagappa, N.; Reddy, K.R.; Raghu, A.V.; Reddy, C.V. Recent progress in metal-doped TiO_2, non-metal doped/codoped TiO_2 and TiO_2 nanostructured hybrids for enhanced photocatalysis. *Int. J. Hydrog. Energy* **2020**, *45*, 7764–7778. [CrossRef]
13. Samuel, E.; Joshi, B.; Kim, M.; Swihart, M.T.; Yoon, S.S. Morphology engineering of photoelectrodes for efficient photoelectrochemical water splitting. *Nano Energy* **2020**, *72*, 104648. [CrossRef]
14. Liu, Q.; Shi, J.; Xu, Z.; Zhang, B.; Liu, H.; Lin, Y.; Gao, F.; Li, S.; Li, G. InGaN Nanorods Decorated with Au Nanoparticles for Enhanced Water Splitting Based on Surface Plasmon Resonance Effects. *Nanomaterials* **2020**, *10*, 912. [CrossRef] [PubMed]
15. Xu, Z.; Zhang, S.; Liang, J.; Lin, J.; Yu, Y.; Li, R.; Gao, F.; Li, G. Surface passivation of InGaN nanorods using H_3PO_4 treatment for enhanced photoelectrochemical performance. *J. Power Sources* **2019**, *419*, 65–71. [CrossRef]
16. Zhang, N.; Liu, S.; Fu, X.; Xu, Y. Synthesis of M@TiO_2 (M= Au, Pd, Pt) core–shell nanocomposites with tunable photoreactivity. *J. Phys. Chem. C* **2011**, *115*, 9136–9145. [CrossRef]
17. Idriss, H. The elusive photocatalytic water splitting reaction using sunlight on suspended nanoparticles: Is there a way forward? *Catal. Sci. Technol.* **2020**, *10*, 304–310. [CrossRef]
18. Zhang, H.; Liang, C.; Liu, J.; Tian, Z.; Wang, G.; Cai, W. Defect-mediated formation of Ag cluster-doped TiO_2 nanoparticles for efficient photodegradation of pentachlorophenol. *Langmuir* **2012**, *28*, 3938–3944. [CrossRef]
19. Wang, W.; Jin, C.; Qi, L. Hierarchical CdS nanorod@SnO_2 nanobowl arrays for efficient and stable photoelectrochemical hydrogen generation. *Small* **2018**, *14*, 1801352. [CrossRef]
20. Guo, C.X.; Xie, J.; Yang, H.; Li, C.M. Au@CdS Core–Shell Nanoparticles-Modified ZnO Nanowires Photoanode for Efficient Photoelectrochemical Water Splitting. *Adv. Sci. Lett.* **2015**, *2*, 1500135. [CrossRef]
21. Cao, S.; Yin, Z.; Barber, J.; Boey, F.Y.; Loo, S.C.J.; Xue, C. Preparation of Au-$BiVO_4$ heterogeneous nanostructures as highly efficient visible-light photocatalysts. *ACS Appl. Mater. Interfaces* **2012**, *4*, 418–423. [CrossRef]
22. Valero, J.M.; Obregón, S.; Colón, G. Active site considerations on the photocatalytic H_2 evolution performance of Cu-doped TiO_2 obtained by different doping methods. *ACS Catal.* **2014**, *4*, 3320–3329. [CrossRef]

23. Yang, J.; Wang, X.; Dai, J.; Li, J. Efficient visible-light-driven photocatalytic degradation with Bi_2O_3 coupling silica doped TiO_2. *Ind. Eng. Chem. Res.* **2014**, *53*, 12575–12586. [CrossRef]
24. Khan, H.; Lone, I.H.; Lofland, S.E.; Ramanujachary, K.V.; Ahmad, T. Exploiting Multiferroicity of $TbFeO_3$ Nanoparticles for Hydrogen Generation through Photo/Electro/Photoelectro-catalytic Water Splitting. *Int. J. Hydrog. Energy* **2022**, *in press*. [CrossRef]
25. Jain, S.K.; Fazil, M.; Naaz, F.; Pandit, N.A.; Ahmed, J.; Alshehri, S.M.; Mao, Y.; Ahmad, T. Silver-doped SnO_2 nanostructures for photocatalytic water splitting and catalytic nitrophenol reduction. *New J. Chem.* **2022**, *46*, 2846–2857. [CrossRef]
26. Wu, M.; Wu, P.; Lin, T.; Lin, T. Photocatalytic performance of Cu-doped TiO_2 nanofibers treated by the hydrothermal synthesis and air-thermal treatment. *Appl. Surf. Sci.* **2018**, *430*, 390–398. [CrossRef]
27. Sulaiman, S.N.A.; Noh, M.Z.; Adnan, N.N.; Bidin, N.; Razak, S.N.A. Effects of photocatalytic activity of metal and non-metal doped TiO_2 for hydrogen production enhancement-a review. *J. Phys. Conf. Ser.* **2018**, *1027*, 012006. [CrossRef]
28. Ko, S.; Banerjee, C.K.; Sankar, J. Photochemical synthesis and photocatalytic activity in simulated solar light of nanosized Ag doped TiO_2 nanoparticle composite. *Compos. B. Eng.* **2011**, *42*, 579–583. [CrossRef]
29. Emran, K.M. Catalytic Activity of Strontium Modified TiO_2 Nanotubes for Hydrogen Evolution Reaction. *Int. J. Electrochem. Sci.* **2020**, *15*, 4218–4231. [CrossRef]
30. Fujishima, A.; Honda, K. Electrochemical photolysis of water at a semiconductor electrode. *Nature* **1972**, *238*, 37–38. [CrossRef]
31. Karunakaran, C.; Abiramasundari, G.; Gomathisankar, P.; Manikandan, G.; Anandi, V. Cu-doped TiO_2 nanoparticles for photocatalytic disinfection of bacteria under visible light. *J. Colloid Interface Sci.* **2010**, *352*, 68–74. [CrossRef] [PubMed]
32. Rohilla, S. Rietveld refinement and structural characterization of $TiO_2/CoFe_2O_4$ nanocomposites. *In IOP Conference Series: Mater. Sci. Eng.* **2020**, *872*, 012171.
33. Yadav, H.M.; Otari, S.V.; Koli, V.B.; Mali, S.S.; Hong, C.K.; Pawar, S.H.; Delekar, S.D. Preparation and characterization of copper-doped anatase TiO_2 nanoparticles with visible light photocatalytic antibacterial activity. *J. Photochem. Photobiol. A* **2014**, *280*, 32–38. [CrossRef]
34. Gogoi, D.; Namdeo, A.; Golder, A.K.; Peela, N.R. Ag-doped TiO_2 photocatalysts with effective charge transfer for highly efficient hydrogen production through water splitting. *Int. J. Hydrog. Energy* **2020**, *45*, 2729–2744. [CrossRef]
35. Tsebriienko, T.; Popov, A.I. Effect of poly (titanium oxide) on the viscoelastic and thermophysical properties of interpenetrating polymer networks. *Crystals* **2021**, *11*, 794. [CrossRef]
36. Jang, D.M.; Kwak, I.H.; Kwon, E.L.; Jung, C.S.; Im, H.S.; Park, K.; Park, J. Transition-metal doping of oxide nanocrystals for enhanced catalytic oxygen evolution. *J. Phys. Chem. C* **2015**, *119*, 1921–1927. [CrossRef]
37. Alshehri, S.M.; Ahmed, J.; Ahamad, T.; Alhokbany, N.; Arunachalam, P.; Al-Mayouf, A.M.; Ahmad, T. Synthesis, characterization, multifunctional electrochemical (OGR/ORR/SCs) and photodegradable activities of $ZnWO_4$ nanobricks. *J. Solgel Sci. Technol.* **2018**, *87*, 137–146. [CrossRef]
38. Behnajady, M.A.; Alizade, B.; Modirshahla, N. Synthesis of Mg-doped TiO_2 nanoparticles under different conditions and its photocatalytic activity. *Photochem. Photobiol.* **2011**, *87*, 1308–1314. [CrossRef]
39. Farooq, U.; Ahmed, J.; Alshehri, S.M.; Ahmad, T. High-surface-area sodium tantalate nanoparticles with enhanced photocatalytic and electrical properties prepared through polymeric citrate precursor route. *ACS Omega* **2019**, *4*, 19408–19419. [CrossRef]
40. Kumar, A.; Patel, A.S.; Mohanty, T. Correlation of photodegradation efficiency with surface potential of silver-TiO_2 nanocomposite thin films. *J. Phys. Chem. C* **2012**, *116*, 20404–20408. [CrossRef]
41. Jung, H.; Yeo, I.; Kim, T.; Ki, H.; Gu, H. Surface plasmon resonance effect of silver nanoparticles on a TiO_2 electrode for dye-sensitized solar cells. *Appl. Surf. Sci.* **2018**, *432*, 266–271. [CrossRef]
42. Barad, H.; Ginsburg, A.; Cohen, H.; Rietwyk, K.J.; Keller, D.A.; Tirosh, S.; Bouhadana, Y.; Anderson, A.Y.; Zaban, A. Hot Electron-Based Solid State TiO_2 | Ag Solar Cells. *Adv. Mater. Interfaces* **2016**, *3*, 1500789. [CrossRef]
43. Zhang, X.; Li, L.; Zhou, Q.; Liang, X.; Liu, D. Facile synthesis of novel gully-like double-sized mesoporous structural Sr-doped ZrO_2-TiO_2 composites with improved photocatalytic efficiency. *J. Solid State Chem.* **2019**, *269*, 375–385. [CrossRef]
44. Hamad, H.; Elsenety, M.M.; Sadik, W.; El-Demerdash, A.; Nashed, A.; Mostafa, A.; Elyamny, S. The superior photocatalytic performance and DFT insights of S-scheme $CuO@TiO_2$ heterojunction composites for simultaneous degradation of organics. *Sci. Rep.* **2022**, *12*, 1–20. [CrossRef] [PubMed]
45. Nayak, S.; Sahoo, B.; Khastgir, D. Flexible Nanocomposites Comprised of Poly (dimethylsiloxane) and High-Permittivity TiO_2 Nanoparticles Doped with La^{3+}/Cu^{+} for Dielectric Applications. *ACS Appl. Nano Mater.* **2019**, *2*, 4211–4221. [CrossRef]
46. Farooq, U.; Chaudhary, P.; Ingole, P.P.; Kalam, A.; Ahmad, T. Development of cuboidal $KNbO_3@\alpha$-Fe_2O_3 hybrid nanostructures for improved photocatalytic and photoelectrocatalytic applications. *ACS Omega* **2020**, *5*, 20491–20505. [CrossRef]
47. Baruah, M.; Ezung, S.L.; Sharma, S.; Sinha, U.B.; Sinha, D. Synthesis and characterization of Ni-doped TiO_2 activated carbon nanocomposite for the photocatalytic degradation of Anthracene. *Inorg. Chem. Commun.* **2022**, *144*, 109905. [CrossRef]
48. Peng, B.; Meng, X.; Tang, F.; Ren, X.; Chen, D.; Ren, J. General synthesis and optical properties of monodisperse multifunctional metal-ion-doped TiO_2 hollow particles. *J. Phys. Chem. C* **2009**, *113*, 20240–20245. [CrossRef]
49. Hamedani, H.A.; Allam, N.K.; Garmestani, H.; El-Sayed, M.A. Electrochemical fabrication of strontium-doped TiO_2 nanotube array electrodes and investigation of their photoelectrochemical properties. *J. Phys. Chem. C* **2011**, *115*, 13480–13486. [CrossRef]
50. Gobal, F.; Faraji, M. Fabrication of nanoporous nickel oxide by de-zincification of Zn–Ni/(TiO_2-nanotubes) for use in electrochemical supercapacitors. *Electrochim. Acta* **2013**, *100*, 133–139. [CrossRef]

51. Wei, X.; Cao, J.; Fang, F. A novel multifunctional Ag and Sr^{2+} co-doped TiO_2@ rGO ternary nanocomposite with enhanced p-nitrophenol degradation, and bactericidal and hydrogen evolution activity. *RSC Adv.* **2018**, *8*, 31822–31829. [CrossRef]
52. Nguyen, D.C.; Doan, T.L.L.; Prabhakaran, S.; Tran, D.T.; Kim, D.H.; Lee, J.H.; Kim, N.H. Hierarchical Co and Nb dual-doped MoS_2 nanosheets shelled micro-TiO_2 hollow spheres as effective multifunctional electrocatalysts for HER, OER, and ORR. *Nano Energy* **2021**, *82*, 105750. [CrossRef]
53. Farooq, U.; Phul, R.; Alshehri, S.M.; Ahmed, J.; Ahmad, T. Electrocatalytic and enhanced photocatalytic applications of sodium niobate nanoparticles developed by citrate precursor route. *Sci. Rep.* **2019**, *9*, 1–17. [CrossRef] [PubMed]
54. Takahiro, N.; Shinagawa, T.; Nishimoto, T.; Takanabe, K. Recent advances in understanding oxygen evolution reaction mechanisms over iridium oxide. *Inorg. Chem. Front.* **2021**, *8*, 2900–2917.
55. Ahmed, J.; Ahamad, T.; Alhokbany, N.; Almaswari, B.M.; Ahmad, T.; Hussain, A.; Al-Farraj, E.S.S.; Alshehri, S.M. Molten salts derived copper tungstate nanoparticles as bifunctional electro-catalysts for electrolysis of water and supercapacitor applications. *ChemElectroChem* **2018**, *5*, 3938–3945. [CrossRef]
56. Haase, F.T.; Rabe, A.; Schmidt, F.P.; Herzog, A.; Jeon, H.S.; Frandsen, W.; Narangoda, P.V.; Spanos, I.; Friedel Ortega, K.; Timoshenko, J.; et al. Role of Nanoscale Inhomogeneities in Co_2FeO_4 Catalysts during the Oxygen Evolution Reaction. *J. Am. Chem. Soc.* **2022**, *144*, 12007–12019. [CrossRef]
57. Grimes, C.A.; Varghese, O.K.; Ranjan, S. Photoelectrolysis. In *Light, Water, Hydrogen*; Springer: Berlin/Heidelberg, Germany, 2008; pp. 115–190.
58. Zhu, C.; Li, C.; Zheng, M.; Delaunay, J. Plasma-induced oxygen vacancies in ultrathin hematite nanoflakes promoting photoelectrochemical water oxidation. *ACS Appl. Mater. Interfaces* **2015**, *7*, 22355–22363. [CrossRef] [PubMed]
59. Ma, Z.; Zhang, Y.; Liu, S.; Xu, W.; Wu, L.; Hsieh, Y.; Liu, P.; Zhu, Y.; Sasaki, K.; Renner, J.N.; et al. Reaction mechanism for oxygen evolution on RuO_2, IrO_2, and RuO_2@ IrO_2 core-shell nanocatalysts. *J. Electroanal. Chem.* **2018**, *819*, 296–305. [CrossRef]
60. Ahmed, J.; Ubaidullah, M.; Ahmad, T.; Alhokbany, N.; Alshehri, S.M. Synthesis of graphite oxide/cobalt molybdenum oxide hybrid nanosheets for enhanced electrochemical performance in supercapacitors and the oxygen evolution reaction. *ChemElectroChem* **2019**, *6*, 2524–2530. [CrossRef]
61. Ahmad, T.; Ganguli, A.K. Synthesis of nanometer-sized particles of barium orthotitanate prepared through a modified reverse micellar route: Structural characterization, phase stability and dielectric properties. *J. Mater. Res.* **2004**, *19*, 2905–2912. [CrossRef]
62. Ahmad, T.; Ganguli, A.K. Reverse micellar route to nanocrystalline titanates ($SrTiO_3$, Sr_2TiO_4, and $PbTiO_3$): Structural aspects and dielectric properties. *J. Am. Ceram. Soc.* **2006**, *89*, 1326–1332. [CrossRef]
63. Jain, S.; Pandit, N.A.; Fazil, M.; Ali, S.A.; Ahmed, J.; Alshehri, S.M.; Mao, Y.; Ahmad, T. Chemical Fabrication, Structural Characterization and Photocatalytic Water Splitting Application of Sr-Doped SnO_2 Nanoparticles. *Nanotechnology* **2022**, *33*, 355706. [CrossRef] [PubMed]
64. Jain, S.K.; Fazil, M.; Pandit, N.A.; Ali, S.A.; Naaz, F.; Khan, H.; Mehtab, A.; Ahmed, J.; Ahmad, T. Modified, Solvothermally Derived Cr-doped SnO_2 Nanostructures for Enhanced Photocatalytic and Electrochemical Water-Splitting Applications. *ACS Omega* **2022**, *7*, 14138–14147. [CrossRef] [PubMed]
65. Fan, X.; Hua, N.; Jia, H.; Zhu, Y.; Wang, Z.; Xu, J.; Wang, C. Synthesis and evaluation of visible-light photocatalyst: Nitrogen-doped TiO_2/Bi_2O_3 heterojunction structures. *Sci. Adv. Mater.* **2014**, *6*, 1892–1899. [CrossRef]
66. Mohamed, I.M.; Dao, V.; Yasin, A.S.; Barakat, N.A.; Choi, H. Design of an efficient photoanode for dye-sensitized solar cells using electrospun one-dimensional GO/N-doped nanocomposite SnO_2/TiO_2. *Appl. Surf. Sci.* **2017**, *400*, 355–364. [CrossRef]
67. Brik, M.G.; Srivastava, A.M.; Popov, A.I. A few common misconceptions in the interpretation of experimental spectroscopic data. *Opt. Mater.* **2022**, *127*, 112276. [CrossRef]
68. Saqib, M.; Althubeiti, K.; Abualnaja, K.M.; Rahman, N.; Khan, R. Effect of Sr and Co co-doping on the TiO_2-diluted magnetic semiconductor for spintronic applications. *J. Mater. Sci.: Mater. Electron.* **2021**, *32*, 28718–28729.
69. Ekoi, E.J.; Gowen, A.; Dorrepaal, R.; Dowling, D.P. Characterisation of titanium oxide layers using Raman spectroscopy and optical profilometry: Influence of oxide properties. *Results Phys.* **2019**, *12*, 1574–1585. [CrossRef]
70. Yang, G.; Jiang, Z.; Shi, H.; Xiao, T.; Yan, Z. Preparation of highly visible-light active N-doped TiO_2 photocatalyst. *J. Mater. Chem.* **2010**, *20*, 5301–5309. [CrossRef]
71. Kumar, M.K.; Bhavani, K.; Srinivas, B.; Kumar, S.N.; Sudhakar, M.; Naresh, G.; Venugopal, A. Nano structured bismuth and nitrogen co-doped TiO_2 as an efficient light harvesting photocatalyst under natural sunlight for the production of H_2 by H_2O splitting. *Appl. Catal. A* **2016**, *515*, 91–100. [CrossRef]
72. Zhang, X.; Song, P.; Cui, X. Nitrogen-doped TiO_2 photocatalysts synthesized from titanium nitride: Characterizations and photocatalytic hydrogen evolution performance. *J. Adv. Oxid. Technol.* **2013**, *16*, 131–136. [CrossRef]
73. Tristantini, D.; Ibadurrohman, M. Photocatalytic hydrogen production from glycerol–water mixture over Pt-N-TiO_2 nanotube photocatalyst. *Int. J. Energy Res.* **2013**, *11*, 1372–1381.
74. Wei, X.; Li, J.; Liu, Z.; Yang, X.; Naraginti, S.; Xu, X.; Wang, X. Visible light photocatalytic mineralization of 17α-ethinyl estradiol (EE2) and hydrogen evolution over silver and strontium modified TiO_2 nanoparticles: Mechanisms and phytotoxicity assessment. *RSC Adv.* **2018**, *8*, 4329–4339. [CrossRef]
75. Chen, D.; Gao, H.; Yao, Y.; Zhu, L.; Zhou, X.; Peng, X.; Zhang, X. Pd loading, Mn+ (n= 1, 2, 3) metal ions doped TiO_2 nanosheets for enhanced photocatalytic H_2 production and reaction mechanism. *Int. J. Hydrog. Energy* **2022**, *47*, 10250–10260. [CrossRef]

76. Nguyen, C.; Dinh, C.; Do, T. Hollow Sr/Rh-codoped TiO_2 photocatalyst for efficient sunlight-driven organic compound degradation. *RSC Adv.* **2017**, *7*, 3480–3487. [CrossRef]
77. Li, C.; Hu, P.; Meng, H.; Jiang, Z. Role of sulfites in the water splitting reaction. *J. Sol. Chem.* **2016**, *45*, 67–80. [CrossRef]
78. Chandra, M.; Bhunia, K.; Pradhan, D. Controlled synthesis of CuS/TiO_2 heterostructured nanocomposites for enhanced photocatalytic hydrogen generation through water splitting. *Inorg. Chem.* **2018**, *57*, 4524–4533. [CrossRef]
79. da Silva, G.C.; Venturini, S.I.; Zhang, S.; Löffler, M.; Scheu, C.; Mayrhofer, K.J.; Ticianelli, E.A.; Cherevko, S. Oxygen evolution reaction on tin oxides supported iridium catalysts: Do we need dopants? *ChemElectroChem* **2020**, *7*, 2330–2339. [CrossRef]
80. Phul, R.; Perwez, M.; Ahmed, J.; Sardar, M.; Alshehri, S.M.; Alhokbany, N.; Khan, M.A.A.; Ahmad, T. Efficient multifunctional catalytic and sensing properties of synthesized ruthenium oxide nanoparticles. *Catalysts* **2020**, *10*, 780. [CrossRef]

catalysts

Review

Photocatalytic Degradation of Recalcitrant Pollutants of Greywater

Mohammad Aslam [1], Dawood Bin Fazal [2], Faizan Ahmad [3], Abdullah Bin Fazal [4], Ahmad Zuhairi Abdullah [1], Mukhtar Ahmed [1], Mohammad Qamar [5] and Mohd Rafatullah [6,*]

1 School of Chemical Engineering, Universiti Sains Malaysia, Nibong Tebal 14300, Penang, Malaysia; mohammadaslam@student.usm.my (M.A.); chzuhairi@usm.my (A.Z.A.); mukhtar.ahmed4848@student.usm.my (M.A.)
2 Department of Petroleum Studies, ZHCET, Aligarh Muslim University, Aligarh 202002, India; dawood.fazal97@gmail.com
3 Department of Post Harvest Engineering and Technology, Aligarh Muslim University, Aligarh 202002, India; faizan.phe@amu.ac.in
4 Department of Geology, Aligarh Muslim University, Aligarh 202002, India; abdullahbinfazal94@gmail.com
5 Interdisciplinary Research Center for Hydrogen and Energy Storage, King Fahd University of Petroleum and Minerals, Dhahran 31261, Saudi Arabia; qamar@kfupm.edu.sa
6 School of Industrial Technology, Universiti Sains Malaysia, USM, George Town 11800, Penang, Malaysia
* Correspondence: mrafatullah@usm.my; Tel.: +60-46532111; Fax: +60-4656375

Abstract: These days, many countries have a water shortage and have limited access to clean water. To overcome this, a new treatment is emerging, namely, the photocatalytic processing of greywater. Photocatalytic processes to remove the organic matter from different greywater sources are critically reviewed. Their efficiency in degrading the organic matter in greywater is scrutinized along with factors that can affect the activity of photocatalysts. Modified TiO_2, ZnO and TiO_2 catalysts show great potential in degrading organic materials that are present in greywater. There are several methods that can be used to modify TiO_2 by using sol-gel, microwave and ultrasonication. Overall, the photocatalytic approach alone is not efficient in mineralizing the organic compounds, but it works well when the photocatalysis is combined with oxidants and Fe^{3+}. However, factors such as pH, concentration and catalyst-loading of organic compounds can significantly affect photocatalytic efficiency.

Keywords: photocatalyst; degradation; greywater; pollutants

1. Introduction

Water is becoming scarce in today's world—over 2 billion people globally experience water shortage (UN, 2018). The lack of water is due to the increased human population, expanded industries, economic activities, long-term droughts in arid regions, and water pollution [1]. Another statement from UNICEF and the World Health Organization (WHO) reported that about 2.2 billion population globally are experiencing limited access to clean and uncontaminated water supply. According to the UN, this causes nearly 1000 children to die every day due to unclean and unsanitised water. In general, food supply, sustainable development, human health, and ecosystems health are essentially dependent on water availability at suitable quality [2]. Since people generally utilize water supply from surface and groundwater, there is a water shortage in these two sources. Thus, an alternative way is required to reduce surface and groundwater consumption. Several practical techniques and technologies have been figured out to solve the water needs [3]. Reusing greywater is an alternative option to optimize water efficiency because this resource is always available. Hence, reusing greywater is a better choice for more sustainable water use [4]. Greywater is defined as any form of domestic wastewater generated from hand basins, kitchen sinks, laundry and baths except wastewater from toilets (toilets and urinals). Reusing greywater is an alternative way to replace surface and groundwater. Domestic water usage is comparatively high [4]. Other than that, the concentration of nutrients, pathogens, harmful chemicals and organic load in greywater is lower than in municipal and industrial effluent [5]. Domestic wastewaters are segregated into two categories, namely, toilet wastewater and greywater [6]. However, the discharged greywaters usually contain surfactants (anionic, cationic and amphoteric), which originate from shampoos, soaps, detergents and dyes that need to be treated [4]. Greywaters also contain salts, food particles, minerals, dissolved organic matter, and personal care products [1]. However, the amount, quality and content in greywater can be affected by different factors such as the country, social and cultural behaviour, lifestyle, personal activities, age distribution, living conditions and cleaning items used at home [6]. Therefore, prior to reuse, greywaters must be treated to remove the harmful substances and pathogens to make them safe for human consumption.

The key issue in greywater treatment is to reduce degradable organic compounds [3]. A new sustainable treatment technology is required with low maintenance costs and high treatment efficiency or robust technology to remove recalcitrant organic compounds from greywater to eliminate the possible environmental and public health hazards. These objectives come with innovative wastewater treatment technologies from advanced oxidation process (AOPs) particularly through a photocatalytic process [3]. Photocatalytic treatment is an evolving technology used to eliminate pollutants present in greywater. Many photocatalyst types are used for removing the contaminants. Recently, semiconductor-based photocatalysts have gained attention [3]. TiO_2 and ZnO semiconductor-based photocatalysts are widely applicable for the treatment. Photocatalysis activity occurs when a light source interacts with the semiconductor surface. The irradiation of UV light on a semiconductor surface will produce highly reactive hydroxyl radicals. The hydroxyl radical will react with the organic molecules to be degraded into water, mineral acids and carbon dioxide [3]. Since the photocatalytic process is operated with the use of solar or UV light, it reduces the electrical power usage and running cost and has fewer operating constraints [6]. This technology can be categorised as highly environmentally friendly and sustainable treatment technology since it is a "zero" waste system [7].

In this study, different types of photocatalytic systems used for the mineralization of organic compounds in greywater are reviewed along with their photocatalytic efficiency in degrading organic materials. Various greywater sources are analyzed by looking into the factors that affect the photocatalyst in degrading the organic compounds. The photocatalytic efficiency in degrading the organic compounds is related to the physical and chemical parameters.

2. Characteristics and Compositions of Greywater

The greywater composition differs widely based on its sources, such as greywater from baths, kitchen or laundry and by the local water quality. Figure 1 shows the characteristics of different sources of greywater. Numerous pollutants are present in greywaters such as acidic and alkaline substances, heavy metals, oil and grease, suspended dissolved solids, fats and synthetic chemicals [8]. Greywaters are also sometimes classified according to high pollutant loads (HGW) or low pollutant loads (LGW). The high pollutant loads category comprises greywaters from kitchen and laundry because they are highly concentrated compared to other greywater sources. In contrast, the low pollutant loads category includes less contaminated greywaters. Still, some researchers consider kitchen greywater as LGW due to highly degradable nutrients and organic matter [9].

2.1. Greywater Quality

Domestic water consumption differs between nations based on geographical location, facilities and livelihoods. This causes greywater generation to be highly varied [10]. Table 1 shows the quantity of greywaters generated in several countries. The greywater volume in-household varies every day with the highest amounts usually generated before or after regular working hours [9]. In the case of geographical location, greywater quantity will often differ seasonally. A high volume of bathroom greywater is generated during hot seasons, while low greywater volume is generated during cold seasons [11]. According to the literature, the typical volume of greywater varies between 90 and 120 L/p/d (litre/person/day) depending on living standards, lifestyles, customs and habits, population structures (age, gender), water installations, and water abundance. On the other hand, greywater volumes can be as low as 20–30 L/p/d in low-income countries with limited water supplies [10].

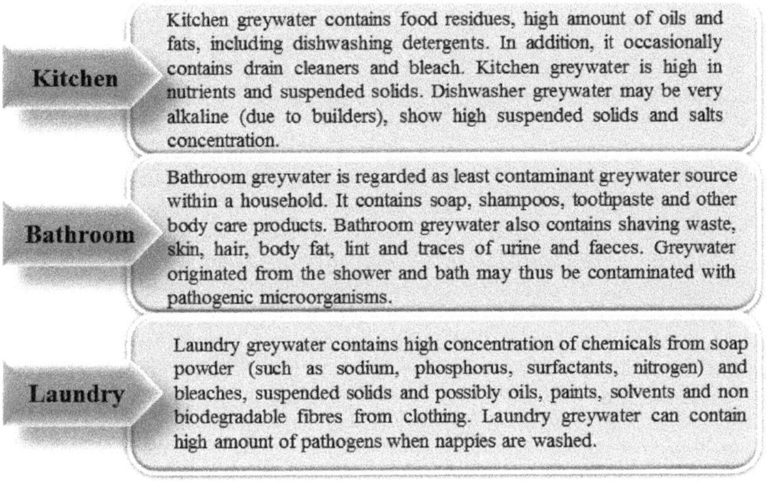

Figure 1. Characteristics of greywater.

2.2. Standard of Greywater

Greywater is produced as result of the people's lifestyles, products they use, and installation's design, so its generation and characteristics are highly variable [12]. Greywater types differ based on where it originates, resulting in a wide range of scenarios and treatment methods. Greywater from bathroom sinks, bathtubs, and showers is called "light greywater," while greywater from laundry facilities, dishwashers, and kitchen sinks is called "black greywater," and is more contaminated [13,14]. Tables 1 and 2 summarize the generation in different countries and quality of different greywater based on literature review [15].

Table 1. Greywater generation reported in different countries.

Location	Generation Rate (Litre/Person/Day)	References
Asia	72–225	[10]
Malaysia	225	[16]
Africa and Middle East	14–161	[10,17,18]
Muscat, Oman	151	[19]
Tucson Arizona, USA	123	[20]
Australia	113	[10]
Switzerland	110	[10]
Vietnam	80–110	[21]
Israel	98	[22]
Nepal	72	[23]
Stockholm	65	[24]
Jordan	50	[25]
Mali	30	[26]
Gauteng, South Africa	20	[27]

Table 2. Physicochemical properties of greywater.

Parameters	Kitchen	Bathroom	Laundry	Light Greywater	D Ark Greywater
Temperature	24.4–30.9 °C	25.8–29.0 °C	22.4–35.0 °C	13.4–29.0 °C	22.4–35.0 °C
pH	5.9–7.4	6.4–8.1	7.1–10	4.90–8.53	5.00–10.33
Total Suspended Solids (mg/L)	134–1300	7–505	68–465	7–793	11–4564
Chemical Oxygen Demand (mg/L)	26–2050	100–633	231–2950	23–1489	58–8071
Biological Oxygen Demand (mg/L)	536–1460	50–300	48–472	20–673	44–3330
Total Nitrogen (mg/L)	11.4–74	3.6–19.4	1.1–40.3	1.3–148.0	0.5–65.0
Total Phosphorus (mg/L)	2.9–>74	0.11–>48.8	ND–>171	0.1–60.0	0.2–187.0
Turbidity (NTU)	210–357	19–375	34–510	13–375	34–510

Although greywater quality varies, an analysis of greywater components by different groups shows that kitchen, and laundry greywater contains more organics, and physical pollutants than bathroom and mixed greywater. In terms of the chemical oxygen demand (COD): biological oxygen demand over 5 days (BOD5) ratios, all types of greywaters are usually biodegradable [15]. When compared to the recommended COD:N:P ratio of 100:20:1 for sewage discharge, bathroom grey water is nitrogen and phosphorus deficient because to the absence of urine and faces [28]. The laundry greywater and mixed greywater are both nitrogen deficient, just like the bathroom greywater. As some detergents are free from phosphorus, concentrations of phosphorus remain low in water arising from the laundry. A kitchen greywater usually contains organic matter, suspended solids, turbidity, and nitrogen. The kitchen greywater, unlike other greywaters, does not lack in nitrogen or phosphorus, and has a COD:N:P ratio that is close to the ratio suggested by Metcalf & Eddy [28]. The wastewater released from kitchens is sometimes excluded from other streams by some authors.

If the greywater is to be treated biologically, however, it is recommended to collect a small quantity of kitchen greywater along with other streams to achieve an optimal COD:N:P ratio as most biodegradable organics and nitrogen particulates come from greywater released from kitchen-sinks and dishwashers. The microorganism contamination of bathroom and laundry greywater is lower than that of other greywater streams, according to the analysis of greywater characteristics by the different categories. Kitchen greywater is more likely to be contaminated by thermally tolerant coliforms than the other greywater streams, presumably due to the presence of easily biodegradable organic substances in large amount. As highlighted by Metcalf and Eddy and Knerr et al., the ratio of C:N:P of mixed greywater is balanced [28,29].

As highlighted by Jefferson et al. [30], a lack of macronutrients and trace nutrients in greywater can limit the biological processes' treatment efficiency. It has been found that the COD: BOD5 ratio in greywater was around 0.50, indicating good biological treatment potential [29,31]. They also claimed that the growth of microorganisms was unaffected by nutrient concentrations. Greywater contained high levels of S, Ca, K, and Al, and trace nutrient concentrations were close to reported requirements as suggested by other researchers [32,33]. The absence of kitchen greywater was clearly the cause of the trace nutrient deficiency as reported by Jafferson et al. [30].

2.3. Greywater Compositions

The water source's quality and household activities significantly influence greywater composition. The cleaning items used in the household and the usage amount affect the number of pollutants substantially. The greywater is a mix of food particles, soaps, oil, fat, textile fibres, and trace amounts of other chemicals. Other than that, there are also large quantities of detergents in greywater [10]. However, greywater is usually composed of biodegradable organics. The quality of greywater can be determined according to its physical properties and chemical properties by measuring their parameters [9]. Table 3 shows that the greywater properties vary according to the country.

Table 3. Range/average of physical and chemical parameters of domestic greywater in several countries.

Parameters	Country													
	Australia [34]	Canada [35]	Egypt [36]	France [37]	Germany [38]	India [38,39]	Israel [38]	Jordan [38]	Malaysia [40]	Palestine [41]	Sweden [38]	UK [42]	USA [38]	Yemen [38]
Total Suspended Solids (mg/L)	74	-	70–202	23–80	-	53.80–788.00	30–298	23–845	19–175	304–4952	-	37–153	17	511
BOD_5 (mg/L)	104	-	220–375	85–155	59	17.10–290.00	74–890	36–1240	1.1–309	407–512	425	8.7–155	86	518
COD (mg/L)	-	278–435	301–557	176–323	109	43.90–733.00	840–1340	58–2263	16–1103	863–1240	890	33–587	-	2000
pH	-	6.7–7.6	6.05–7.96	6.46–7.48	7.6	5.90–8.34	6.3–8.2	6.4–9	6.5–7.2	5.8–8.26	7.8	6.6–7.8	6.4	6
Total Phosporus (mg/L)	3	0.24–1.02	8.4–12.1	-	1.6	0.01–3.84	1.9–48	0.69–51.58	4.5	5.8–15.16	4.2	0.4–0.9	4	-
Total Nitrogen (mg/L)	5.3	-	-	-	15.2	17.00–28.82	10–34.3	6.44–61	-	111–322	75	4.6–10.4	13.5	-
Total Coliforms (MPN/100 mL)	-	-	-	1.7×10^8–1.4×10^9	-	-	-	250–1.0×10^7	-	-	-	1.8×10^3–2.2×10^7	-	-
Faecal Coliforms (MPN/100 mL)	-	4.7×10^4– 8.3×10^5	-	4.0×10^3–5.7×10^6	1.4×10^5	5.0×10^1–1.2×10^2	3.5×10^4–4.0×10^6	1.3×10^1–3.0×10^5	0–1.9×10^6	-	1.7×10^5	1.0×10^1–2.2×10^5	-	1.9×10^7
E. coli (MPN/100 mL)	-	-	-	-	-	-	5.0×10^4	2.0×10^5	0–6.7×10^3	-	-	1.0×10^1–3.9×10^5	5.4×10^2	-

2.3.1. Physical Parameters

The greywater quality can be determined through its physical properties by measuring the physical parameters that comprise of temperature, turbidity and suspended solids (SS) [43].

Temperature

The temperature of greywater usually varies between 18 and 30 °C, according to Oteng-Peprah et.al. [43]. The utilization of warm water in household activities such as cooking activities causes greywater temperature to be high. The high temperature can lead to microbial growth and deposition of carbonates such as $CaCO_3$ and other mineral compounds [43].

Suspended Solids

In [43], studies show that greywater contains suspended solids with a concentration in a range of 190–537 mg/L but which can also reach high as 1500 mg/L. The source for the highest concentration of suspended solids is from kitchen and laundry greywaters. Hair, fibre from clothes, food particles, oil and soil from the washing of vegetables and fruits contribute to the high content of solids in greywater [10].

Turbidity

Greywater usually has a turbidity level between 19 and 444 NTU, according to Oteng-Peprah et al. [43]. The household activities are the primary influence causing the water to be turbid. Greywater from the laundry and kitchen is more likely to be turbid due to the presence of more suspended particles in those greywaters. [43].

2.3.2. Chemical Parameters

The greywater quality can be determined through its chemical properties by measuring the chemical parameters such as electrical conductivity, pH, COD and BOD [9].

pH

The greywater can be identified by measuring the pH, whether acidic or alkaline. According to [43], greywater generally has pH from 5 to 9. Chemicals such as bleaching agent, disinfectants, and fabric softeners will increase greywater's pH value due to the presence of alkalinity. Greywater from laundry will mostly contribute to a high pH value [43].

Ionic Conductivity

The ionic conductivity in greywater is identified in the range of 14 and 3000 μS/cm [43]. The phosphates, sodium, and potassium from the detergents simultaneously increase dissolved solids and increase ionic conductivity [8]. Areas with a shortage of water and groundwater sources may have water with high ionic conductivity because of the dissolved materials [9].

Biological Oxygen Demand (BOD)

BOD is used to determine the amount of dissolved oxygen in greywater required by microorganisms to breakdown the organic matter. The BOD_5/COD ratios are used to assess the degradability of greywater. The bacteria present in greywater are defined through the rate of the decomposition of organic matter. The typical greywater BOD_5/COD ratios range from 0.31 to 0.71 [43].

Chemical Oxygen Demand (COD)

The COD values of greywaters usually range from 38 to 5000 mg L^{-1}, depending on the country. The majority of the COD values are affected by the sources that originate from laundry activities [10].

3. Classification of Advanced Oxidation Processes (AOPs)

Figure 2 depicts various types of AOPs [44]. AOP also can be generally classified into two categories, namely, homogeneous and heterogeneous processes [45]. However, photocatalysis varies from other AOPs because it uses reusable catalysts with low energy UV-A light and does not require any strong oxidants.

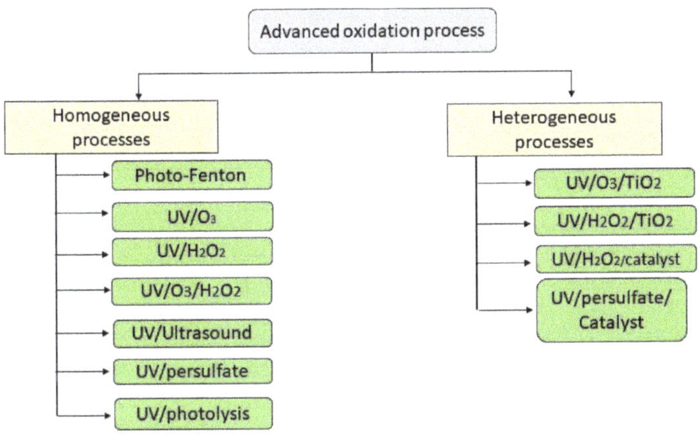

Figure 2. Classification of Advanced Oxidation Processes (AOPs).

3.1. Homogeneous AOP

Homogeneous processes make use of light as a source of energy. UV rays, sonolysis (ultrasonic), electrical energy, and microwave irradiation are some of the most common methods for pollutant degradation. Among all the AOPs, the homogenous ones are the majority. Homogeneous AOPs involve the widely known (photo)Fenton (considered as (photo)catalysis due to the Fe^{+3}/Fe^{+2} cycle), use of H_2O_2/UV, O_3/UV, electrochemical oxidation, photolysis, etc. Another reported homogeneous AOP is homogeneous photocatalysis when using organic molecules acting as photocatalyst (also described as photosensitizers). The main advantage of these substances remains in their ability to absorb UV-visible light, taking advantage of solar light. These substances may be naturally in the greywater and may promote the degradation of the organic matter. Some examples of homogeneous photocatalysts/photosensitizers are humic/fulvic acids [46,47] or organic dyes [48] such as riboflavin [49], rose Bengal [50,51] or triphenilpyrilium salts [52,53]. These compounds may generate reactive oxygen species (such as singlet oxygen) but also can promote a photo-induced electron transfer, removing an electron from the contaminants and thus, generating oxidized radicals derived from the contaminants (photoredox process). Subsequently, these radicals may continue decomposing to reach mineralization.

3.2. Heterogeneous AOP

Heterogeneous semiconductor-mediated photocatalytic processes are capable of effectively degrading a variety of toxic contaminants into biodegradable substances, which are later converted to CO_2 and H_2O [54]. The word 'heterogeneous' corresponds to a dual-phase system, in which the catalyst is in the solid phase and the contaminants are in the liquid phase. Two or more phases, as well as a light source (UV/solar radiation) and a catalyst, are used in the heterogeneous process. Semiconductors such as TiO_2, ZnO, and zinc sulphide (ZnS) are used as catalysts in the heterogeneous process. The charge carriers (electrons and holes) are produced in the process that initiates the redox reactions that cause contaminants to degrade on the catalyst surface [55,56].

Although numerous technologies, ozonation [57] and photo-Fenton [58] for instance, are used as advanced oxidation treatment processes for water ablution, most of these processes are expensive, time-consuming, inadequate, and involving significant inputs of energy [59]. Among these technologies, heterogeneous photocatalysis, on the other hand, is seen as a promising, effective, environmentally acceptable, and cost-efficient technique to address the challenges associated with the removal of dangerous chemicals present in the environment [60,61]. To enhance the breakdown of sustained contaminants, photocatalytic methods were created on the basis of the formation of redox species. The adsorption of incident photons by a photocatalyst, and generation of electrons and holes in the conduction and valence band, is the basis of the photocatalytic reaction process (Figure 3). It is hypothesized that when TiO_2 NPs were exposed to UV irradiation, more electrons were energized from the valence band (VB) to the conduction band (CB), resulting in the formation of more positively charged holes (by h+) on the surface. The electrons move to the conduction band (CB), and leave vacancies in the valence band, thereby producing excitons e^- and h^+ pairs as presented in Equation (1). The activated electrons are picked up by the surface-adsorbed oxygen molecules, producing super-oxide anions $O_2^{\bullet-}$ (Equation (5)). At the same time, the produced holes oxidise H_2O and OH molecules present on the photocatalyst's surface, forming hydroxyl-radicals (OH•) (Equations (5) and (6)). OH•radicals are exceptional oxidising agents with very high standard potentials (Eo (OH•/H_2O) = 2.80 V/SHE), capable of oxidising any organic contaminants. If a photocatalytic process is allowed to happen for a long time, the final products will be water, carbon dioxide, and simple mineral acids [(Equation (1)). Equation (9) depicts the overall photocatalytic reactivity towards organic pollutants [62,63].

$$\text{Catalyst} + \text{irradiation} \rightarrow h^+_{VB} + e^-_{CB} \quad (1)$$

$$TiO_2 + h\nu(UV) \rightarrow TiO_2^* + h^+ + e^- \quad (2)$$

$$TiO_2\left(h^+\right) + H_2O \rightarrow TiO_2 + h^+ + OH^\bullet \quad (3)$$

$$TiO_2\left(e^-\right) + O_2 \rightarrow TiO_2 + O_2^{-\bullet} \quad (4)$$

$$O_2 + e^- \rightarrow O_2^{\bullet-} \quad (5)$$

$$H_2O + h^+ \rightarrow OH^\bullet \quad (6)$$

$$OH^- + h^+ \rightarrow H^+ + OH^- \quad (7)$$

$$R - H + OH^\bullet \rightarrow RCOO^\bullet \rightarrow CO + H_2O + \text{Organic} - \text{ions} \quad (8)$$

$$\text{Organic pollutants} \xrightarrow{\text{semiconductor} + h\nu} \text{intermediate(s)} \rightarrow CO_2 + H_2O \quad (9)$$

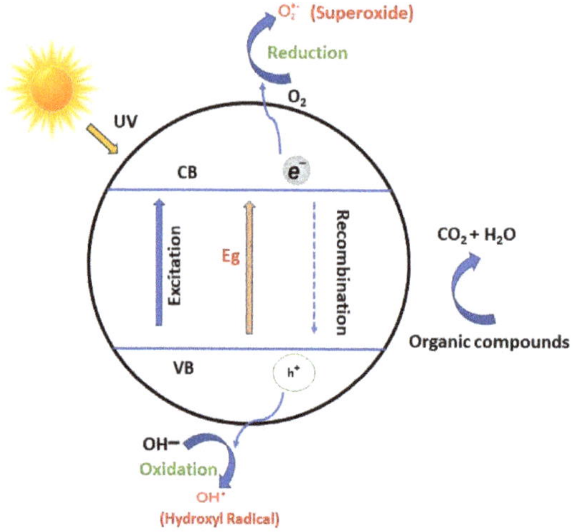

Figure 3. Basic mechanism of semiconductor–mediated photocatalytic process.

4. Heterogeneous Photocatalyst

Photocatalytic research is primarily concerned with the advancement of solar energy utilization. Solar batteries [64,65], solar heat [66], and photocatalysis [67] are three different ways to use solar energy technology. The primary technique among them is the conversion of solar-energy to chemical-energy. The near-ultraviolet (UV) region of the sun spectrum (wavelength less than 400 nm) is used to photo-excite a semiconducting catalyst in exposure to water and oxygen in the heterogeneous solar photocatalytic process. Many semiconductor catalysts can absorb light at wavelengths greater than 400 nm (Fe_2O_3, CdS, etc.) [68,69]. The synthesis in order to initiate a chemical reaction of chemical energy is referred to as this conversion. A photocatalyst is a catalyst that accelerates the solar photo-reaction. The following conditions must be met to become a photocatalyst: (i) They can participate directly into reaction, via charge/energy transfer, etc., and can be recovered back after the catalytic cycle; and (ii) other mechanism routes from existing photo-reactions and a faster reaction-rate are required [67]. Photodegradation is a relatively new and reliable method for degrading organic pollutants in greywater [70–73]. It is categorized under the advanced oxidation process (AOP) since it can tackle an issue such as removing the pollutants from polluted water like waste and greywater. There are several AOPs such as photocatalytic, Fenton-like processes, ozonation, etc. [74].

Photocatalysis has the ability to degrade deadly organic chemicals in wastewater to innocuous and safe components like carbon dioxide and water [75]. Sunlight can power the photocatalytic process, which works under mild settings. As a result, the amount of electricity used and consequently the operational expenses are greatly reduced [6]. Because solar light comprises around 5% UV and 42% visible light, which is abundant and free [76], photocatalytic degradation using sunlight will be cost-effective. When a semiconductor, such as ZnO, is exposed to light energy, it absorbs photons with a wavelength greater than its band gap, resulting in the formation of an electron (e^-)–hole (h^+) pair. The electron functions as a reducing agent while the produced h^+ acts as an oxidizing agent [77].

Under visible light irradiation, the mechanism of heterogeneously accelerated degradation of pollutants by semi-conductor materials occurs with and without PDS. A semiconductor stimulated by visible light creates electrons and holes, which can combine with O_2 and H_2O to produce oxidative free radicals (O_2• and •OH) [78]. The pollutants that are targeted by O_2•/•OH degrade into low-toxic products. Nonetheless, the high rate of photogenerated carrier recombination reduces the pollutants' degrading efficiency.

5. TiO_2 and ZnO Photocatalyst

5.1. TiO_2 Catalyst

Photocatalysts that are non-toxic, inert, photostable, photoactive, and affordable are required. It also should not endanger people or the environment. Finally, the photocatalyst must be excited when exposed to visible and near-ultraviolet light [76,77]. As a result of the rapid growth of nanotechnology, several photosensitive nano semiconductor metal oxides have been developed for use as photocatalysts, including Bismuth (III) oxide (Bi_2O_3), titanium-dioxide (TiO_2), iron (III) oxide (Fe_2O_3), and zinc-oxide (ZnO). TiO_2 is a highly effective photocatalyst for removing organic pollutants present in aqueous as well air environments. Because the use of the TiO_2 for large scale industrial scale water and wastewater treatment is excessively expensive, efforts are being made to develop viable alternatives [79]. TiO_2 is a moderate oxidant-catalyst, which is usually considered as a limitation. However, when the concentration and quantity of pollutants rise, the process becomes more complex, requiring solutions to difficulties such slow kinetics, catalyst deactivation, low photo efficiencies, and unpredictable processes [68].

TiO_2 has been utilized as a photocatalyst for the treatment of greywater originated from different resources. Findings of some studies related to the treatment of greywater with TiO_2 are presented in Table 4. Jin and colleagues conducted a study to identify the optimum photocatalytic conditions for the removal of RB5 dye from greywater. The optimum condition of photocatalytic degradation was used in laundry wastewater to determine efficiency removal of the RB5 dye [1]. Tsoumachidou et al. [6] prepared synthetic greywater to be used in the study that resembles the combination of a washing machine, shower and hand basin water. The study was conducted by applying artificial and solar light in mineralization of the synthetic greywater. Sanchez et al. [4] carried out research on photocatalytic oxidation of greywater from two sources, using titanium dioxide particles. The study was conducted in greywater from two different hotels and a household source, specifically including the first rinse of laundry water and the last rinse of laundry water. They used different catalyst concentrations to remove dissolved organic carbon (DOC) in both hotel and laundry greywater. Birben et al. and Boyjoo et al. [80,81] conducted photocatalytic degradation in synthetically-prepared greywater samples with different composition to identify photocatalytic efficiency and for shower water under artificial light; the TOC concentration was examined along with the time to reduce the TOC concentration for each. The synthetic greywater was prepared with seven different compositions that contained low organic matter (OM) and low anion-content (L1), low organic matter with high anionic species (L2), low organic matter with low anionic species and bacteria (L3), high organic matter with low anionic species (H1), high organic matter with high anionic species (H2) and high organic matter load greywater with low anionic species and bacteria (H3) [80].

5.2. Modification of TiO_2 Catalyst

TiO_2 can only absorb light in the UVA and UVB regions, meaning that just a small portion of the solar spectrum is utilized. Modifying the structure and composition of catalysts, adequate electron acceptors are required to avoid electron–hole recombination [68]. In addition to bare TiO_2, modified TiO_2 is also utilized for the treatment of greywater collected from the residential apartment. Some findings are summarized in Table 5. Behera and colleagues conducted studies of modified TiO_2 catalyst coated on gravel catalyst under solar light and solar photocatalytic-treatment in greywater under modified titanium-dioxide catalyst. The NP-TiO_2 shows high adsorption of light compared to TiO_2 [82]. Various modified titanium-dioxide catalysts were used in this study, namely, microwaved TiO_2 (M-TiO_2), ultrasonicated TiO_2 (U-TiO_2), nitrogen doped (N-doped) titanium-dioxide with P25 (NP-TiO_2) and N-doped titanium-dioxide with titanium isopropoxide (NT-TiO_2) [83].

Table 6 highlights the studies on removal of COD in greywater for reclamation of greywater using pilot scale solar photocatalytic tubular reactors. In addition, hydrogen production along with the treatment of the waste-activated sludge was shown in a newly designed photocatalytic reactor using glass-tubes coated with AgX/TiO_2 [84]. By running the reactors in the batch recycle mode under solar irradiation, the activity of the Ag/TiO_2 was investigated by monitoring degradation of organics present in greywater [85]. TiO_2-coated Al_2O_3 is utilized to process domestic and agricultural wastewaters [86]. Photocatalytic treatment of metoprolol is performed using B-doped TiO_2 [87].

Table 4. Conditions and efficiency of TiO_2 catalyst for photocatalytic treatment of greywater.

Source of Greywater	Type of Catalyst	Method /Supplier	Nature of Lamp	Power (W)	Intensity (kilo lux)/(W/m^2)	Time (h)/(min), Reactor Design with Dimensions (L)	λ (nm)	Findings	References
Laundry water	TiO_2	Sigma-Aldrich	-	-	-	150 min, UV-photoreactor	-	The optimum conditions were = pH = 5, photocatalyst amount = 0.1 gL^{-1} without compressed air sparging and initial Reactive Black 5 (RB5) concentration of a 1 ppm. Reaction time = 150 min, RB5 removal = 97%. Lesser removal (76%) of Reactive Black 5 from real greywater was observed after 330 min. The monitoring findings revealed 60% O&G, 54% COD, 69% BOD5, and 41% removal of TN.	[1]
Simulated greywater (Shower, hand basin, washing machine)	TiO_2 P-25	Evonik	UV-A lamp	9 Watt/78	Natural Sunlight	250–300 min, Thermostated pyrex cell (0.5 L)	350–400 nm	In photo-Fenton-mediated titania photocatalytic process, ~72% DOC removal was observed in the bench-scale treatment after 210 min, whereas under the same photocatalytic conditions but under solar light in pilot reactor, the DOC removal reached decreased to ~64%.	[6]
Hotel greywater Laundry greywater	TiO_2 P-25	Evonik Degussa (Aeroxide® P 25)	Hg lamp TQ 150Z1 (Heraeus Nobelight)	-	-	160–400 min, Batch cylindrical glass photoreactor (1.0 L)	200–700 nm	Treatment of greywater showed removal of 65% DOC after 150 min. Anionic surfactants were completely removed.	[4]
Synthetic greywater	TiO_2 P-25	Evonik P-25	Black light fluorescent lamp (BLF)	125 W	-	0–180 min	300–420 nm	The best dissolved organic carbon removal rates were obtained from greywater containing low OM and low anion content (L1).	[80]
Shower water	TiO_2	Sigma-Aldrich	UV mercury lamp	Primarc Ltd. (PM 3426, 800 W)	-	6.5 h, Pilot-scale reactor (31 L)		Under optimum conditions, approximately 57% removal of total organic carbon (TOC) was achieved after 6 h: initial solution pH = 3, photocatalyst amount = 0.07 gL^{-1}, flow rate of air = 1.8 Lmin^{-1}, circulation rate of solution = 4.4 Lmin^{-1}.	[81]

Table 5. Studies showing the photocatalytic treatment of greywater with modified TiO_2.

Source of Greywater	Type of Catalyst	Material/ Method	Nature of Lamp	Power (W)	Intensity (kilo lux)/ (W/m²)	Time (h)/(min)	λ (nm)	Reactor Design with Dimensions	Findings	References
Residential apartment	Gravel-NP-TiO_2	Sol-gel method	Solar light	-	18.3, 45.7 kilo lux	6 h	-	Tray-type reactor	Significant removal was achieved by solar photocatalytic process: TOC removal was 93.7 % with 0.393 h^{-1} removal rate. Approximately 50% TKN was removed with photocatalytic oxidation to nitrate. 43% removal of nitrate was observed due to photocatalytic reduction. Toxicity of the treated greywater(for 30 min incubation) was reduced from the 13.6 to 4%.	[82]
Simulated greywater	M-TiO_2 U-TiO_2 NP-TiO_2 NT-TiO_2	Microwave Ultrasonication Modified sol-gel	Visible light	Tungsten-halogen lamp 150 W	-	3 h 6 h	400–700 nm	Pencil-type immersion photo-reactor Solar photocatalytic reactor	The performance of different photocatalysts for the mineralisation was found to be NT-TiO_2 > UeTiO_2 > NP-TiO_2M-TiO_2 > P-25. The maximum mineralisation (~75%) was observed after 3 h in presence of NT-TiO_2. Then activity for nitrate degradation was determined to be NT-TiO_2 > UeTiO_2~NP-TiO_2 > M-TiO_2 > P-25. The nitrate degradation was ~92.5 %. As compared to other photocatalytic systems, the NT-TiO_2 was found to be most energy-efficient (~31.86 kWh (kgCOD)1).	[83]

Table 6. Removal of COD from greywater by solar photocatalytic treatment using doped TiO_2 catalyst.

Source	Catalyst	Reactor	Light Source	Findings	Reference
Bathroom and kitchen Greywater	Ag deposited TiO_2 catalyst	tubular reactors	sunlight	In presence of bare and Ag-modified TiO_2, the COD removal was ~32% and ~69%, respectively. The experiments were conducted under sunlight at neutral pH for 3 h. The findings suggest that silver deposition boosted the effectiveness of TiO_2 photocatalysis by serving as electron sink and facilitating interfacial electron transport, minimizing the charge carrier recombination- and producing more ROS in Ag coated titanium dioxide as compared to bare titanium-dioxide.	[85]
Greywater	Silver and silver compounds (AgX) doped TiO_2 film	fluidized tubular photocatalytic reactor (SFTPR)	sunlight	The AgX/TiO_2 filmcoated reactor had a substantially higher rate of waste activated sludge (WAS) degradation, as measured by COD elimination, than the titanium-dioxide filmcoated reactor, with 69.1 percent and 45.3 percent respectively, in 72 h.	[84]
Greywater	TiO_2 (coated on α-alumina)	Photoreactor	sunlight	Chen et al. used TiO_2-coated –Al_2O_3 to treat wastewater originated from domestic and agricultural uses (COD was 36.27.4) before transferring them to a bench-scale wetland system.	[86]
Greywater	Boron-modified TiO_2	solar simulator	Sunlight	A total of 35% COD reduction was achieved in the UW trials. The consequent mineralization, on the other hand, was lower [12%], indicating that chemicals in the effluent are resistant to mineralization.	[87]

5.3. *ZnO Photocatalyst Degradation of Organic Compounds*

In photocatalytic processes, ZnO is the most potential substitute for TiO_2, which has been most widely used photocatalyst. This is because of its band gap of (3.37 eV), low cost and low technological requirements [88]. With an exciton binding energy of 60 meV, ZnO has a strong capacity to absorb UV irradiation [89]. Zinc oxide is designated as a "GRAS" substance by the Food and Drug Administration (FDA) of the United States of America [90]. Zinc oxide does not permeate or can only penetrate the surface of skin, which indicates its nontoxicity towards humans and the environment [91]. Due to a variety of qualities, ZnO is an ideal alternative to TiO_2 as a photocatalyst. According to new research, ZnO is more effective than TiO_2 at photocatalytically destroying organic molecules under visible light. As result, in the presence of sunlight, zinc oxide is the best photocatalyst for removing organic molecules from photodegraded materials [92]. Poulios et al. [93] investigated how ZnO and TiO_2 photodegrade Auramine O in aqueous solution. When compared to TiO_2, evidence shows that ZnO degrades it more quickly (Degussa P25). Kaneva et al. [94] discovered that ZnO NPs degrade pollutants more quickly than TiO_2 NPs. The higher activity was attributed to efficient absorption by ZnO NPs and bigger light quanta as compared TiO_2 [95]. ZnO produces H_2O_2 more efficiently than TiO_2 which facilitates the removal of xenobiotic organic compounds. ZnO possessed high mineralization rate and a greater number of active centers per surface area.

In another study [94], it has been demonstrated that zinc oxide has higher photocatalytic efficacy than TiO_2 due to separation of the high mobility, production, photoinduced electrons and holes. As a result, there is an increasing trend to integrate water that has been treated with visible light in photocatalyst applications. Yashni et al. [96] performed the study of photocatalytic degradation in artificial bathroom greywater under solar light using zinc oxide nano-particles (ZnO NPs). The optimum amounts of ZnO NPs, pH and BR51 concentration were identified by evaluating the efficacy of degradation when comparing different amount of zinc oxide nano-particles, using different pH levels and BR51 concentrations. ZnO NPs loadings of 50, 100, 150 and 200 mg were investigated for degradation. A constant volume (100 mL) of artificial bathroom greywater was exposed to direct solar irradiation for 5.5 h to determine whether the degradation of BR51 in artificial bathroom greywater is due to photocatalytic degradation. The BR51 was degraded to 74.48% at the loadings of 100 mg ZnO NPs. The higher the zinc oxide nanoparticles loading from 100 mg to 150 mg and 200 mg, the lower the degradation rate of BR51. The photocatalytic treatment of a range of pollutants from greywater is shown in Table 7.

Table 7. ZnO-supported photocatalyst matrix is used to photocatalytically degrade organic pollutants in greywater.

Organic Pollutant in Greywater	Type of Catalyst	Amount of Catalyst (g)/Time of Irradiation (h)	Catalyst Synthesis Method	Findings	References
MB	CdS/ZnO composites	3/3	Adsorption, interaction of successive ionic layers. Deposition in a chemical bath	Methylene blue degradation is estimated to be around 91%.	[95]
MO, Methyl green	ZnO / PVP composites	0.1/4	Coprecipitation	Both MO and methyl-green degraded at a rate of 82.7 percent and 99.5 percent.	[97]
Amoxycillin	ZnO ultrathin layers	0.2/2	Deposition of Sol-gel, and spin-coatings	The efficacy of degradation increased by about 65%.	[98]
Color	Nb_2O_5-ZnO-composite	-/4	Chemical solution method with no smooth surfactants	Nb_2O_5/ZnO had ability to elutriate POME with a colour-removal rate of 100%.	[99]
Caffeine	ZnO/ZEO-composite	25/2	Impregnation method	UV light was able to remove nearly all of the caffeine.	[100]
Crystal violet, Methyl red and Basic blue	ZnO, SnO_2 and TiO_2	-/-	NA	When compared to TiO_2 and SnO_2, ZnO had highest photocatalytic activity.	[101]
Crystal violet	ZnO-modified polymer nanocomposite	0.1/5	Chemical precipitation, free radical polymerization	The elimination efficiency was 94 percent in presence of sunlight and 84 percent in the absence of the sunlight, respectively.	[102]
RhB	Mg/ZnO nano-particles	-/2	Sol-gel	The Mg/ZnO nano-particles photocatalytic activitywas controlled by their maximum content, with best removal attained at 2 weight percent Mg.	[103]
Orthonitrophenol	ZnO nano-particles	0.05/5	Microwave assisted combustion	98 percent of the orthonitrophenol was removed.	[104]
MB	SnO_2 doped ZnO nano-particles	1/2	Smooth chemical method	SnO_2 doping boosts the photocatalytic activity of ZnO.	[105]
Aniline	Hybrid chitosanphthalocyanine- TiO_2	0.04/1	SolGel	Using chitosan-phthalocyanine-TiO_2 as a hybrid photocatalyst resulted in more degradation.	[106]
Anthracene	ZnO nano-particles	1/5	Corriandrum Sativum by green synthesis method	Anthracene was photocatalytically degraded at a rate of 96%.	[107]
MB	Poly(Ethylenedioxy-Terthiophene)/ZnO Composites	-/5	Ball mill	Using Poly(Ethylenedioxy-Terthiophene)/ZnO Composite produced by ball-milling and exposed to UV radiation, nearly 100 percent elimination was achieved.	[108]
Malachite green	ZnO nano-particles with activated carbon	0.005/1	Sol-gel method	The high-adsorption capacity of ZnO NPs loaded on activated carbon (322.58 mg g-1) enabled the removal of malachite green within twenty min of adsorption.	[109]
Benzene	Pt-ZnO hydroxyapatite nano-particles	-/-	Template and ultrasonication	The photocatalytic activity of Pt–ZnO NPs altered with hydroxyapatite for benzene removal from aqueous solution was significantly improved.	[110]
Dichloro-benzene Methyl orange	ZnO and Fe-doped ZnO (ZnO/Fe) nanowires	0.1/4	Hydro-thermal	ZnO/Fe nano-wires outperform ZnO in terms of photocatalytic activity.	[111]
RhB	ZnO nano-powder	0.02/0.13	Solution-combustion	Application of the ZnO catalyst resulted in the most dye decolorization (more than 95%).	[79]
MB	Surface-decorated ZnO nano-particles	-/0.8	Solution-combustion	ZnO NPs-mediated colour removal from waste-water while lowering chemical oxygen requirement by 62 percent.	[112]
MB	Ag/ZnO nano-particles	0.15/8	Laser induction	MB has been degraded by 92 percent.	[113]
MB	ZnO/RGO	-/0.6	Hydro-thermal	RGO was mixed with ZnO nano-particles to increase colour removal.	[114]
RhB	Zinc-oxide modified Titanate nanotubes (ZnO–TNTs)	0.2/0.8	Smooth chemical method	ZnO-TNTs nanocomposite outperformed both pure TNTs and ZnO for removal of RhB to.	[115]

MB = Methylene blue, MO = Methyl orange, RhB = Rhodamine B, RGO = reduced graphene oxide.

To suppress the recombination of electron-hole pairs and retain their reducing and oxidizing ability, special strategies are required to design optimal photocatalysts [98]. A simple and facile chemical method is to synthesize SnO_2 doped ZnO nanostructures which have been observed in the presence of polyethylene glycol (PEG) as a surfactant for highly efficient photocatalysis. The structural investigation indicated that the XRD patterns reveal highly crystalline ZnO nanoparticles. The FE-SEM images show that the synthesized SnO_2 doped ZnO has aggregated layers with cave-like structures. The newly prepared SnO_2 doped ZnO nanostructures have been evaluated for photodegradation of methylene blue (MB) under visible light. The photodegradation of MB proceeds much more rapidly in the presence of SnO_2 doped ZnO compared to the undoped ZnO nanoparticles. The photocatalytic performance was in the order of 0.5% SnO_2/ZnO > 1.0% SnO_2/ZnO > 0.2% SnO_2/ZnO > undoped ZnO, suggesting that doping of SnO_2 improves the photocatalytic activity of ZnO. These results indicate that SnO_2 doped ZnO nanostructures are very promising to fabricate highly efficient photocatalysts [105].

Heterojunction modification is necessary to address the high recombination rates of photogenerated electron-hole pairs as well as their inadequate reduction and oxidation capabilities in a single photocatalyst. The type-II and Z-scheme heterojunctions are the two most common types of heterojunctions to gain attention. This direct Z-scheme heterojunction photocatalyst has a low fabrication cost that is comparable to standard type-II heterojunction systems. It also has other benefits, such as the ability to tailor its redox potential to specific photocatalytic activities. Furthermore, because of the electrostatic attraction between electrons and holes, charge transfer on the direct Z-scheme heterojunction photocatalyst is physically better than that on the type-II heterojunction photocatalyst [116]. Figure 4 shows the various types of heterojunction photocatalyst used for degradation of organic pollutants present in greywater.

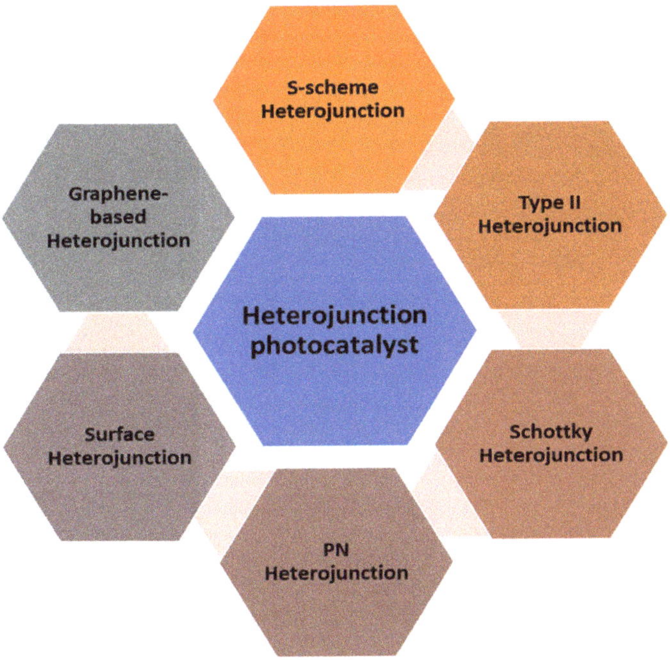

Figure 4. Various types of heterojunction photocatalyst material.

6. Factors Controlling Photocatalytic Reaction

Photocatalysis is influenced by various factors such as pH, catalyst quantity, concentration, and content of organic contaminants. It is critical to use enough catalyst in the process to degrade organic contaminants [117,118]. The above-mentioned increase in catalyst concentration can result in light scattering, a screening effect, and lower specific activity. As a result, the catalyst surface is no longer available for photon absorption and pollutant adsorption, resulting in a reduction in reaction rate [119,120]. An important characteristic for consideration is the pH because different effluents must be handled at recommended pH levels. pH dictates the photocatalyst's surface and charge properties [121]. Other factors that have significant impact on the photocatalytic degradation and reaction kinetics are concentrations and nature of organic pollutants. Highly concentrated solutions of organic pollutants in the aqueous media can adsorb significantly onto the surface of catalysts and block the incident photons from interacting with the surface. As a result, lower photonic efficiency is obtained. Some examples about the effect of different parameters on the photocatalytic activity and/or degradation of organic pollutants are described below [70,71].

The concentration of photocatalyst in the treatment system is critical during the degradation process. The optimum dose is acquired by adjusting the catalyst loading amount in order to cut costs and energy while increasing photocatalytic efficiency and performance [122]. Several researchers have found that increasing the quantity of photocatalytic activity in a treatment system increases the number of photons absorbed on the catalyst's surface [123,124]. As a result, the creation of electron–hole pairs increased, as did the quantity of hydroxyl radicals (OH•).

The photocatalytic degradation rates are strongly influenced by the initial pH of the aqueous solution. This is due to the pH of the solution, which might affect the adsorption of contaminants on the photocatalyst's surface [125]. In general, pH is critical because it affects the surface charge of the photocatalyst, TiO_2. When the pH of the solution is at the catalyst particle's point of zero charge (PZC) of the photocatalyst, the adsorption of contaminants (such as dyes) is at its lowest [126,127]. In general, and in the instance of TiO_2 as a photocatalyst, the contact in between catalyst particles and the water contaminate(s) would be low at the PZC of TiO_2. The lack of an electrostatic force is the reason for this. When operating at pH PZC(TiO_2), however, the photocatalyst's surface charges become positively charged, causing an electrostatic attraction force towards negatively charged molecules. The adsorption of the organic contaminant(s) onto the surface of the activated TiO_2 will be intensified if the organic molecules present have an anionic charge, and under such conditions [128]. When operating at pH > PZC(TiO_2), on the other hand, the photocatalyst surface is negatively charged, repelling anionic molecules in water [63].

7. Conclusions

In conclusion, semiconductor-based photocatalysts are widely used to mineralize the organic compounds in greywater due to high reusability and being chemically inert; the photocatalysts widely used in mineralising the organic compounds in greywater are TiO_2, modified TiO_2 and ZnO. A few improvements are needed for the photocatalytic process to decompose the organic pollutants in the greywater efficiently. One of the methods for studying the optimum conditions for photocatalytic degradation is to include different light intensities because light intensity affects photocatalytic performance in degrading organic compounds. However, the photocatalytic process able to remove the organic compound efficiently with oxidant (hydrogen peroxide) and Fe^{3+} causes the high generation of OH radicals that enhance the mineralization rate. This process is called a photon-Fenton process. Other than that, photocatalytic efficiency is also determined by solution pH, pollutant concentration, amount of catalyst, and nature of the greywater. In contrast, different light sources such as solar and UV light also show a significant difference in the removal of organic materials whereby more degradation occurs under UV light. The higher the amount of catalyst, the higher the degradation rate, but sometimes a high amount of catalyst does not show any improvement in degradation because the greywater becomes more turbid and reduces the penetration of light causing low degradation. The pH value also affects the TiO_2 surface since it can alter the surface charge causing low degradation. Therefore, the optimum pH and catalyst loading are required for efficient photocatalytic degradation.

A study on the morphology and characteristic of photocatalysts also should be conducted. Further study is needed regarding new or unexplored catalysts for the photocatalytic process.

Author Contributions: Conceptualization, M.A. (Mohammad Aslam), D.B.F., F.A., A.B.F. and M.R.; Supervision, A.Z.A., F.A. and M.R.; Writing—original draft preparation, M.A. (Mohammad Aslam), D.B.F. and M.A. (Mukhtar Ahmed); Writing—review and editing, F.A., A.B.F., A.Z.A., M.Q. and M.R.; Funding acquisition, M.R. All authors have read and agreed to the published version of the manuscript.

Funding: The authors would like to express their appreciation to the Ministry of Higher Education, Malaysia, for the Fundamental Research Grant Scheme with Project Code: FRGS/1/2019/STG07/USM/02/12.

Institutional Review Board Statement: Not applicable.

Informed Consent Statement: Not applicable.

Data Availability Statement: Not applicable.

Acknowledgments: The authors are grateful to the Universiti Sains Malaysia and Aligarh Muslim University for providing constant support during this review.

Conflicts of Interest: The authors declare that they do not have any conflict of interests.

References

1. Chong, M.N.; Cho, Y.J.; Poh, P.E.; Jin, B. Evaluation of Titanium dioxide photocatalytic technology for the treatment of reactive Black 5 dye in synthetic and real greywater effluents. *J. Clean. Prod.* **2015**, *89*, 196–202. [CrossRef]
2. Hariharan, C. Photocatalytic degradation of organic contaminants in water by ZnO nanoparticles: Revisited. *Appl. Catal. A Gen.* **2006**, *304*, 55–61. [CrossRef]
3. Bhadiyadra, J.G.; Vaghani, M.V. A review on applicability of photocatalyst titanium dioxide for treatment of greywater. *Int. J. Eng. Res. Appl.* **2015**, *5*, 102–105.
4. Sanchez, M.; Rivero, M.J.; Ortiz, I. Photocatalytic oxidation of grey water over titanium dioxide suspensions. *Desalination* **2010**, *262*, 141–146. [CrossRef]
5. Gulyas, H.; Argáez, Á.S.O.; Kong, F.; Jorge, C.L.; Eggers, S.; Otterpohl, R. Combining activated carbon adsorption with heterogeneous photocatalytic oxidation: Lack of synergy for biologically treated greywater and tetraethylene glycol dimethyl ether. *Environ. Technol.* **2013**, *34*, 1393–1403. [CrossRef]
6. Tsoumachidou, S.; Velegraki, T.; Antoniadis, A.; Poulios, I. Greywater as a sustainable water source: A photocatalytic treatment technology under artificial and solar illumination. *J. Environ. Manag.* **2017**, *195*, 232–241. [CrossRef]
7. Bodzek, M.; Rajca, M. Photocatalysis in the treatment and disinfection of water. Part I. Theoretical backgrounds. *Ecol. Chem. Eng. S* **2012**, *19*, 489–512. [CrossRef]
8. Rakesh, S.S.; Ramesh, P.T.; Murugaragavan, R.; Avudainayagam, S.; Karthikeyan, S. Characterization and treatment of grey water: A review. *IJCS* **2020**, *8*, 34–40. [CrossRef]
9. Boyjoo, Y.; Pareek, V.K.; Ang, M. A review of greywater characteristics and treatment processes. *Water Sci. Technol.* **2013**, *67*, 1403–1424. [CrossRef]
10. Morel, A. *Greywater Management in Low and Middle-Income Countries*; (No. 628.2 G842g); Swiss Federal Institute of Aquatic Science and Technology: Dubenforf, Switzerland, 2006.
11. Oh, K.S.; Leong, J.Y.C.; Poh, P.E.; Chong, M.N.; Von Lau, E. A review of greywater recycling related issues: Challenges and future prospects in Malaysia. *J. Clean. Prod.* **2018**, *171*, 17–29. [CrossRef]
12. Eriksson, E. Potential and Problems Related to Reuse of Water in Households. Ph.D. Thesis, Environment & Resources DTU, Technical University of Denmark, Lyngby, Denmark, 2002.
13. Birks, R.; Hills, S. Characterisation of indicator organisms and pathogens in domestic greywater for recycling. *Environ. Monit. Assess.* **2007**, *129*, 61–69. [CrossRef] [PubMed]
14. Sushmitha, M.B.; Chanakya, H.N.; Khuntia, H.K. Efficient grey water treatment and reuse options for India—A Review. In *Waste Water Recycling and Management*; Springer: Singapore, 2019; pp. 143–149. [CrossRef]

15. Li, F. Treatment of Household Grey Water for Non-Potable Reuses. Ph.D. Thesis, Hamburg University of Technology, Hamburg, Germany, 2009.
16. Martin, C. *Ecological Sanitation Greywater Demonstration Project at Hui Sing Garden*; Urban Environmental Management System Project Report; Natural Resources and Environment Board: Kuching, Malaysia, 2005.
17. Halalsheh, M.; Dalahmeh, S.; Sayed, M.; Suleiman, W.; Shareef, M.; Mansour, M.; Safi, M. Grey water characteristics and treatment options for rural areas in Jordan. *Bioresour. Technol.* **2008**, *99*, 6635–6641. [CrossRef] [PubMed]
18. Al-Hamaiedeh, H.; Bino, M. Effect of treated grey water reuse in irrigation on soil and plants. *Desalination* **2010**, *256*, 115–119. [CrossRef]
19. Jamrah, A.; Al-Futaisi, A.; Prathapar, S.; Harrasi, A.A. Evaluating greywater reuse potential for sustainable water resources management in Oman. *Environ. Monit. Assess.* **2008**, *137*, 315–327. [CrossRef]
20. Casanova, L.M.; Gerba, C.P.; Karpiscak, M. Chemical and microbial characterization of household greywater. *J. Environ. Sci. Health Part A Toxic Hazard. Subst. Environ. Eng.* **2001**, *34*, 395–401. [CrossRef]
21. Büsser, S.; Pham, T.N.; Antoine, M.; Nguyen, V.A. *Characteristics and Quantities of Domestic Wastewater in Urban and Peri-Urban Households in Hanoi*; Annual Report of FY 2006, The Core University Program between Japan Society for the Promotion of Science (JSPS) and Vietnamese Academy of Science and Technology (VAST); Core University Program Office, Ike Laboratory, Div. of Sustainable Energy and Environmental Engineering, Osaka University: Osaka, Japan, 2007; pp. 395–397.
22. Friedler, E. Quality of individual domestic greywater streams and its implication for on-site treatment and reuse possibilities. *Environ. Technol.* **2004**, *25*, 997–1008. [CrossRef]
23. Shrestha, R.R.; Haberl, R.; Laber, J. Constructed wetland technology transfer to Nepal. *Water Sci. Technol.* **2001**, *43*, 345–360. [CrossRef]
24. Ottoson, J.; Stenstrom, T.A. Faecal contamination of greywater and associated microbial risks. *Water Res.* **2003**, *37*, 645–655. [CrossRef]
25. Faraqui, N.; Al-Jayyousi, O. Greywater reuse in urban agriculture for poverty alleviation. A case study in Jordan. *Water Int.* **2002**, *27*, 387–394. [CrossRef]
26. Alderlieste, M.C.; Langeveld, J.G. Wastewater planning in Djenne, Mali. A pilot project for the local infiltration of domestic wastewater. *Water Sci. Technol.* **2005**, *51*, 57–64. [CrossRef]
27. Adendorff, J.; Stimie, C. Food from used water-making the previously impossible happen. *Water Wheel. S. Afr. Water Res. Comm. (WRC)* **2005**, *4*, 26–29.
28. Metcalf and Eddy, Inc. *Wastewater Engineering—Treatment, Disposal and Reuse*, 3rd ed.; Tchobanoglous, G., Burton, F.L., Eds.; Series in Water Resources and Environmental Engineering; McGraw-Hill: New York, NY, USA, 1991.
29. Knerr, H.; Engelhart Hansen, J.; Sagawe, G. Separated grey- and blackwater treatment by the KOMPLETT water recycling system—A possibility to close domestic water cycle. In Proceedings of the Sanitation Challenge: New Sanitation Concepts and Models of Governance, Wageningen, The Netherlands, 19–21 May 2008; pp. 260–269.
30. Jefferson, B.; Burgess, J.E.; Pichon, A.; Harkness, J.; Judd, S.J. Nutrient addition to enhance biological treatment of greywater. *Water Res.* **2001**, *35*, 2702–2710. [CrossRef]
31. Hernandez Leal, L.; Zeeman, G.; Temmink, H.; Buisman, C. Characterisation and biological treatment of greywater. *Water Sci. Technol.* **2007**, *56*, 193–200. [CrossRef] [PubMed]
32. Palmquist, H.; Hanæus, J. Hazardous substances in separately collected grey-and blackwater from ordinary Swedish households. *Sci. Total Environ.* **2005**, *348*, 151–163. [CrossRef] [PubMed]
33. Burgess, J.E.; Quarmby, J.; Stephenson, T. Role of micronutrients in activated sludge-based biotreatment of industrial effluents. *Biotechnol. Adv.* **1999**, *17*, 49–70. [CrossRef]
34. Fowdar, H.S.; Hatt, B.E.; Breen, P.; Cook, P.L.; Deletic, A. Designing living walls for greywater treatment. *Water Res.* **2017**, *110*, 218–232. [CrossRef]
35. Finley, S.; Barrington, S.; Lyew, D. Reuse of domestic greywater for the irrigation of food crops. *Water Air Soil Pollut.* **2009**, *199*, 235–245. [CrossRef]
36. Abdel-Shafy, H.I.; Al-Sulaiman, A.M. Assessment of physico-chemical processes for treatment and reuse of greywater. *Egypt. J. Chem.* **2014**, *57*, 215–231. [CrossRef]
37. Hourlier, F.; Masse, A.; Jaouen, P.; Lakel, A.; Gerente, C.; Faur, C.; Le Cloirec, P. Formulation of synthetic greywater as an evaluation tool for wastewater recycling technologies. *Environ. Technol.* **2010**, *31*, 215–223. [CrossRef]
38. Ghaitidak, D.M.; Yadav, K.D. Characteristics and treatment of greywater—A review. *Environ. Sci. Pollut. Res.* **2013**, *20*, 2795–2809. [CrossRef]
39. Ramprasad, C.; Smith, C.S.; Memon, F.A.; Philip, L. Removal of chemical and microbial contaminants from greywater using a novel constructed wetland: GROW. *Ecol. Eng.* **2017**, *106*, 55–65. [CrossRef]
40. Wurochekke, A.A.; Harun, N.A.; Mohamed, R.M.S.R.; Kassim, A.H.B.M. Constructed wetland of Lepironiaarticulata for household greywater treatment. *APCBEE Procedia* **2014**, *10*, 103–109. [CrossRef]
41. Al-Atawneh, N.; Mahmoud, N.; van der Steen, P.; Lens, P.N. Wastewater characteristics in partially sealed cesspit: Case study from Beit Dajan, Palestine. *Linnaeus Eco-Tech* **2014**, 1–17. [CrossRef]
42. Williams, R.F.; Avery, L.; Winward, G.; Jeffrey, P.; Smith, C.S.; Liu, S.; Memon, F.A.; Jefferson, B. Constructed wetlands for urban grey water recycling. *Int. J. Environ. Pollut.* **2008**, *33*, 93. [CrossRef]

43. Oteng-Peprah, M.; Acheampong, M.A.; DeVries, N.K. Greywater characteristics, treatment systems, reuse strategies and user perception—A review. *Water Air Soil Pollut.* **2018**, *229*, 255. [CrossRef]
44. Poyatos, J.M.; Muñio, M.M.; Almecija, M.C.; Torres, J.C.; Hontoria, E.; Osorio, F. Advanced oxidation processes for wastewater treatment: State of the art. *Water Air Soil Pollut.* **2010**, *205*, 187–204. [CrossRef]
45. Antonopoulou, M.; Kosma, C.; Albanis, T.; Konstantinou, I. An overview of homogeneous and heterogeneous photocatalysis applications for the removal of pharmaceutical compounds from real or synthetic hospital wastewaters under lab or pilot scale. *Sci. Total Environ.* **2021**, *765*, 144163. [CrossRef]
46. Zhan, M.; Yang, X.; Xian, Q.; Kong, L. Photosensitized degradation of bisphenol A involving reactive oxygen species in the presence of humic substances. *Chemosphere* **2006**, *63*, 378–386. [CrossRef]
47. Carlos, L.; Martire, D.O.; Gonzalez, M.C.; Gomis, J.; Bernabeu, A.; Amat, A.M.; Arques, A. Photochemical fate of a mixture of emerging pollutants in the presence of humic substances. *Water Res.* **2012**, *46*, 4732–4740. [CrossRef]
48. Marin, M.L.; Santos-Juanes, L.; Arques, A.; Amat, A.M.; Miranda, M.A. Organic photocatalysts for the oxidation of pollutants and model compounds. *Chem. Rev.* **2012**, *112*, 1710–1750. [CrossRef]
49. Barbieri, Y.; Massad, W.A.; Díaz, D.J.; Sanz, J.; Amat-Guerri, F.; García, N.A. Photodegradation of bisphenol A and related compounds under natural-like conditions in the presence of riboflavin: Kinetics, mechanism and photoproducts. *Chemosphere* **2008**, *73*, 564–571. [CrossRef] [PubMed]
50. Escalada, J.P.; Pajares, A.; Gianotti, J.; Massad, W.A.; Bertolotti, S.; Amat-Guerri, F.; García, N.A. Dye-sensitized photodegradation of the fungicide carbendazim and related benzimidazoles. *Chemosphere* **2006**, *65*, 237–244. [CrossRef] [PubMed]
51. Martinez-Haya, R.; Miranda, M.A.; Marin, M.L. Type I vs. Type II photodegradation of pollutants. *Catal. Today* **2018**, *313*, 161–166. [CrossRef]
52. Martinez-Haya, R.; Gomis, J.; Arques, A.; Marin, M.L.; Amat, A.M.; Miranda, M.A. Time-resolved kinetic assessment of the role of singlet and triplet excited states in the photocatalytic treatment of pollutants at different concentrations. *Appl. Catal. B Environ.* **2017**, *203*, 381–388. [CrossRef]
53. Martinez-Haya, R.; Gomis, J.; Arques, A.; Amat, A.M.; Miranda, M.A.; Marin, M.L. Direct detection of the triphenylpyrylium-derived short-lived intermediates in the photocatalyzed degradation of acetaminophen, acetamiprid, caffeine and carbamazepine. *J. Hazard. Mater.* **2018**, *356*, 91–97. [CrossRef]
54. Turchi, C.S.; Ollis, D.F. Photocatalytic degradation of organic water contaminants: Mechanisms involving hydroxyl radical attack. *J. Catal.* **1990**, *122*, 178–192. [CrossRef]
55. Ibhadon, A.O.; Fitzpatrick, P. Heterogeneous photocatalysis: Recent advances and applications. *Catalysts* **2013**, *3*, 189–218. [CrossRef]
56. Fujishima, A.; Zhang, X.; Tryk, D.A. Heterogeneous photocatalysis: From water photolysis to applications in environmental cleanup. *Int. J. Hydrogen Energy* **2007**, *32*, 2664–2672. [CrossRef]
57. Fu, L.; Wu, C.; Zhou, Y.; Zuo, J.; Song, G.; Tan, Y. Ozonation reactivity characteristics of dissolved organic matter in secondary petrochemical wastewater by single ozone, ozone/H2O2, and ozone/catalyst. *Chemosphere* **2019**, *233*, 34–43. [CrossRef]
58. Ahmad, M.; Chen, S.; Ye, F.; Quan, X.; Afzal, S.; Yu, H.; Zhao, X. Efficient photo-Fenton activity in mesoporous MIL-100 (Fe) decorated with ZnO nanosphere for pollutants degradation. *Appl. Catal. B Environ.* **2019**, *245*, 428–438. [CrossRef]
59. Mishra, A.; Mehta, A.; Basu, S. Clay supported TiO_2 nanoparticles for photocatalytic degradation of environmental pollutants: A review. *J. Environ. Chem. Eng.* **2018**, *6*, 6088–6107. [CrossRef]
60. Lee, C.M.; Palaniandy, P.; Dahlan, I. Pharmaceutical residues in aquatic environment and water remediation by TiO_2 heterogeneous photocatalysis: A review. *Environ. Earth Sci.* **2017**, *76*, 611. [CrossRef]
61. Ye, S.; Yan, M.; Tan, X.; Liang, J.; Zeng, G.; Wu, H.; Wang, H. Facile assembled biochar-based nanocomposite with improved graphitization for efficient photocatalytic activity driven by visible light. *Appl. Catal. B Environ.* **2019**, *250*, 78–88. [CrossRef]
62. Akerdi, A.G.; Bahrami, S.H. Application of heterogeneous nano-semiconductors for photocatalytic advanced oxidation of organic compounds: A review. *J. Environ. Chem. Eng.* **2019**, *7*, 103283. [CrossRef]
63. Chong, M.N.; Jin, B.; Chow, C.W.; Saint, C. Recent developments in photocatalytic water treatment technology: A review. *Water Res.* **2010**, *44*, 2997–3027. [CrossRef]
64. Nottrott, A.; Kleissl, J.; Washom, B. Energy dispatch schedule optimization and cost benefit analysis for grid-connected, photovoltaic-battery storage systems. *Renew. Energy* **2013**, *55*, 230–240. [CrossRef]
65. Lewis, N.S. Toward cost-effective solar energy use. *Science* **2007**, *315*, 798–801. [CrossRef]
66. Bouadila, S.; Kooli, S.; Lazaar, M.; Skouri, S.; Farhat, A. Performance of a new solar air heater with packed-bed latent storage energy for nocturnal use. *Appl. Energy* **2013**, *110*, 267–275. [CrossRef]
67. Lee, S.Y.; Park, S.J. TiO_2 photocatalyst for water treatment applications. *J. Ind. Eng. Chem.* **2013**, *19*, 1761–1769. [CrossRef]
68. Malato, S.; Fernández-Ibáñez, P.; Maldonado, M.I.; Blanco, J.; Gernjak, W. Decontamination and disinfection of water by solar photocatalysis: Recent overview and trends. *Catal. Today* **2009**, *147*, 1–59. [CrossRef]
69. Corrigan, N.; Shanmugam, S.; Xu, J.; Boyer, C. Photocatalysis in organic and polymer synthesis. *Chem. Soc. Rev.* **2016**, *45*, 6165–6212. [CrossRef] [PubMed]
70. Singh, P.; Ojha, A.; Borthakur, A.; Singh, R.; Lahiry, D.; Tiwary, D.; Mishra, P.K. Emerging trends in photodegradation of petrochemical wastes: A review. *Environ. Sci. Pollut. Control Ser.* **2016**, *23*, 22340–22364. [CrossRef] [PubMed]

71. Singh, P.; Vishnu, M.C.; Sharma, K.K.; Borthakur, A.; Srivastava, P.; Pal, D.B.; Tiwary, D.; Mishra, P.K. Photocatalytic degradation of Acid Red dye stuff in the presence of activated carbon-TiO_2 composite and its kinetic enumeration. *J. Water Process Eng.* **2016**, *12*, 20–31. [CrossRef]
72. Zhou, C.; Luo, J.; Chen, Q.; Jiang, Y.; Dong, X.; Cui, F. Titanate nanosheets as highly efficient non-light-driven catalysts for degradation of organic dyes. *Chem. Commun.* **2015**, *51*, 10847–10849. [CrossRef]
73. Sood, S.; Umar, A.; Mehta, S.K.; Kansal, S.K. Highly effective Fe-doped TiO_2 nanoparticles photocatalysts for visible-light driven photocatalytic degradation of toxic organic compounds. *J. Colloid Interface Sci.* **2015**, *450*, e213–e223. [CrossRef]
74. Lee, K.M. Degradation of Dyes Using Zinc Oxide as The Photocatalyst. Ph.D. Thesis, Universiti Putra Malaysia, Selangor, Malaysia, 2008.
75. Meephon, S.; Rungrotmongkol, T.; Puttamat, S.; Praserthdam, S.; Pavarajarn, V. Heterogeneous photocatalytic degradation of diuron on zinc oxide: Influence of surface-dependent adsorption on kinetics, degradation pathway, and toxicity of intermediates. *J. Environ. Sci.* **2019**, *84*, 97–111. [CrossRef]
76. Lam, S.M.; Sin, J.C.; Abdullah, A.Z.; Mohamed, A.R. Degradation of wastewaters containing organic dyes photocatalysed by zinc oxide: A review. *Desalination Water Treat.* **2012**, *41*, 131–169. [CrossRef]
77. Ghosh, M.; Manoli, K.; Shen, X.; Wang, J.; Ray, A.K. Solar photocatalytic degradation of caffeine with titanium dioxide and zinc oxide nanoparticles. *J. Photochem. Photobiol. A Chem.* **2019**, *377*, 1–7. [CrossRef]
78. Saray, A.M.; Zare-Dehnavi, N.; Jamali-Sheini, F.; Yousefi, R. Type-II p (SnSe)-n(g-C3N4) heterostructure as a fast visible-light photocatalytic material: Boosted by an efficient interfacial charge transfer of p-n heterojunction. *J. Alloys Compd.* **2020**, *829*, 154436. [CrossRef]
79. Nagaraja, R.; Kottam, N.; Girija, C.R.; Nagabhushana, B.M. Photocatalytic degradation of Rhodamine B dye under UV/solar light using ZnOnanopowder synthesized by solution combustion route. *Powder Technol.* **2012**, *215–216*, 91–97. [CrossRef]
80. Birben, N.C.; Uyguner-Demirel, C.S.; Bekbolet, M. Photocatalytic removal of microbiological consortium and organic matter in greywater. *Catalysts* **2016**, *6*, 91. [CrossRef]
81. Boyjoo, Y.; Ang, M.; Pareek, V. Photocatalytic treatment of shower water using a pilot scale reactor. *Int. J. Photoenergy* **2012**, *2012*, 578916. [CrossRef]
82. Priyanka, K.; Remya, N.; Behera, M. Greywater treatment using modified solar photocatalyst- degradation, kinetics, pathway and toxicity analysis. *Sep. Purif. Technol.* **2020**, *251*, 117319. [CrossRef]
83. Priyanka, K.; Remya, N.; Behera, M. Comparison of titanium dioxide based catalysts preparation methods in the mineralisation and nutrients removal from greywater by solar photocatalysis. *J. Clean. Prod.* **2019**, *235*, 1–10. [CrossRef]
84. Liu, C.; Lei, Z.; Yang, Y.; Zhang, Z. Preliminary trial on degradation of waste activated sludge and simultaneous hydrogen production in a newly-developed solar photocatalytic reactor with AgX/TiO_2-coated glass tubes. *Water Res.* **2013**, *47*, 4986–4992. [CrossRef]
85. Saran, S.; Arunkumar, P.; Manjari, G.; Devipriya, S.P. Reclamation of grey water for non-potable purposes using pilot-scale solar photocatalytic tubular reactors. *Environ. Technol.* **2019**, *40*, 3190–3199. [CrossRef] [PubMed]
86. Lazar, M.A.; Varghese, S.; Nair, S.S. Photocatalytic water treatment by titanium dioxide: Recent updates. *Catalysts* **2012**, *2*, 572–601. [CrossRef]
87. Cavalcante, R.P.; Dantas, R.F.; Wender, H.; Bayarri, B.; González, O.; Giménez, J.; Machulek, A., Jr. Photocatalytic treatment of metoprolol with B-doped TiO_2: Effect of water matrix, toxicological evaluation and identification of intermediates. *Appl. Catal. B Environ.* **2015**, *176*, 173–182. [CrossRef]
88. Duarah, R.; Karak, N. Hyperbranched polyurethane/reduced carbon dot-zinc oxide nanocomposite-mediated solar-assisted photocatalytic degradation of organic contaminant: An approach towards environmental remediation. *Chem. Eng. J.* **2019**, *370*, 716–728. [CrossRef]
89. Ta, Q.T.H.; Cho, E.; Sreedhar, A.; Noh, J.S. Mixed-dimensional, three-level hierarchical nanostructures of silver and zinc oxide for fast photocatalytic degradation of multiple dyes. *J. Catal.* **2019**, *371*, 1–9. [CrossRef]
90. Jiang, J.; Pi, J.; Cai, J. The Advancing of Zinc Oxide Nanoparticles for Biomedical Applications. *Bioinorg. Chem. Appl.* **2018**, *2018*, 1062562. [CrossRef] [PubMed]
91. Mohammed, Y.H.; Holmes, A.; Haridass, I.N.; Sanchez, W.Y.; Studier, H.; Grice, J.E.; Roberts, M.S. Support for the safe use of zinc oxide nanoparticle sunscreens: Lack of skin penetration or cellular toxicity after repeated application in volunteers. *J. Investig. Dermatol.* **2019**, *139*, 308–315. [CrossRef] [PubMed]
92. Sakthivel, S.; Neppolian, B.; Shankar, M.V.; Arabindoo, B.; Palanichamy, M.; Murugesan, V. Solar photocatalytic degradation of azo dye: Comparison of photocatalytic efficiency of ZnO and TiO_2. *Sol. Energy Mater. Sol. Cells* **2003**, *77*, 65–82. [CrossRef]
93. Poulios, I.; Kositzi, M.; Pitarakis, K.; Beltsios, S.; Oikonomou, I. Photocatalytic oxidation of Auramine O in the presence of semiconducting oxides. *Int. J. Environ. Pollut.* **2000**, *28*, 33–44. [CrossRef]
94. Kaneva, N.V.; Dimitrov, D.T.; Dushkin, C.D. Effect of nickel doping on the photocatalytic activity of ZnO thin films under UV and visible light. *Appl. Surf. Sci.* **2011**, *257*, 8113–8120. [CrossRef]
95. Pannee, A.; Chradda, B.; Wichien, S. Photocatalytic Activity under Solar Irradiation of Silver and Copper Doped Zincoxide: Photodeposition Versus Liquid Impregnation Methods. *J. Appl. Sci.* **2012**, *12*, 1809–1816. [CrossRef]
96. Yashni, G.; Al Gheethi, A.; Mohamed, R.M.S.R.; Arifin, S.N.H.; Shanmugan, V.A.; Kassim, A.H.M. Photocatalytic degradation of basic red 51 dye in artificial bathroom greywater using zinc oxide nanoparticles. *Mater. Today Proc.* **2020**, *31*, 136–139. [CrossRef]

97. Lam, S.M.; Kee, M.W.; Sin, J.C. Influence of PVP surfactant on the morphology and properties of ZnO micro/nanoflowers for dye mixtures and textile wastewater degradation. *Mater. Chem. Phys.* **2018**, *212*, 35–43. [CrossRef]
98. Boussatha, N.; Gilliot, M.; Ghoualem, H.; Martin, J. Formation of nanogranular ZnO ultrathin films and estimation of their performance for photocatalytic degradation of amoxicillin antibiotic. *Mater. Res. Bull.* **2017**, *99*, 485–490. [CrossRef]
99. Chin, Y.H.; Sin, J.C.; Lam, S.M. A facile route for fabrication of hierarchical porous Nb_2O_5/ZnO composites with enhanced photocatalytic degradation of palm oil mill effluent. *Mater. Lett.* **2018**, *216*, 8–11. [CrossRef]
100. Sacco, O.; Vaiano, V.; Matarangolo, M. ZnO supported on zeolite pellets as efficient catalytic system for the removal of caffeine by adsorption and photocatalysis. *Sep. Purif. Technol.* **2017**, *193*, 303–310. [CrossRef]
101. Shinde, D.R.; Tambade, P.S.; Chaskar, M.G.; Gadave, K.M. Photocatalytic degradation of Dyes in Water by Analytical Reagent Grade Photocatalysts—A comparative study. *Drink. Water Eng. Sci. Discuss.* **2017**, *10*, 1–16. [CrossRef]
102. Mohamed, S.K.; Hegazy, S.H.; Abdelwahab, N.A.; Ramadan, A.M. Coupled adsorption-photocatalytic degradation of crystal violet under sunlight using chemically synthesized grafted sodium alginate/ZnO/graphene oxide composite. *Int. J. Biol. Macromol.* **2017**, *108*, 1185–1198. [CrossRef] [PubMed]
103. Zarei, N.; Behnajady, M.A. Synthesis, characterization, and photocatalytic activity of sol–gel prepared Mg/ZnO nanoparticles. *Desalination Water Treat.* **2016**, *57*, 16855–16861. [CrossRef]
104. Assi, N.; Mohammadi, A.; Manuchehri, Q.S.; Walker, R.B. Synthesis and characterization of ZnO nanoparticle synthesized by a microwave-assisted combustion method and catalytic activity for the removal of ortho-nitrophenol. *Desalination Water Treat.* **2015**, *54*, 1939–1948. [CrossRef]
105. Faisal, M.; Ibrahim, A.A.; Harraz, F.A.; Bouzid, H.; Al-Assiri, M.S.A.A. Ismail, SnO2 doped ZnO nanostructures for highly efficient photocatalyst. *J. Mol. Catal. A Chem.* **2015**, *397*, 19–25. [CrossRef]
106. Hamdia, A.; Boufib, S.; Bouattour, S. Phthalocyanine/chitosan-TiO_2 photocatalysts: Characterization and photocatalytic activity. *Appl. Surf. Sci.* **2015**, *339*, 128–136. [CrossRef]
107. Hassan, S.S.M.; el Azab, W.I.M.; Ali, H.R.; Mansour, M.S.M. Green synthesis and characterization of ZnO nanoparticles for photocatalytic degradation of anthracene. *Adv. Nat. Sci. Nanosci. Nanotechnol.* **2015**, *6*, 045012. [CrossRef]
108. Abdiryim, T.; Ali, A.; Jamal, R.; Osman, Y.; Zhang, Y. A facile solid-state heating method for preparation of poly (3,4-ethelenedioxythiophene)/ZnO nanocomposite and photocatalytic activity. *Nanoscale* **2014**, *9*, 89. [CrossRef]
109. Ghaedi, M.; Ansari, A.; Habibi, M.H.; Asghari, A.R. Removal of malachite green from aqueous solution by zinc oxide nanoparticle loaded on activated carbon: Kinetics and isotherm study. *J. Ind. Eng. Chem.* **2014**, *20*, 17–28. [CrossRef]
110. Mohamed, R.M.; Aazam, E. Synthesis and characterization of Pt-ZnO-hydroxyapatite nanoparticles for photocatalytic degradation of benzene under visible light. *Desalination Water Treat.* **2013**, *51*, 6082–6090. [CrossRef]
111. Zhang, Y.; Ram, M.K.; Stefanakos, E.K.; Goswami, D.Y. Enhanced photocatalytic activity of iron doped zinc oxide nanowires for water decontamination. *Surf. Coat. Technol.* **2013**, *217*, 119–123. [CrossRef]
112. Rezaee, A.; Masoumbeigi, H.; Soltani, R.D.C.; Khataee, A.R.; Hashemiyan, S. Photocatalytic decolorization of methylene blue using immobilized ZnO nanoparticles prepared by solution combustion method. *Desalination Water Treat.* **2012**, *44*, 174–179. [CrossRef]
113. Whang, T.-J.; Hsieh, M.-T.; Chen, H.-H. Visible-light photocatalytic degradation of methylene blue with laserinduced Ag/ZnO nanoparticles. *Appl. Surf. Sci.* **2012**, *258*, 2796–2801. [CrossRef]
114. Zhou, X.; Shi, T.; Zhou, H. Applied surface science hdrothermal preparation of ZnO-reduced graphene oxide hybrid with high performance in photocatalytic degradation. *Appl. Surf. Sci.* **2012**, *258*, 6204–6211. [CrossRef]
115. Wang, L.S.; Xiao, M.W.; Huang, X.J.; Wu, Y.D. synthesis, characterization, and photocatalytic activities of titanate nanotubes surface-decorated by zinc oxide nanoparticles. *J. Hazard. Mater.* **2009**, *161*, 49–54. [CrossRef]
116. Low, J.; Yu, J.; Jaroniec, M.; Wageh, S.; Al-Ghamdi, A.A. Heterojunction photocatalysts. *Adv. Mater.* **2017**, *29*, 1601694. [CrossRef]
117. Akpan, U.G.; Hameed, B.H. Parameters affecting the photocatalytic degradation of dyes using TiO_2-based photocatalysts: A review. *J. Hazard. Mater.* **2009**, *170*, 520–529. [CrossRef]
118. Diyauddeen, B.H.; Daud, W.M.A.W.; Abdul Aziz, A.R. Treatment technologies for petroleum refinery effluents: A review. *Process Saf. Environ. Protect.* **2011**, *89*, 95–105. [CrossRef]
119. Gao, H.T.; Si, C.D.; Zhou, J.; Liu, G.J. Sound assisted photocatalytic degradation of formaldehyde in fluidized bed reactor. *J. Taiwan Inst. Chem. Eng.* **2011**, *42*, 108–113. [CrossRef]
120. Gayaa, U.I.; Abdullah, A.H. Heterogeneous photocatalytic degradation of organic contaminants over titanium dioxide: A review of fundamentals, progress and problems. *J. Photochem. Photobiol. C Photochem. Rev.* **2008**, *9*, 1–12. [CrossRef]
121. Turolla, A.; Santoro, D.; de Bruyn, J.R.; Crapulli, F.; Antonelli, M. Nanoparticle scattering characterization and mechanistic modelling of UVeTiO_2 photocatalytic reactors using computational fluid dynamics. *Water Res.* **2016**, *88*, 117–126. [CrossRef] [PubMed]
122. Saggioro, E.M.; Oliveira, A.S.; Pavesi, T.; Maia, C.G.; Ferreira, L.F.V.; Moreira, J.C. Use of titanium dioxide photocatalysis on the remediation of model textile wastewaters containing azo dyes. *Molecules* **2011**, *16*, 10370–10386. [CrossRef] [PubMed]
123. Kaur, A.; Kansal, S.K. Bi2WO6 nanocuboids: An efficient visible light active photocatalyst for the degradation of levofloxacin drug in aqueous phase. *Chem. Eng. J.* **2016**, *302*, 194–203. [CrossRef]
124. Khaki, M.R.D.; Shafeeyan, M.S.; Raman, A.A.A.; Daud, W.M.A.W. Application of doped photocatalysts for organic pollutant degradation—A review. *J. Environ. Manag.* **2017**, *198*, 78–94. [CrossRef]

125. Jallouli, N.; Elghniji, K.; Trabelsi, H.; Ksibi, M. Photocatalytic degradation of paracetamol on TiO_2 nanoparticles and TiO_2/cellulosic fiber under UV and sunlight irradiation. *Arab. J. Chem.* **2017**, *10*, S3640–S3645. [CrossRef]
126. Cavigli, L.; Bogani, F.; Vinattieri, A.; Faso, V.; Baldi, G. Volume versus surface-mediated recombination in anatase TiO_2 nanoparticles. *J. Appl. Phys.* **2009**, *106*, 053516. [CrossRef]
127. Kumar, A.; Pandey, G. A review on the factors affecting the photocatalytic degradation of hazardous materials. *Mater. Sci. Eng. Int. J.* **2017**, *3*, 106–114. [CrossRef]
128. Gogniat, G.; Thyssen, M.; Denis, M.; Pulgarin, C.; Dukan, S. The bactericidal effect of TiO_2 photocatalysis involves adsorption onto catalyst and the loss of membrane integrity. *FEMS Microbiol. Lett.* **2006**, *258*, 18–24. [CrossRef]

catalysts

MDPI

Review

Semiconductors Application Forms and Doping Benefits to Wastewater Treatment: A Comparison of TiO_2, WO_3, and g-C_3N_4

Eryk Fernandes *, João Gomes and Rui C. Martins

University of Coimbra, CIEPQPF—Chemical Engineering Processes and Forest Products Research Center, Department of Chemical Engineering, Faculty of Sciences and Technology, Rua Sílvio Lima, Polo II, 3030-790 Coimbra, Portugal
* Correspondence: eryk@eq.uc.pt; Tel.: +351-239798723; Fax: +351-239798

Abstract: Photocatalysis has been vastly applied for the removal of contaminants of emerging concern (CECs) and other micropollutants, with the aim of future water reclamation. As a process based upon photon irradiation, materials that may be activated through natural light sources are highly pursued, to facilitate their application and reduce costs. TiO_2 is a reference material, and it has been greatly optimized. However, in its typical configuration, it is known to be mainly active under ultraviolet radiation. Thus, multiple alternative visible light driven (VLD) materials have been intensively studied recently. WO_3 and g-C_3N_4 are currently attractive VLD catalysts, with WO_3 possessing similarities with TiO_2 as a metal oxide, allowing correlations between the knowledge regarding the reference catalyst, and g-C_3N_4 having an interesting and distinct non-metallic polymeric structure with the benefit of easy production. In this review, recent developments towards CECs degradation in TiO_2 based photocatalysis are discussed, as reference catalyst, alongside the selected alternative materials, WO_3 and g-C_3N_4. The aim here is to evaluate the different techniques more commonly explored to enhance catalyst photo-activity, specifically doping with multiple elements and the formation of composite materials. Moreover, the possible combination of photocatalysis and ozonation is also explored, as a promising route to potentialize their individual efficiencies and overcome typical drawbacks.

Keywords: contaminants of emerging concern; catalysts doping; composite materials; visible photocatalysis; ozonation

1. Introduction

Society and industry are subject to constant evolution and increasing complexity. Consequently, the consumption of crucial resources for their expansion also progressively grows. Water, as the most vital of these resources, becomes an endangered commodity, making the discussion regarding its renewability against the growing usage and degradation a high-interest topic. Fresh and drinkable water scarcity is caused by different factors, such as climate change, loss of biodiversity, and pollution, which can be due to the direct irregular disposal of sewage, effluents, and other contaminated sources into water bodies, or by the inefficient treatment of those in wastewater treatment plants (WWTPs) [1]. Nowadays, an estimated 1.2 billion people live in locations facing water scarcity and 780 million have no access to basic water services, resulting in a variety of diseases associated with contaminated water consumption [2].

The WWTPs are currently responsible for the provision of proper sanitary living conditions for more than 5.2 billion people, but as industries evolve, a large number of different products and chemicals are consumed by the population, eventually making their way into these facilities, which are not designed to treat them [3,4]. There is a great number of these pollutants, but a certain group is particularly important, the contaminants of emerging concern (CECs), which are a variety of chemicals, such as pesticides, pharmaceutical and personal care products (PPCPs), hormones, and stabilizers [5]. Due to their ineffectiveness, WWTPs become a major source of CECs, which may accumulate in the environment and have already been detected, at ng L^{-1} and μg L^{-1} ranges, in surface and groundwater, as well as in remote places, such as high-altitude rivers and the Antarctic Peninsula [6–9].

Different treatments and techniques have been researched by the scientific community to solve the problem associated with CECs, but a feasible and broad-range solution is yet to be found [10]. Advanced oxidation processes (AOPs) are presented as suitable alternatives for the degradation of a wide range of pollutants, based upon the formation of reactive oxidative species (ROS), most importantly hydroxyl radicals (·OH). AOPs include an array of treatments, vastly explored towards CECs abatements, such as ultraviolet-, ozone-, photocatalysis-, Fenton-, and sulfate-based processes [11–14] (Figure 1). Ozone-based water disinfection treatments are well-established, but also currently one of the most applied AOP for micropollutants removal, being already implemented as tertiary wastewater treatment in some countries [15].

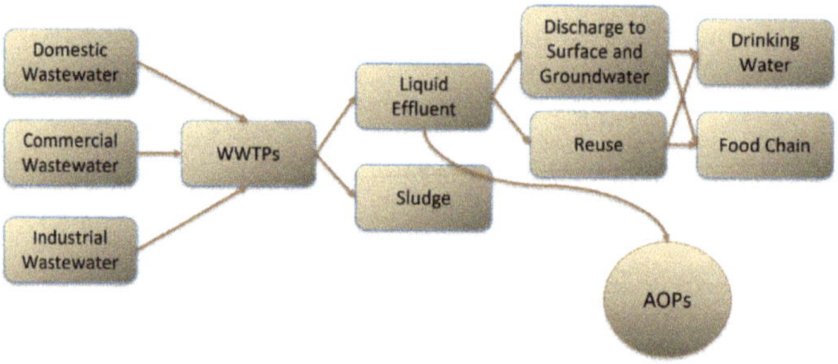

Figure 1. Water treatment scheme and routes of contaminants introduction in the environment.

Ozone is a very strong oxidant, with a redox potential of 2.08 eV, capable of directly reacting with microorganisms and different organic compounds, but it may also be responsible for indirect reactions, producing hydroxyl radicals, which then can interact with the microcontaminants [16]. Nevertheless, this process faces some disadvantages, namely the low solubility of ozone in water, the high energy requirement for its production, low mineralization, and the formation of byproducts potentially more toxic than the parent compounds [17,18]. To enhance the single ozone process, a very common strategy is to integrate or couple it with other AOPs, for instance, O_3/UV, O_3/H_2O_2, O_3/Cl, and O_3/Photocatalysis [11,19,20]. Photocatalytic ozonation is then a promising combined technique, boosting the treatment efficacy and suppressing the individual processes' disadvantages. Photocatalysis is based upon the activation of a catalyst through photon absorption, which then promotes multiple radical's production reactions that attack a large range of contaminants. The contaminants may also be eliminated directly by the catalyst through adsorption [14]. The photocatalysis, when in combination with ozone, can promote a higher decomposition of the gas in water, diminishing ozone demand and increasing the production of ROS. Besides, ozone may also enhance photocatalysis by acting as an electron receiver, reducing electron-hole recombination, which is one of the major disadvantages of this process [21].

Photocatalytic systems are very adaptable and, currently, there is a great variety of catalysts available to be used, the most common being TiO_2. Notwithstanding, TiO_2 presents some disadvantages that challenge its large-scale application, but as these materials are easily tunable, recent studies present multiple adaptations to enhance photocatalysts' performances, such as doping using metal and non-metal elements, immobilization onto other materials, as well as the coupling of different catalysts and materials with distinct natures. Another route is the exploration of alternative catalysts, other than TiO_2, that present superior performances over visible/solar radiation, lowering the process overall cost, which nowadays corresponds to a high-interest field of exploration [22].

Regarding other catalysts, visible-light-driven (VLD) materials became the aim of an increasing number of studies, presenting a better and easier alternative regarding their activation. g-C_3N_4 is a currently very significant catalyst, possessing a metal-free polymeric structure, with simple and very adaptable synthesis methods. Another example, WO_3, also possesses lower bandgap energy and broader absorption of solar radiation while sharing the metal oxide characteristic of TiO_2, which allows an easier correlation between possible known mechanisms and adaptations that can be made. These alternative materials, although with interesting light absorption properties, still face some drawbacks, some also found in TiO_2, such as high electron-hole recombination, low specific surface areas, and improper energy band positions, which hinder the potential of the redox reactions responsible for contaminants elimination.

This current review focuses on the critical analysis of current TiO_2 data as a reference catalyst, and its comparison with VLD materials, g-C_3N_4 and WO_3. Graphitic carbon nitride was selected due to its high recent relevance among this group of alternative catalysts, possessing a high density of recent scientific studies and distinct composition, while WO_3 also has appealing features and structures, and may represent an important connection between the broad existent knowledge of TiO_2 and metal oxides, with visible light active catalysts. Advantageously, these materials may also be involved in modification techniques to boost their efficiencies and overcome disadvantages, as will be further discussed, focusing on doping techniques and composite formation.

2. General Features of Catalysts for Photo-based Treatment Processes

Titanium dioxide is certainly the most applied catalyst for photocatalytic water treatment and possesses different crystalline phases, anatase, rutile, and brookite, with anatase having a higher general photocatalytic activity [23]. Degussa P25 is a commercial TiO_2 catalyst, vastly used due to its optimized characteristics, presenting a mixed anatase-rutile crystal phase.

The benchmarked P25 catalyst was applied by Gomes et al. [24] for the degradation of a complex solution of insecticides, namely azoxystrobin, buprofezin, imidacloprid, procymidone, simazine, terbutryn, and thiamethoxam. Under solar radiation for 120 min, Degussa P25 TiO_2 completely removed 100 $\mu g\ L^{-1}$ of almost all pesticides, excluding thiamethoxam (~90%) and procymidone (~50%). The authors also evaluated the scale-up effect, in a pilot-scale 120 L photoreactor, P25 achieved lower degradation yields, expected due to the larger volume, but still favorable, with removal rates higher than 70% with a photon flux of approximately 8 $kJ\ L^{-1}$. Photocatalytic ozonation was also assessed, and the higher production of ROS improved the degradation of the insecticides by more than 80%, while single ozonation had considerably lower removal rates for some contaminants, such as Procymidone (~50%) and Imidacloprid (~50%). However, the high electronic density groups characteristics of the contaminants that promote a fast reaction with molecular ozone still allow a noticeably efficient treatment by the single process.

Even with broad usage, TiO_2 faces compromising disadvantages that challenge its full-scale application. Primarily, its high bandgap energy, ~3.2 eV for anatase and ~3.0 eV for rutile TiO_2, makes its photoactivation possible only under UV radiation, with $\lambda \leq 390$ nm, hampering its activation through sunlight as the UV portion represents only 4–6% [14]. Moreover, TiO_2 also presents a high recombination rate of e^-/h^+ and a weak separation of photocarriers, reducing its photocatalytic activity [25]. Thus, other catalysts have been constantly developed and investigated to overcome the disadvantages of TiO_2 and fulfill the need for a feasible material. In the last years, a few catalysts have been the focus of an increasing number of publications, e.g., g-C_3N_4 and WO_3 materials, which will be more extensively discussed in this review.

Graphitic carbon nitride (g-C_3N_4) is a metal-free semiconductor catalyst, with high photochemical stability and photoelectric properties, that can be obtained through the thermal polymerization of numerous low-price nitrogen-rich organic precursors. It emerges as a promising visible-light-driven photocatalyst, as it possesses a typical absorption edge at 450–470 nm which corresponds to a bandgap energy of approximately 2.7 eV [26]. The graphitic C_3N_4 is the most stable among the multiple allotropic forms of C_3N_4 (e.g., cubic, beta, alpha), due to its particular 2D structure formed by triazine or heptazine rings.

As mentioned, g-C_3N_4 can be synthesized through the polymerization of different precursors, which will intrinsically affect the properties of the photocatalyst. Nguyen et al. [27] evaluated this parameter using dicyandiamide, melamine, urea, and thiourea as g-C_3N_4 precursors. Their physical and optical properties were considerably distinct, with urea-based g-C_3N_4 having noticeably better values, a specific surface area (S_{BET}) more than three times higher than the others (78.9 m^2 g^{-1}), and lower bandgap energy (2.72 eV). The authors also conducted a chemical oxidation treatment over the catalysts, a method used for the exfoliation of g-C_3N_4, improving their properties, such as the surface area, hydrophilicity, and the addition of reactive functional groups (e.g., hydroxyl). The exfoliation treatment resulted in generally higher S_{BET}, pore density, and a slower recombination rate of the photogenerated electron-hole pairs, indicated by the photoluminescence spectra and electrochemical impedance spectroscopy (EIP). However, the E_{bg} of the treated g-C_3N_4 catalyst was also higher (3.4–3.6 eV), attributed to the exfoliation of the bulk structure into thin layers and the quantum confinement effect. Nonetheless, when applied for methylene blue (MB) photodegradation under visible light, the exfoliated catalyst had a significantly better performance, with urea- and thiourea-based catalysts achieving a complete abatement in 180 min. It is important to mention that the use of thiourea as a precursor also inflicts the addition of sulfur in the catalyst structure, which will be further discussed in the next sections.

The g-C_3N_4 exfoliation may also be obtained through other methods, such as ultrasonication in water or alcohol solutions, or thermal treatment [26]. Fernandes et al. [28] compared Degussa P25 TiO_2 with dicyandiamide-based g-C_3N_4, subjected to a posterior thermal treatment, for the removal of methyl-, ethyl-, and propylparaben (0.08 mM), individually and in the mixture. Under visible light, the polymeric catalyst achieved complete elimination of parabens within 20 min, individually, and 30 min when in a mixture, while P25 TiO_2 obtained the same results in 120 min. The g-C_3N_4 also proved capable to maintain its stability and efficiency when tested in real water matrices, tap and river water. The thermal exfoliation may also partially remove amino groups and thus introduce defects in the catalyst structure, which reduce electron and hole recombination [29].

The combination of g-C_3N_4 photocatalysis and ozonation is still very minimally investigated, yet some studies have been conducted to demonstrate a propitious tertiary wastewater treatment. The more negative conduction band of g-C_3N_4, in comparison to TiO_2, makes it easier for the excited electrons to be captured by ozone. This system was detailed by Orge et al. [29], applying a dicyandiamide as a catalyst precursor with a thermal post-treatment for exfoliation, for the elimination of oxamic acid (OMA), a common by-product of the oxidation of nitrogenous compounds, using a LED system. The exfoliated catalyst, which had an increase of more than 10 times its surface area, presented a much faster removal rate and full elimination of OMA in 120 min, compared with the same system using a benchmark TiO_2 or single ozonation, which resulted in near 80% and 5% removals. Thus, the beneficial synergetic effect was proposed to be due to the interaction between ozone and the graphitic layers of g-C_3N_4. However, in the catalyst's reusing evaluation, the authors indicated that ozone may also deactivate the catalyst, causing a decrease in its activity, pointed to be due to the insertion of defects on the g-C_3N_4 structure and oxygenated groups at its surface.

The benefits of g-C_3N_4 have also been explored for disinfection proposes in recent years. Liu et al. [30] achieved a 4.80 and 4.24 log reduction of *E. coli* and *Staphylococcus aureus*, respectively, using ultrathin urea-based g-C_3N_4 with nitrogen vacancies under visible light, which kept its disinfection efficiency even after five runs. The presence of nitrogen vacancies coupled with a higher specific surface area was shown to be determinant of the catalyst's bactericidal activity. Bacteriophage MS2 virus (*Emesvirus zinderi*) elimination using g-C_3N_4 was assessed by Li et al. [31], achieving a nearly 8-log reduction of the pathogen under 360 min. The photogenerated electrons and superoxide radicals were indicated as the main ones responsible for the virus inactivation, through the oxidation of their proteins.

Tungsten trioxide (WO_3) is a transition metal oxide that also appears as a promising photocatalyst to be used in chemical and biological CECs removal. The high photostability, corrosion resistance, low-cost fabrication, and bandgap of 2.5–2.8 eV are some of its benefits for photocatalytic usage [26]. Bulk WO_3 possesses a cubic perovskite structure and may be obtained in multiple crystalline forms, such as cubic, hexagonal, and monoclinic, the last being more stable at room conditions, formed between 17 °C and 330 °C.

The synthesis methods and procedures are determinants of the type of structure and the characteristics of the catalyst. The morphology and phases of WO_3 are especially relevant as it possesses multiple possible structures, affected by the synthesis method or post-treatments. Nagy et al. [32] found a direct influence of the pH of the solution during catalyst production on its obtained morphology, with a lower pH resulting in a cuboid shape (0.10), changing into nanorod (0.51), nanoneedles (1.52), and nanowires (2.01) with the respective increase of this parameter. Crystallinity may also be intentionally altered, resulting in different single or mixed phases, which present variations in the final product performance. Besides the pH, the addition of some compounds can alter WO_3 structures, such as polyethylene glycol and citric acid (Figure 2) [33]. This easy customization of the structure characteristics is a key advantage of WO_3, and their balanced effect can considerably increase the catalyst photoactivity for targeted applications.

Figure 2. SEM images of WO_3 morphology alteration by variation of citric acid addition, (**a**) nanoflakes, (**b**) nanosheets, and (**c**) hollow spheres (Retrieved with permission from Ref. [33] Copyright (2020) Elsevier).

Different forms of WO_3 catalysts, synthesized through sol-gel and hydrothermal methods, have also been applied for photocatalytic ozonation by Mena et al. [34] for N,N-diethyl-meta-toluamide (DEET) abatement under visible light. Sol-gel monoclinic WO_3 calcinated at higher temperatures (600–700 °C) presented faster removals, eliminating DEET in 10 min. In comparison, over 2 h, the photocatalytic oxidation of DEET achieved only a 22% removal. The presence of ozone in the reaction medium can significantly decrease the typical high recombination rate of the photogenerated species.

The tungsten trioxide catalyst has also been tested for the treatment of more complex effluents. Razali et al. [35] used a WO_3 catalyst synthesized through the coprecipitation method, for the color and suspended solid removal of palm oil mill effluent (POME), retrieved from different ponds of the conventional treatment. The applied catalyst was able to obtain up to 65% and 91% color and suspended solids removal, respectively. The number of active surface sites and their adsorptive characteristics are appointed to be determinants over the treatment process.

Even with a higher visible-light photoactivity, the alternative catalysts still face compromising drawbacks for their application in more rounded contaminant treatment technologies. The lower specific surface areas, high electron-hole recombination, the limited number of active sites, and low electron reduction potential are some examples of those disadvantages that still need to be circumvented [26].

Among different modifications, catalyst doping, and the production of different composites based on the commented catalysts appear as practical techniques to overcome these disadvantages. In the following sections, a recent overview of the general aspect of studies regarding the application of doped TiO_2 and composites of TiO_2 and the alternative catalysts for the photocatalytic degradation of contaminants will be given.

3. Catalyst Doping

3.1. TiO_2

TiO_2 is the most used semiconductor catalyst for the photocatalytic elimination of contaminants, but still presents some typical characteristics that hinder the process scale-up. Its high bandgap energy typically implies the use of external UV radiation sources, with $\lambda \leq 390$ nm, hampering its activation through sunlight as the UV portion represents only 4–6% [14]. The high recombination rate of e^-/h^+ and a weak separation of photocarriers also reduces its photocatalytic activity [25]. Among different modifications, catalyst doping thus appears as a practical technique to overcome these disadvantages. In this section, an overview of the general aspect of recent studies regarding the application of doped TiO_2 for the photocatalytic degradation of contaminants will be presented.

3.1.1. Transition Metals Doping

Both metal and non-metal elements may be incorporated in the catalyst structure and are capable of altering light absorption capacity and, more importantly, visible light, preventing the recombination of electron-hole [36]. Regarding metal doping, transition metals are more commonly applied, due to their partially filled *d* states, which promote the creation of intra-bandgap energy states and the absorption shift [37]. Table 1 summarizes the results and main conditions of the application of metal-doped TiO_2 found in the literature.

In reference to the catalyst synthesis, the sol-gel method is one of the main alternatives and allows better control of the product characteristics, such as porosity, structure, composition, and homogeneity, which is especially important in the incorporation of dopants. Moreover, a calcination step post-synthesis is usually applied as a simple technique to transform an amorphous structure into a crystalline (Anatase, Rutile, and Brookite). Karuppasamy et al. [38] attested that, during the investigation of the doping effect of different transition metals (Zn, Cu, and Zr) on TiO_2, the presence of doping elements may also alter the crystalline phase formation, reducing the temperature at the calcination stage needed to achieve the different structures, which was confirmed by other studies [39,40]. Regarding the different elements used and their photocatalytic activity, Zn-TiO_2 presented the best performance in methylene blue elimination under visible light. The interactions between Zn correspondent electronic states with the TiO_2 conduction band may provoke a red shift in the bandgap, improving the material light absorption between 400 nm and 700 nm.

The amount of dopant is also a crucial parameter to be studied, as excess material can lead to the formation of metallic oxides on the catalyst surface during thermal treatment, which may hinder the catalyst's exposure to photon irradiation. Lee et al. [39] applied hollow TiO_2 spheres doped with copper for the degradation of phenol under visible irradiation. To remove the copper oxides (CuO and Cu_2O), as detected in XRD and TEM analysis, an acid treatment was applied, capable of dissolving these species. The untreated catalyst presented minimal photodegradation, even lower than bare TiO_2, while acid-treated Cu-TiO_2 completely removed the contaminant in 4 h.

Table 1. Application of metal-doped TiO_2 for the photocatalytic removal of contaminants.

Catalyst	Precursor		Method	E_{bg} (eV)	Radiation Source	Contaminant	Removal (Time)	Ref.
	Ti	Dopant						
Ag-TiO_2	-	Silver Nitrate	Precipitation	3.65	UV	Rhodamine B, 5 mg L^{-1}	97% (20 min)	[41]
	-	Silver Nitrate	Ion-exchange	-	UV-Visible	Methyl Orange, 1 mg L^{-1}	80% (150 min)	[42]
						S. aureus, 10^6 CFU mL^{-1}	100% (2 h)	
						E. coli, 10^6 CFU mL^{-1}	100% (1 h)	
Au-TiO_2	TBOT	Gold (III) Chloride	Solvothermal	3.70	UV-Visible	Methylene Blue and Diuron 0.03 mM each	65% MB (180 min) and 95% DIU (120 min)	[43]
Ce-TiO_2	-	Cerium Nitrate	EDTA-Citrate	2.50	Solar	Ciprofloxacin and Norfloxacin, 10 mg L^{-1} each	93.2% CPR and 93.6% NOR (180 min)	[44]
						E. coli, 10^8 CFU mL^{-1}	95.0% (180 min)	
Cu-TiO_2	TTIP	Copper Nitrate	Sol-gel	3.70	Visible	Methylene Blue, 0.05 M	27.5% (60 min)	[38]
	TBOT	Copper Nitrate	Sol-gel and Ion-exchange	2.38	Visible	Phenol, 5 mg L^{-1}	100% (4 h)	[39]
Eu-TiO_2	TTIP	Europium Oxide	Sol-gel	2.86	Visible	Methylene Blue and Methyl Orange, 5 mg L^{-1}	72.1% MB and 71.8% MO (180 min)	[45]
Fe-TiO_2	TTIP	Iron Acetylacetonate	Sol-gel	2.80	Visible	Acid Orange Azo Dye, 10 mg L^{-1}	80% (60 min)	[46]
	TEOT	Iron Nitrate	Sol-gel	-	UV-Visible	Nitrobenzene, 2.45×10^{-4} M	97.3% (240 min)	[47]
La-TiO_2	TTIP	Lanthanum Nitrate	Electrospinning	2.68	Visible	Ciprofloxacin and Methylene Blue, 10 mg L^{-1} each	91% MB and 99.5% CIP (300 min)	[48]
Mn-TiO_2	TTIP	Manganese Acetate	Microwave-assisted Hydrothermal	1.65	UV	Prozac®, 10 mg L^{-1}	95% (30 min)	[40]
Pr-TiO_2	TTIP	Praseodymium Nitrate	Sol-gel	3.00	Visible	Acid Orange Azo Dye, 10 mg L^{-1}	53% (60 min)	[46]
Zn-TiO_2	TTIP	Zinc Nitrate	Sol-gel	2.83	Visible	Methylene Blue, 0.05 M	99.6 (60 min)	[38]
Zr-TiO_2	TTIP	Zirconium Nitrate	Sol-gel	3.30	Visible	Methylene Blue, 0.05 M	81.9% (60 min)	[38]

Various factors are determinants of the performance and characteristics of doped catalysts, but regarding the element's properties, both the ionic radius and electronegativity are very important, due to their role over the element's solubility in TiO_2 lattice [49]. If the radius and electronegativity of the metal ion are close to Ti^{4+}, doping tends to be substitutional, meaning that the substitution of the existing Ti^{4+} ions for the dopant will occur, but if the ion radius is lower than the existing elements, interstitial doping may occur, as they will occupy the interstitial sites of the catalyst lattice. Moreover, if the radius of the doping element is higher, it will not be able to penetrate the catalyst structure and will be mostly present at its surface [50].

Different analytical techniques may be applied to better detect the possible doping types. Through the study of a Mn-TiO_2 catalyst, Moreira et al. [40] detected a negative shift of the typical 144 cm^{-1} E_g mode of anatase TiO_2 in Raman spectroscopy results, up to 5 cm^{-1} with the increase of Mn%. This mode is representative of the symmetrical stretching of O-Ti-O bonds, and the presented shift was suggested to be indicative of the Mn^{3+} substitution of Ti^{4+}. Moreover, Lee et al. [39] conducted an XPS analysis of Cu-TiO_2, and the results indicated a peak shift of the binding energy in both O 1s and Ti 2p spectrums, which was associated with the incorporation of Cu over the TiO_2 lattice, through substitution of Ti^{4+} by Cu^{2+} atoms, forming Cu-O-Ti bonds, also corroborated by

Cu 2p spectra, resulting in the modification of the electron density and charge distribution. Regarding interstitial and surface doping, Kayani et al. [51] studied vanadium doping TiO_2 and suggested that, in the XRD analysis, the reduction in the intensity of the anatase phase diffraction peak (101) is an indication of the interstitial doping of vanadium, which was expected due to the lower ionic radius of V^{5+} in comparison to Ti^{4+}.

The efficiency of photocatalytic systems may also be improved by combination with other AOPs, such as ozonation, where the incorporation of ozone can increase radicals' formation and minimize photocatalytic drawbacks [52]. Ozone, as an excellent electron acceptor, can diminish electron-hole recombination by adsorbing onto the catalyst and retrieving the formed electrons, producing the ozonide radical ($\cdot O_3^-$) and engaging in multiple reactions, including the formation of hydroxyl and other radicals (Equations (1)–(3)).

$$O_{3\ (ads)} + e^- \rightarrow \cdot O_3^- \quad (1)$$

$$\cdot O_3^- + H^+ \rightarrow \cdot O_3H \quad (2)$$

$$\cdot O_3H \rightarrow O_2 + \cdot OH \quad (3)$$

Considering that one of the main advantages of catalyst doping is the reduction of E_{bg} and the resultant easier activation, their use in photocatalytic ozonation induces a higher electronic excitation, and thus an even higher radical generation. The ozone present in the medium may also directly react with the contaminants, especially those with high electron density, such as aromatic chemicals [53].

Catalyst doping is also an approach to turn photocatalytic ozonation into a more feasible technology by not only improving ozone decomposition but compensating the ozone production cost by using solar irradiation. Mecha et al. [54] demonstrated that, when applied for the photocatalytic ozonation treatment of phenol in SWW, Cu-TiO_2, Fe-TiO_2, and Ag-TiO_2 had an up to almost 40% lower energy requirement compared to the undoped catalyst, due to their increased activity. Moreover, notably, Fe-TiO_2 also maintained an appreciable activity under UV irradiation, indicating a broader active spectrum. Still, photocatalytic ozonation is not very much studied for the doped and codoped TiO_2 catalysts and there is plenty of space for developing this interesting and efficient technology. Moreover, the possibility of sunlight radiation usage, which is a natural resource available for a great part of the year, in such an approach is considered.

3.1.2. Noble and Rare-Earth Metals Doping

Noble metals (e.g., Au, Ag, Pd, Pt) represent a high-interest group of transition metals currently being vastly explored for TiO_2 doping due to their light-gathering capability and their role as electron trappers. These metals present, in general, a larger ionic radius compared to Ti^{4+}, which difficult their penetration into the catalyst lattice, being deposited mostly on the surface. Saber et al. [43] detected multiple peaks corresponding to the presence of an Au crystalline phase in Au-TiO_2, prepared by solvothermal method, that can be indicative of surface modifications. Additionally, visible light absorption was considerably increased, again attributed to surface plasmon resonance.

Noble metal doping is known to produce a surface plasmon resonance effect, which induces a higher photoactivity of the catalyst under visible irradiation. This effect can be identified through diffuse reflectance spectroscopy (UV-Vis DRS), with a broader shoulder-like peak attributed to the heteroatom addition and the resulting oscillation of the conduction band electrons on its surface during photoirradiation [41]. Moreover, Ellouzi et al. [41] identified such a phenomenon, which resulted in considerable faster degradation of RhB, with an observed rate constant of 0.1827 min^{-1} using a Ag-TiO_2 catalyst, while non-doped TiO_2 led to an apparent constant rate of 0.1034 min^{-1}. The lack of general and more specific legislation and regulation challenges the classification of CECs, but different synthetic dyes are also considered by many researchers as examples of CECs, due to their potential toxicity, with some studies already appointing their carcinogenic and mutagenic characteristics and ubiquitous behavior [55].

The use of noble metals in catalyst doping, especially silver, us also known to potentialize the antimicrobial activity of catalysts for disinfection purposes. In the medium, the metal particles present on the surface or even some that are released from the catalyst may interact with the cell membrane of microorganisms, damaging it and causing it to rupture, with the consequent oxidization of its proteins and genetic material (Figure 3) [42].

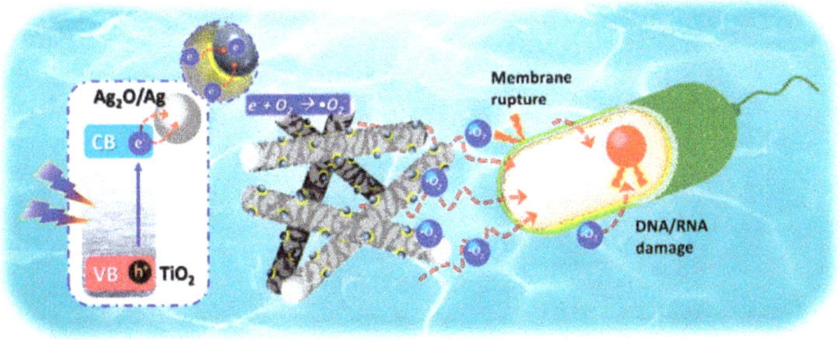

Figure 3. The mechanism for bacterial disinfection using Ag-TiO_2 (Retrieved with permission from Ref. [42] Copyright (2019) Elsevier).

Wu et al. [42] proved the capacity of Ag-TiO_2 nanofibers to inactivate *S. aureus* and *E. coli* bacteria. In 1 h, the number of bacterial colonies, initially ~10^6 CFU mL^{-1}, was substantially reduced when in contact with Ag-TiO_2 under visible light and eliminated in the next hour, while no significant inactivation was found for the undoped catalyst. A synergic effect between Ag_2O, AgO, and metallic silver, present in the catalyst surface, and the photogenerated radical species can considerably boost the disinfection capacity of the material, as it is able to rupture the bacteria cell membranes and degenerate their genetic material.

The surface plasmon resonance phenomenon and other positive effects of noble metals can also be an important ally in photocatalytic ozonation. However, specific properties of each element can greatly affect the final product performance and need to be evaluated for a better selection. For instance, the high electronegativity of Au (2.54) entails a higher capacity to withhold the photogenerated electrons. Although it could contribute to the mitigation of electron-hole recombination, it also impedes ozone reduction and ozonide radical formation, reducing the overall efficiency of the ozone-based process [56]. Thus, other elements with lower electronegativities can produce a more balanced effect, such as Ag (1.93), and boost contaminants removal while decreasing ozone consumption.

Rare-earth metals (e.g., Eu, Ce, Pr) also possess a high capacity to enhance the solar (visible and NIR) photocatalytic activity of TiO_2, in addition to its UV activity [23]. These elements are particularly interesting due to the partially filled 4f states, and the f-f electronic transitions present within. These electronic features combined with the interactions between the f and other orbitals, and with the TiO_2 CB and VB are responsible for the formation of new electronic states and other beneficial doping effects regarding luminescence properties [50]. The 5s and 5p orbitals also promote a shielding effect of the 4f orbital, producing a narrowing of the transition emission bands [57]. As their ionic radius is overall larger than Ti^{4+}, the surface doping and formation of their respective oxide layers are expected, but multiple doping forms may occur simultaneously [57].

These metals can also considerably increase the number of defects over the catalyst surface, which may act as photogenerated species trapping sites. Moreover, even though, due to their ionic radius and symmetric difference, the impregnation of such elements occurs typically on the material surface, some studies have detected their presence within the catalyst structure [58].

The rare-earth metal doping can also influence the crystalline phase transition of TiO_2, which is known to inhibit the transition to rutile from the anatase phase, which possesses a higher photoactivity [58]. More specifically, Pascariu et al. [48] detected an increase from 13% to 70% in the contribution of the anatase phase as a result of La^{3+} doping (Figure 4). Additionally, due to the crystalline phase composition, a higher number of surface defects, and OH groups when La^{3+} was applied, the doped catalyst performed a higher degradation of contaminants, ciprofloxacin, and methylene blue, under visible light. An optimum concentration of dopant was also found, with a further increase of lanthanum concentration higher than 0.1% possibly leading to the blocking of active sites and an excess of trap sites, with a shorter distance between them, having a controversial result, increasing electron-hole recombination [59,60].

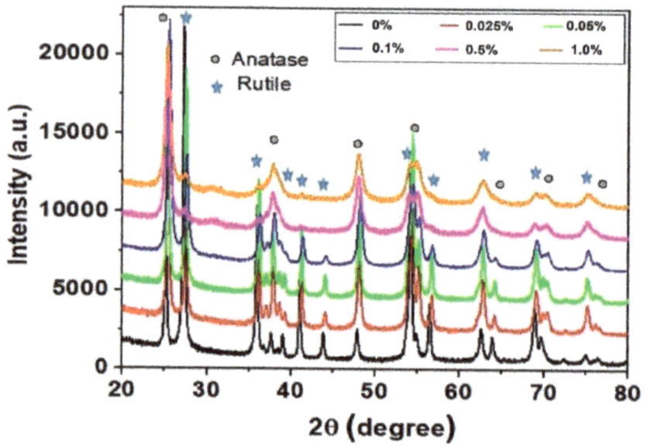

Figure 4. Crystalline phase transformation through La^{3+} at different concentrations (0–1.0%) (Retrieved with permission from Ref. [48] Copyright (2022) Elsevier).

Overall, noble and rare-earth metals are able to produce significant alterations in the catalyst light absorption characteristics and its charge separation. However, in a more practical application, other circumstances must be considered which may hinder their use. These elements possess typically high price precursors and, even if used in low concentration, in comparison to other earth-abundant materials, will elevate the material and consequently the final process cost, which is counter to what is needed to broaden the photocatalysis application. Moreover, they present intricate or distinct impregnations, necessitating a case-by-case analysis of their introduction to evaluate whether it will in fact produce the desired effect in more specific materials and morphologies. The high heterogenicity of studies involving such elements, with different impregnations, irradiation sources, contrasting results, and other parameters, challenge a proper selection of a more promising element.

3.1.3. Non-Metals Doping

Non-metal elements (e.g., N, S, B, C) are vastly explored for TiO_2 doping, with facile production methods, and capable of enhancing the catalyst stability and photoactivity, forming new energy levels with the consequent bandgap shortening, and inducing the formation of oxygen vacancies [37]. Contrarily to metals, in non-metal doping, it is expected that the introduced elements will have an influence over the valence band through interactions with O 2p states, even though cationic interactions may also take place [25]. Table 2 summarizes the application of different non-metals in TiO_2 doping.

Nitrogen doping is one of the most explored techniques, due to its promising red shift of the absorption edge and a similar ionic radius to oxygen, facilitating its substitution, although it may be inserted in the catalyst structure in different forms (Figure 5) [61]. Because of the broad exploration, different preparation methods have been applied, such as pulsed laser deposition [62], thermal annealing [63], hydrothermal [64,65], solvothermal [66], and sol-gel [67–69].

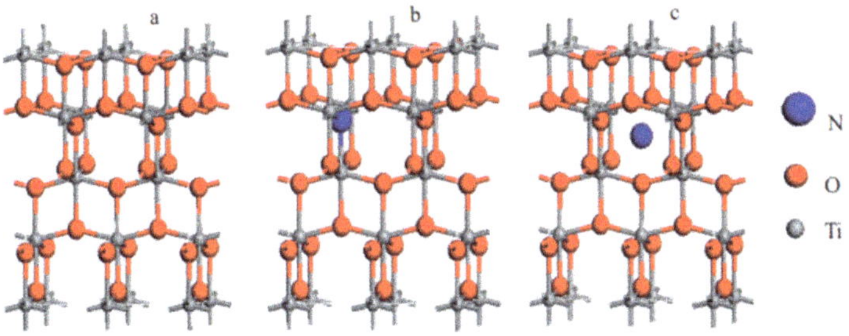

Figure 5. Schematics of different nitrogen incorporation over TiO_2 lattice: (**a**) pure TiO_2, (**b**) substitutional, and (**c**) interstitial doping (Retrieved with permission from Ref. [70] Copyright (2009) John Wiley and Sons).

The sol-gel method is the most widely applied, due to its facile, flexible, and controllable operation, allowing different modifications of the basic processes to achieve the desired characteristics [50]. Using such a method, Assayehegn et al. [71] synthesized multiple N-TiO_2 catalysts applying different ratios of guanidium chloride (GUA), an environmentally friendly N precursor, to Ti, and indicated that the incorporation of N precursor showed a direct effect over the crystalline phase of N-TiO_2 formation post calcination. This can be caused by the perturbance of N^{3-} ions on the lattice orientation and density of charges, due to its higher negative charges and ionic radius compared to O^{2-}. Moreover, the found optimum mixed crystalline phase, 42% anatase and 58% rutile, is pointed out to have a beneficial synergic effect with nitrogen doping, which was also concluded by other research groups [39,68]. When applied for MB degradation under visible light, the best N-TiO_2 had an apparent reaction rate constant of 0.0325 min^{-1}, which is almost 17 times higher than the undoped catalyst, and represented 97% of MB removal within 100 min.

Dopant precursor is a key parameter, especially due to the large variety of compounds to be used. Bakre et al. [68] used semicarbazide, *N,N'*-dimethyl urea, and urea as nitrogen precursors in the sol-gel synthesis of N-TiO_2. Regarding the crystalline phase composition, urea led to a pure anatase phase catalyst, while the other dopants resulted in rutile-anatase mixed phases. Following the characterization of the N-TiO_2 catalysts, the urea catalyst presented a higher absorption on the visible range, seconded by *N,N'*-dimethyl urea. FTIR spectroscopy results showed proper nitrogen incorporation over all catalysts lattice due to an indication of N-Ti-N bonds existence, possibly interstitially as suggested by N1s XPS spectra binding energies peaks. Additionally, other substances containing nitrogen have been vastly applied for doping, such as ammonia, ammonium hydroxide, ethylenediamine, and hydrazine hydrate [66,72–74].

Table 2. Application of non-metal doped TiO_2 for photocatalytic removal of contaminants.

Catalyst	Precursor		Method	E_{bg} (eV)	Radiation Source	Contaminant	Removal (Time)	Ref.
	Ti	Dopant						
B-TiO_2	* P25	Boric Acid	EDTA-Citrate	2.87	Solar	Ciprofloxacin and Norfloxacin, 10 mg L^{-1}	93.2% CPR and 93.0% NOR (180 min)	[44]
						E. coli, 10^8 CFU mL^{-1}	99.9% (180 min)	
	TBOT	Boric Acid	Sol-gel	3.01	Visible	Catechol, 10 mg L^{-1}	100% (60 min)	[75]
	TTIP	Boric Acid	Sol-gel	2.98	UV	Diclofenac, 50 mg L^{-1}	98% (180 min)	[67]
F-TiO_2	$TiCl_4$	Ammonium Fluoride	Nebulizer Spray	2.79	Visible	Malachite Green, 300 mg L^{-1}	90% (60 min)	[76]
I-TiO_2	TBOT	Iodic Acid	Sol-gel	3.18	Solar	Methylene Blue, 4.8 mg L^{-1}	30% (60 min)	[77]

Table 2. *Cont.*

Catalyst	Precursor		Method	E_{bg} (eV)	Radiation Source	Contaminant	Removal (Time)	Ref.
	Ti	Dopant						
$N\text{-}TiO_2$	TTIP	Urea	Sol-gel	2.27	UV	Diclofenac, 50 mg L^{-1}	95% (180 min)	[67]
	TBOT	Guanidinium Chloride	Sol-gel	2.91	Visible	Methylene Blue, 10 mg L^{-1}	97% (100 min)	[71]
	TTIP	Urea	Sol-gel	-	Solar	Methylene Blue, 10 mg L^{-1}	99% (100 min)	[68]
	TTIP	N,N′-dimethyl urea	Sol-gel	-			98% (120 min)	
	TTIP	Semicarbazide	Sol-gel	-			98% (80 min)	
	TBOT	Urea	Sol-gel	2.58	Visible	*Microcystis aeruginosa*, 3×10^6 cells mL^{-1}	99.1% (12 h)	[78]
	TTIP	Ammonium Hydroxide	Sol-gel	2.31	Visible	*E. coli*, 10^5 CFU mL^{-1}	~100% (12 h)	[79]
						S. aureus, 10^5 CFU mL^{-1}	~100% (12 h)	
						Mycobacterium avium, 10^5 CFU mL^{-1}	~100% (12 h)	
						Candida albicans, 10^5 CFU mL^{-1}	99.9% (12 h)	
$S\text{-}TiO_2$	TTIP	Sulfuric Acid	Flame Spray Pyrolysis	2.78	Visible	Acetaldehyde, 0.5 mM	65% (300 min)	[80]
	TTIP	Hydrogen Sulfide	Chemical Vapor Deposition	-	Visible	Methyl Orange, 5 mg L^{-1}	38% (300 min)	[81]

* Commercial Degussa P25 TiO_2.

Sulfur emerges as an interesting dopant, presenting in both cationic and anionic species, as in the form of S^{6+} and S^{4+} it can substitute Ti^{4+}, although the O^{2-} substitution by S^{2-} can also occur. Through the study of sulfur-doped TiO_2 films, Bento et al. [81] suggested that the cationic substitution by S^{6+} would be more advantageous for photocatalytic purposes in comparison to the anionic, due to the formation of impurity energy levels located above the TiO_2 valence band, decreasing the E_{bg} and improving activation by visible light. Furthermore, the authors verified the formation of $SO_4{}^{2-}$ on the surface of the catalyst, known to occur during S doping, which may act as electron trap centers and improve radicals' formation. The replacement of O^{2-} by S^{2-} is may be less chemically favorable due to the larger ionic radius of the sulfur anion, hampering its penetration into the catalyst lattice and inducing higher thermodynamic energy for the S-Ti-S bond to be formed [80].

Boron has been found to promote a higher reduction of Ti^{4+} into Ti^{3+}, which induces the formation of oxygen vacancies, a broader catalyst spectrum response, and improved photocatalytic activity. Yadav et al. [67] identified Ti^{3+} correspondent peaks in interstitially boron-doped TiO_2 $Ti2p_{2/3}$ XPS spectra, indicating its existence in the catalyst lattice. Moreover, the full-width half maxima (FWHM) was used as a correlation of the presence of Ti^{3+}, and when compared to a $N\text{-}TiO_2$ catalyst, it was found that $B\text{-}TiO_2$ presented a higher density of Ti at a less oxidized state. In addition, both $N\text{-}TiO_2$ and $B\text{-}TiO_2$ were compared regarding diclofenac degradation, and $B\text{-}TiO_2$ presented a higher photocatalytic activity, with an apparent constant rate value 30% higher than $N\text{-}TiO_2$. Zhang et al. [75] also confirmed the benefits of boron doping regarding Ti^{4+} reduction and suggested that, besides its role as electron trapping sites, Ti^{3+} altered the adsorptive properties of the catalyst. The formed $O\text{-}Ti^{3+}$ bonds may weaken the water adsorption and enhance the formation of cooperative hydrogen bonds on the catalyst surface, promoting a higher interaction with contaminants, such as phenolic compounds.

Halogens, such as chlorine, iodine, and fluorine, possess attractive properties, such as high electronegativity and the capacity to occupy both titanium and oxygen sites [82]. Ravidhas et al. [76] successfully synthesized $F\text{-}TiO_2$ thin films and suggested that fluorine, the most electronegative element, could substitute oxygen atoms in the catalyst structure, leading to the formation of Ti-F-Ti bonds and the reduction of Ti^{4+}, promoting an adjustment of the electron density of titanium and an increased concentration of oxygen vacancies.

This incorporation of impurities into the structure also caused a decrease of the E_{bg} down to 2.90 eV, from an initial 3.30 eV of the undoped catalyst. F-TiO_2 also presented an improved degradation of malachite green dye under visible and solar radiation, attributed to its higher surface area and number of active sites, oxygen vacancies, and lower E_{bg}. The disturbance of charges was also found by Hwang et al. [77] after iodine doping of TiO_2. In the I 3d XPS spectra, multiple valence states were found, with peaks corresponding to I^{5+}, I^{7+}, and I^- being detected in coexistence. The authors pointed out that I^{5+} had substituted Ti^{4+} atoms, due to their similar radius, with the ensuing formation of Ti^{3+} and Ti^{2+} due to compensation for the disproportionate charges. Meanwhile, the larger ionic radius of I^- induces its dispersion on the TiO_2 surface.

Apart from chemical contaminants, photocatalytic processes using non-metal doped catalysts have been applied for microorganisms' elimination, as a facile disinfection method that does not produce secondary pollutants [83]. The disinfection of a series of pathogenic contaminants, *S. aureus*, *E. coli*, *Mycobacterium avium*, and *Candida albicans*, was studied by Tzeng et al. [79] using a N-TiO_2 catalyst under visible light. Starting with a 10^5 CFU mL^{-1} individual solutions of each species, in 12 h treatment, N-TiO_2 obtained a final survival rate, C/C_0, of approximately 10^{-1}, 10^{-2}, 10^{-4}, and 10^{-5} of, respectively, *C. albicans*, *M. avium*, *E. coli*, and *S. aureus*. The order of the disinfection yields is directly related to the complexity of the cells, as *C. albicans* is a unicellular fungus, it presents a higher structural complexity and contains substances less prominent to oxidative radicals' attacks, whereas *S. aureus* is a Gram-positive bacteria, which has no lipopolysaccharide layer.

The precursors may also deeply affect the disinfection characteristics of the material. Using urea and N_2 as nitrogen precursors, Zhou et al. [78] demonstrated that even with the catalyst only calcinated in the presence of N_2 having the highest incorporation of nitrogen, the urea doped catalyst calcinated in the same atmosphere presented a removal of 99% of chlorophyll-a, related to the *Microcystis aeruginosa* algae cells degradation, in 12 h.

The doping process may also lead to the formation of other compounds within the catalyst structure, such as the elements' respective oxides, which may provide additional mechanisms for chemical and biological contaminant removal. For instance, regarding boron doping, the formation of a B_2O_3 can enhance the catalyst bacterial inactivation ability, as it can dissolve in water and lead to the formation of boric acid [44,84].

The typical low cost of non-metal precursors can be an interesting aspect to reduce photocatalytic ozonation costs. As said previously, the combined process may present an increased cost due to ozone production energy requirement and possible use of artificial radiation sources, in the case of typical TiO_2. Thus, the lower cost associated with non-metal doping and the produced benefits, such as the increase in photocatalytic activity under solar radiation, may render the photocatalytic ozonation process more feasible. In previous studies, Fernandes et al. [18,21] developed multiple nitrogen-doped catalysts for application in the photocatalytic ozonation removal of a mixture of parabens. The doped catalyst, especially ammonia doped, led to an improved degradation of all five contaminants (methyl-, ethyl-, propyl-, butyl-, and benzylparaben), completely removing them faster and with a lower consumption of ozone. As ozone has low solubility in water, the photocatalyst may enhance its solubilization and decomposition, allowing a more efficient consumption, leading to a lower concentration needed through the production of other reactive species.

3.1.4. Co-Doping

The application of two or more dopants can be highly beneficial, as it can enhance the catalyst performance through multiple mechanisms. Multiple combinations of elements are possible and have been explored, even with different classifications, such as metal and non-metal co-doping. The co-doped Fe-Pr-TiO_2 catalyst, studied by Mancuso et al. [46], presented an improved performance regarding AO dye removal compared to the single and undoped catalyst, with 87% removal of the AO and 80% TOC removal in 60 min under visible light. In the studied case, the co-doping was accounted to considerably decrease the E_{bg} (2.7 eV) and enhance the formation of oxygen vacancies, confirmed in photoluminescence spectra, acting as electron trappers, and reducing the e^-/h^+ recombination. The authors also investigated in another study the co-doping of Fe and N, and again it showed a superior performance as it can benefit from both the substitution of Ti^{4+} by Fe^{3+}, acting as electron acceptors and inhibiting e^-/h^+ recombination, and the nitrogen replacement of oxygen sites, producing Ti^{3+} and oxygen vacancies [59].

Jahdi et al. [85] explored the combination of fluorine with palladium for the degradation of sulfamethoxazole (SMX), using a hydrothermal-produced F-Pd-TiO_2 catalyst. The addition of F was suggested to control the crystalline phase growth, while Pd significantly improved the photocatalytic activity under higher wavelengths, with its increasing concentration (0–10%) leading to E_{bg} down to 0.54 eV. The best catalyst achieved 98.4% of SMX removal under direct sunlight in 40 min. It should be noticed though that the presence of fluorine was found to lead to the formation of white fluorine-based polymeric substances floating in the treated solution.

Different precursors may be used to incorporate more than one element simultaneously [86]. Thiourea is commonly used as a source of carbon, sulfur, and nitrogen, and the three dopants' simultaneous incorporation may promote an even more significant narrowing of the bandgap of TiO_2 through the formation of mid-gap levels due to interactions with the O 2p orbitals. Additionally, Khedr et al. [86] indicated that sulfate and carbonaceous species can also be formed and may improve the photocatalytic activity, acting as electron trapping sites, inducing visible-light absorption, and enhancing radicals' formation. In this specific study, N-C-S-TiO_2 had substantially better performance over ibuprofen degradation under visible light irradiation, almost fully eliminating the contaminant under 240 min of reaction, against a 12.2% degradation reached by the undoped catalyst. Table 3 presents some key studies involving the use of co-doped TiO_2 catalysts.

Table 3. Application of co-doped TiO_2 for photocatalytic removal of contaminants.

Catalyst	Precursor		Method	E_{bg} (eV)	Radiation Source	Contaminant	Removal (Time)	Ref.
	Ti	Dopant						
C-Co-TiO_2	TTIP	Glucose and Cobalt Chloride	Co-precipitation	2.81	Visible	Phenol, 100 mg L^{-1}	100% (120 min)	[87]
C-N-TiO_2	TBOT	Extrapallial Fluid of Mussels	Chemical Deposition	2.90	Visible	High-density Polyethylene, 0.4% *w*/*v*	72.0% (50 h)	[88]
C-N-S-TiO_2	$Ti_2(SO_4)_3$	Thiourea	Hydrothermal	2.90	Visible	Ibuprofen, 20 mg L^{-1}	100% (5 h)	[86]
Cu-N-TiO_2	TTIP	Urea and Copper (III) Nitrate	Sol-gel	-	Visible	Methylene Blue, 12.5 mg L^{-1}	56.3% (90 min)	[89]
F-N-TiO_2	TBOT	Ammonium Fluoride and 2-nitrophenol	Solvothermal	-	Solar	2-nitrophenol, 10 mg L^{-1}	~98.1% (75 min)	[90]
		Ammonium Fluoride and 4-nitrophenol		-	Solar	4-nitrophenol, 10 mg L^{-1}	~93.9% (75 min)	
F-Pd-TiO_2	TBOT	Trifluoroacetic Acid and Palladium Chloride	Microwave-assisted Hydrothermal	0.54	Solar	Sulfamethoxazole, 30 mg L^{-1}	98.4% (40 min)	[85]
Fe-Eu-TiO_2	TTIP	Europium Oxide and Iron Nitrate	Sol-gel	2.78	Visible	Methylene Blue and Methyl Orange, 5 mg L^{-1}	97.9% MB and 99.7% MO (180 min)	[45]
Fe-N-TiO_2	TTIP	Urea and Iron Acetylacetonate	Sol-gel	2.70	Visible	Acid Orange Azo Dye, 10 mg L^{-1}	90% (60 min)	[59]
Fe-Pr-TiO_2	TTIP	Praseodymium Nitrate and Iron Acetylacetonate	Sol-gel	2.70	Visible	Acid Orange Azo Dye, 10 mg L^{-1}	87% (60 min)	[46]

3.2. WO_3

The WO_3 photocatalysts appear as an interesting alternative to typical TiO_2, with good visible-light activity, but also chemical and electronic properties [91]. However, some drawbacks are still found, principally the high electron-hole recombination rate and a more positive position of the CB, which compromises the superoxide production [50]. The introduction of heteroatoms to the WO_3 structure has not been widely explored compared with other catalysts, but some studies have been conducted showing that this adaptation may be responsible for promoting the enhancement of the catalyst performance and its typical disadvantages. Studies regarding the application of doped WO_3 catalysts are summarized in Table 4.

Table 4. Application of doped WO_3 for photocatalytic removal of contaminants.

Catalyst	Precursor W	Precursor Dopant	Method	E_{bg} (eV)	Radiation Source	Contaminant	Removal (Time)	Ref.
Ag-WO_3	Sodium Tungstate	Silver Nitrate	Hydrothermal, HCl	2.63	Solar	Acetaminophen, 5 mg L^{-1}	75.4% (120 min)	[92]
Cd-WO_3	-	Cadmium Nitrate	Ion-exchange	1.85	Visible	Methylene Blue, 10 mg L^{-1}	75.5% (80 min)	[93]
Co-WO_3	Sodium Tungstate	Cobalt Chloride	Co-precipitation	-	Visible	Methyl Red, 10 mg L^{-1}	90% (120 min)	[94]
Cu-WO_3	-	Copper Nitrate	Precipitation	2.60	Visible	Tetracycline, 50 mg L^{-1}	96.7% (120 min)	[95]
Fe-WO_3	Ammonium Paratungstate	Iron Chloride	Sol-gel	2.39	Visible	Methylene Blue, 10 mg L^{-1}	95% (120 min)	[96]
Gd-WO_3	Sodium Tungstate	Gadolinium Nitrate	Hydrothermal	2.64	Visible	Rhodamine B, 20 mg L^{-1}	94% (100 min)	[97]
Mn-WO_3	Tungstic Acid	Manganese Chloride	Microwave-assisted Precipitation	2.00	Visible	Sulfamethoxazole, 1 mg L^{-1}	100% (70 min)	[98]
C-WO_3	Sodium Tungstate	Carbonized Glucose	Hydrothermal	-	UV-Visible	Rhodamine B, 20 mg L^{-1}	95% (180 min)	[99]
I-WO_3	Ammonium Paratungstate and *Spondias mombin* leaves extract	Ammonium Iodide	Hydrolysis and Precipitation	2.17	Solar	Dyeing Wastewater, TOC = 576.8 mg L^{-1} COD = 991 mg L^{-1}	88.2% TOC 89.1% COD (240 min)	[100]
P-WO_3		Ammonium Phosphate		2.41			86.8% TOC 86.6% COD (240 min)	
P-I-WO_3		Ammonium Phosphate and Ammonium Iodide		2.02			93.4% TOC 95.1% COD (240 min)	
S-WO_3	Sodium Tungstate	Thiourea	Hydrothermal	-	Visible	Methyl Orange, 20 mg L^{-1}	97% (3 h)	[101]

3.2.1. Transition Metals Doping

WO_3 doping can considerably interfere with basic catalyst characteristics, such as crystallinity, morphology, and optical properties, which will then produce variations in their photocatalytic activity. Cu-WO_3 was found to have a lower E_{bg} (2.78 to 2.60 eV) and higher crystallinity compared to the studied undoped catalyst by Quyen et al. [95]. Besides, changes in its surface properties were also detected, with the incorporation of Cu being mainly at this level, promoting the formation of a more porous structure, meaning a higher surface area and number of active sites capable of enhancing contaminants' interaction and degradation. The presence of the heteroatom also promoted a higher presence of the W^{5+} state, which may imply the equivalent formation of oxygen vacancies due to electronic rebalance, which may increase the catalyst electroconductivity and pollutants adsorption.

Mehmood et al. [94] also presented a broad study of the incorporation of another metallic element, cobalt, over the WO_3 lattice. The Co atoms were appointed to be doped substitutionally, by the replacement of W^{6+} by Co^{2+}, which promoted an expansion of the lattice due to the slightly higher ionic radius of Co^{2+}. Thus, excessive addition of the dopant could degrade the catalyst crystallinity due to the collapse of its structure. The dopant also supported the creation of localized states within the catalyst bandgap, reducing the E_{bg}. The Co-WO_3 presented incredibly better performance for the degradation of a methyl red solution, increasing the removal rate from less than 5% to 90% under 2 h. Additionally, the use of this catalyst was also applied for the inactivation of cancerous cells, which reduced the viability of MCF-7 breast cancer and Hep-2 liver cancer cells down to 60%, possibly through the attack of their mitochondria and rupture of the cell wall.

The doped catalysts must also be subjected to lixiviation tests, especially when more dangerous metal elements are used, as their continuous reutilization may be responsible for the release of this species into the medium. The use of cadmium (Cd) as a metal dopant has been indicated to diminish e^-/h^+ pairs recombination and increase the catalyst crystallinity and porosity, resulting in a higher photocatalytic degradation [93]. However, although the dopant may substitute W atoms on the catalyst structure, implying stronger bounds, Cd lixiviation can still occur, especially at acidic conditions. It must be noted that US EPA regulations, for example, establish a maximum of 0.005 ppm of cadmium in drinking water, with serious negative health effects when in contact with higher concentrations.

3.2.2. Noble and Rare-Earth Metals Doping

The use of different precursors during the photocatalyst and doping synthesis can alter the mechanisms and the configuration of the formed product. Palharim et al. [92] studied the alteration caused by the addition of HCl or HNO_3 during Ag-WO_3 synthesis, using $AgNO_3$ as the dopant precursor. It was found that HNO_3 promoted the incorporation of the dopant metallic silver, while HCl reacted with $AgNO_3$ to form AgCl, being detected in the catalyst structure. This promoted differences in the catalyst, as an increase in Ag concentration led to lower E_{bg} in HNO_3/Ag-WO_3 due to the particle agglomeration, and the opposite occurred for HCl/Ag-WO_3. Regardless of the acid used, Ag-WO_3 had better removal rates of acetaminophen under visible light than the undoped catalyst, because of the combination of factors such as the localized surface plasmon resonance, characteristic of noble metal doping, the insertion of new energy level within the catalyst bandgap and the reduction of E_{bg}. Moreover, HCl doped presented a better performance, possibly due to the interaction between Ag-WO_3 and the AgCl clusters formed, and the contribution towards decreasing the WO_3 conduction band to more negative values, allowing oxygen reduction.

The f-orbitals of rare-earth metals can also be beneficial for the improvement of WO_3 performance. Tahir and Sagir [102] conducted a wide study of the application of these metals by synthesizing lanthanum-, erbium- and gadolinium-doped WO_3 through a hydrothermal method. As these rare-earth metals possess considerably higher ionic radius, the substitution of W atoms is not favored, and the dopants were found to occupy interstitial sites. The catalyst crystallite size decreased with the incorporation of the heteroatoms, having advantages regarding the reduction of photogenerated species recombination, and the specific surface areas were also greatly increased, especially for Gd-WO_3, 106 $m^2\ g^{-1}$ compared to 30.7 $m^2\ g^{-1}$ of the undoped catalyst. Moreover, the higher adsorption locations and catalytic center of Gd-WO_3, combined with lower recombination of electron-hole, presented a faster and almost complete removal of multiple single contaminants (methotrexate, tetracycline, methyl orange, methylene blue, and crystal violet).

3.2.3. Non-Metals Doping

The combination of the non-metal elements and WO_3 can also promote variations of the CB and VB of the catalyst, hinder the recombination factor of photoinduced charges, and increase photocatalytic activity under visible light. The S-WO_3 nanowires catalyst was investigated by Han et al. [101] using thiourea as the sulfur precursor. The non-metal element was able to substitute W^{6+} in the form of S^{6+} in the WO_3 lattice, as opposed to the anionic replacement of O^{2-}, due to the higher energy form of W-S bonds than W-O. The doping method also accounted for the formation of surface hydroxyl groups because of oxygen adsorption, which may act as electron trapping centers. Sulfur also was responsible for creating intermediate energy levels above the VB of WO_3, enhancing the catalyst's photoresponse. Even if not mentioned by the authors, the use of thiourea as a precursor could also involve the parallel incorporation of nitrogen over the structure.

Similar effects were found regarding the positive effects of non-metallic heteroatoms by Zheng et al. [99], in interstitially carbon-doped WO_3. The location of the C atoms may denote its interaction with WO_3 and consequent distortion of the WO_6 octahedron structure and an induced increased dipole, possibly more conducive for the transmission of photocarriers and photocatalytic activity. Comparably with other non-metals, new C 2p induced levels were formed in the interior of the bandgap of WO_3, acting as transfer stations to the photocarriers.

3.2.4. Co-Doping

As dopants may have different mechanisms, their combination can severely upgrade the photocatalytic activity through the synergic combined effect, being able to attain higher removal rates, even at complex effluents. Tijani et al. [100] proposed a green synthesis of I-P-WO_3, by applying a plant extract in bulk catalyst production. The I^- and P^{3+} atoms acted as structure-directing agents, affecting the morphology of the rod-like catalyst, with P conducing to more elongated shapes and I with shorter and more uniform. Alterations of lattice parameters indicated distortions of the catalyst structure, which may be a result of the partial substitution of W^{6+} and O^{2-}, by respectively P^{3+}, which has a smaller ionic radius than W, and I^- that has a higher ionic radius compared to O. This substitution is pointed out to have detrimental effects as it may occur the collapsing of the WO_3 structure. However, due to the low concentration of the incorporation, it only resulted in beneficial decreased grain sizes and enhanced surface areas, up to 416.3 $m^2\ g^{-1}$. The optical properties of I-P-WO_3 were also improved, decreasing the E_{bg} from 2.61 to 2.02 eV, due to acceptor energy levels below the CB of WO_3. The codoped catalyst was applied for the treatment of dyeing wastewater under sunlight, achieving high color and odor removal, as well as 93.4%, 95.1%, and 92.0% removals of total organic carbon (TOC), chemical, and biological oxygen demand (COD and BOD), respectively, accounting for values below the regulated international standards. Thus, codoping may lead to the formation of a more feasible photocatalyst regarding more complex applications.

3.3. *g-C_3N_4*

Even with a typically lower E_{bg} compared to TiO_2, g-C_3N_4 still exhibits a fast recombination rate of the photogenerated e^-/h^+ pairs and a limited light absorption range, which diminishes the visible and solar radiation conversion. Thus, the doping of g-C_3N_4 may contribute to charge separation and concurrently a shift of light absorption. As g-C_3N_4 has gotten much attention recently, investigation of the incorporation of multiple elements over the catalyst structure has been reported.

3.3.1. Metals Doping

The doping using metallic elements is known to promote the formation of oxygen vacancies and considerably improve the catalyst activity. The existence of various unbounded electrons in the hexazine structures of g-C_3N_4 makes it susceptible for metallic elements to be inserted in its lattice [103]. The results regarding the use of metal-doped g-C_3N_4 are present in Table 5. For the removal of paracetamol under visible light, doping using cobalt (II) nitrate was made over a melamine-based g-C_3N_4 through a calcination method [104]. The detected Co^{2+} and Co^{3+} in the catalyst and the O 1s XPS spectra indicated the formation of oxygen vacancies. Moreover, the doping affected the crystallinity of g-C_3N_4 and its morphology, presenting a mixed structure of cobalt oxide closely attached to carbon nitride layers. Optical characterization also showed indirect recombination of charge carriers and smaller charge transfer resistance, which indicates a higher charge transport kinetics and separation. Ultimately, the Co-doped g-C_3N_4 presented a reaction rate almost three times higher than the undoped catalyst, 0.0382 min^{-1}, completely removing 1 mg L^{-1} of paracetamol within 120 min. No leaching of Co ions was detected after the oxidation experiments.

Pham et al. [105] studied Ni-doped g-C_3N_4 for volatile organic compounds (VOCs), nitrobenzene, and toluene, removal under natural sunlight. The presence of Ni promoted a red shift of the absorption edge, suggesting a higher photoactivity over the visible range, and E_{bg} was estimated as 2.25 eV, noticeably lower than the undoped g-C_3N_4, 2.76 eV. The doped catalyst also presented a higher specific surface area and pore volume, which promote a higher interaction with the different compounds involved in the oxidative reactions. The doping treatment finally improved VOCs elimination, with reaction rate constants of 0.0170 and 0.0173 min^{-1} for, respectively, toluene and nitrobenzene, compared with 0.0049 and 0.0072 min^{-1} of the undoped catalyst.

The incorporation of iron over g-C_3N_4 can also improve the oxidation potential and may act as electron trap centers. With a relatively larger ionic radius, Fe atoms tend to be doped into the interlayers of the catalyst, forming strong bonds with the N atoms present in the aromatic rings, which decreases the E_{bg}. However, an excess amount of Fe may also cause the collapse of the sheet structure. The dopant exists mainly as Fe^{3+}, and as the reduction potential of Fe^{3+}/Fe^{2+} lies between the CB and VB of g-C_3N_4, the photogenerated e^- may easily react with Fe^{3+}, reducing it into Fe^{2+} and diminishing e^-/h^+ recombination. The instability of Fe^{2+} makes it prone to suffer oxygen reduction and return to the Fe^{3+} form [103].

Although not often, the use of alkaline and alkaline-earth metals (such as K, Mg, and Ca) for g-C_3N_4 doping has been explored. Studies regarding the use of these elements in more surveyed catalysts present mixed results, with possibly poisoning effects, but also an increase in photoactivity and surface area [50]. Tripathi and Narayanan [106] tested potassium doping of melamine-based g-C_3N_4. The presence of K during the thermal polymerization was suggested to inhibit the formation of hydrogen bonds between the polymeric melon strings of the intra-layer infrastructure, reducing the layer's stacking, but also the E_{bg}, from 2.7 to 2.5 eV, with changes of the VB energy and Fermi level towards more positive potentials. This condition may promote an increase in the oxidation capacity of the K-doped catalyst. Furthermore, under natural sunlight, the doped catalyst obtained 56.5% elimination of phenol present in real wastewater.

3.3.2. Noble and Rare-Earth Metals Doping

Noble metals, such as Ag and Au, are electron-rich transition metal elements that can capture photogenerated electrons and provide the excess energy of plasmonic states, provoking a shift of the Fermi levels to less positive potentials due to an excess of negative charges, reducing electron-hole recombination and boosting its catalytic performance. Tri et al. [107] explored Ag-doped g-C_3N_4 for the treatment of tetracycline, an antibiotic, in hospital wastewater under solar irradiation. Besides the faster production and higher separation of the e^-/h^+, Ag significantly altered the optical properties of the catalyst, decreasing its E_{bg} down to 2.19 eV, and possibly increasing the number of active sites and surface area, culminating in an optimum removal of 96.8% of tetracycline under 120 min.

Bawazeer et al. [108] also proposed a facile ball mill-assisted synthesis method for Au-doped g-C_3N_4, without the use of solvents. Even with no significant modification of the E_{bg} and a decrease in specific surface area, the presence of gold atoms boosted the catalyst performance over Arsenazo III dye decolorization, eliminating the contaminant within 50 min and with a reaction rate constant 2.5 times higher. The role of Au as recombination centers and a higher charge photogeneration due to the surface plasmon resonance is appointed to be responsible for the improved efficiency.

Due to their unfilled f-orbitals, rare-earth metals may act as trapping sites, distributing the photogenerated carriers, but can also form complexes with Lewis bases by bonding with their respective functional groups, promoting a higher interaction with organic pollutants at the catalyst surface. An Er-doped g-C_3N_4 was produced by Li et al. [109], with the dopant existing in the Er^{3+} state. The foreign element promoted a higher degree of polymerization and crystallization of the g-C_3N_4 network, but when in excess may also weaken tri-s-triazine bonds. The doped catalyst showed great efficiencies towards the degradation of three different contaminants, i.e., RhB, tylosin, and tetracycline, under simulated solar irradiation, with kinetics rate constants, respectively, 3.6, 1.6, and 1.7 times higher than the bulk g-C_3N_4. Additionally, Er^{3+} presented a detrimental effect over the specific surface area, which indicates that changes in the optoelectronic properties may play more important roles in the photodegradation performance.

Table 5. Application of metal-doped g-C_3N_4 for photocatalytic removal of contaminants.

Catalyst	Precursor		Method	E_{bg} (eV)	Radiation Source	Contaminant	Removal (Time)	Ref.
	C_3N_4	Dopant						
Ag-g-C_3N_4	Melamine and Urea	Silver Nitrate	Thermal Polymerization and Ion-exchange	2.19	Solar	Tetracycline, 20 mg L^{-1}	96.8% (120 min)	[107]
	Melamine	Silver Nitrate	Thermal Polymerization	2.46	Visible	Oxytetracycline, 10 mg L^{-1} and	98.6% (120 min)	[110]
						Antibiotics Wastewater: Oxytetracycline, 101.5 mg L^{-1}; Tetracycline, 85.3 mg L^{-1}; Gatifloxacin, 89.4 mg L^{-1}	94.5% OTC, 81.8% TC, 67.3% GFA (120 min)	
Au-g-C_3N_4	Melamine	Au NPs	Thermal Polymerization	2.86	Visible	Arsenazo, 4 mg L^{-1}	100% (60 min)	[108]
Co-g-C_3N_4	Melamine	Cobalt Nitrate	Thermal Polymerization	2.38	Visible	Paracetamol, 10 mg L^{-1}	96.3% (120 min)	[104]
Er-g-C_3N_4	Melamine	Europium Nitrate	Thermal Polymerization	2.50	Solar	Tetracycline, 25 mg L^{-1}	85% (90 min)	[109]
						Tylosin, 25 mg L^{-1}	70% (90 min)	
						Rhodamine B, 5 mg L^{-1}	90% (30 min)	
	Cyanuric Acid and Melamine	Europium Chloride	Thermal Polymerization	2.70	Visible	Rhodamine B, 10 μg L^{-1}	100% (90 min)	[111]
Fe-g-C_3N_4	Melamine	Iron Nitrate	Thermal Polymerization	2.73	Visible	Rhodamine B, 10 mg L^{-1}	100% (150 min)	[103]
K-g-C_3N_4	Melamine	Potassium Chloride	Thermal Polymerization and Ion-exchange	2.50	Solar	Phenolic Effluent, Ph = 980 mg L^{-1} and COD = 6300 mg L^{-1}	56.5% (300 min)	[106]
	Thiourea	Potassium Bromide	Thermal Polymerization	2.15	Visible	Nitric Oxide, 600 ppb	36.8% (30 min)	[112]
Ni-g-C_3N_4	Melamine	Nickel Nitrate	Thermal Polymerization	2.25	Solar	Toluene and Nitrobenzene, 5 mg L^{-1}	85.8% TOL 98.6% NBZ (120 min)	[105]
V-g-C_3N_4	Melamine	Ammonium Metavanadate	Thermal Polymerization	2.63	Solar	Tetracycline, 10 mg L^{-1}	81.9% (240 min)	[113]

3.3.3. Non-Metals Doping

Non-metal and metalloid elements have a larger record of use in g-C_3N_4 doping and are able to overcome the disadvantageous surface characteristics of the catalyst and improve its electronic and optical properties. Table 6 presents the results of the use of non-metal doped g-C_3N_4.

Zhang et al. [114] synthesized a mesoporous g-C_3N_4 which presented a high specific surface area, 91.1 m^2 g^{-1}, due to its porous structure, providing a larger number of active sites. However, due to the quantum size effect, the UV-Vis absorption spectrum suffered a blue shift as bandwidth decreased with the concurrent decrease of particle size. It was found that oxygen doping, besides the reduction of the photo-induced carriers' recombination, may compensate for the absorption blue shift and enhance photocatalytic activity under simulated solar irradiation. The oxygen-doped mesoporous catalyst attained incredibly faster quasi-first-order kinetic constants for RhB and MO degradation, 64 and 24 times higher than simple mesoporous g-C_3N_4.

Significant surface alterations may also be found when doping with oxygen, as investigated by Praus et al. [115]. The authors presented a facile synthesis of the doped catalyst, by the combination of melamine and cyanuric acid, an oxygen-rich compound, in the thermal polymerization step. The modification increased the catalyst specific surface area from 14 to up to 41 m^2 g^{-1} and a lowering of protonated nitrogen groups at the catalyst surface occurred, by means of substitution by -OH and -O- groups. Under LED lights, the O-doped g-C_3N_4 presented faster kinetic rates regarding ofloxacin and RhB, i.e., ~30 mol L^{-1} min^{-1} and ~3 min^{-1}, respectively. The conduction and valence bands potentials at pH 7 were also calculated against normal hydrogen electrode (NHE), with the best doped catalyst presenting values of −0.93 and 1.69 eV, respectively. As the redox potential of hydroxyl radicals, $E^0(\cdot OH/H_2O)$, is 2.74 eV, which means the oxidation of water to form commented radicals by the positive holes at the VB is not energetically favored, the direct contaminant oxidation by h^+ and superoxide radicals are the main degradation mechanisms (Figure 6). An interesting approach to overcome the low production of ·OH due to the respective low VB potential would be the incorporation of ozone in the reaction. The ozonide racial redox potential, $E^0(\cdot O_3^-/O_3)$, is within the catalyst band potential, 0.89 eV, which can improve the production of radicals including •OH through consecutive reactions (Equations (1)–(3)) [116].

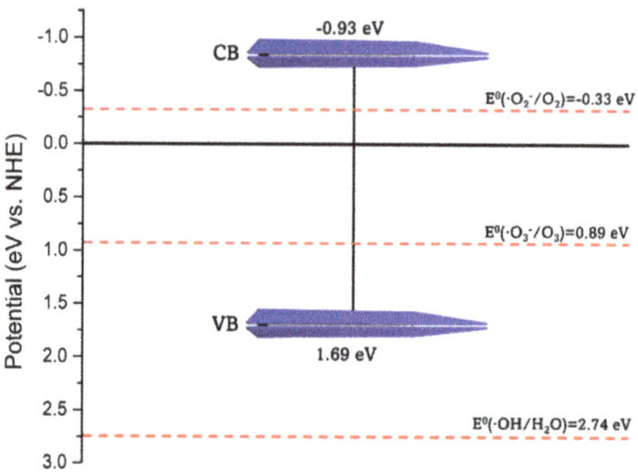

Figure 6. Representation of the energy diagram of Ag-g-C_3N_4 CB and VB potentials [115].

Other elements, such as silicon and phosphorous, have been indicated to substantially alter the bandgap energy and/or position of CB and VB of g-C_3N_4 catalysts, due to orbitals interaction. Wang et al. [117] produced P-doped g-C_3N_4 nanobelts and showed a significant decrease of the CB compared to bulk g-C_3N_4, –0.92 to –0.04 eV, with P successfully substituting C atoms, and even with a less negative CB, e^-/h^+ recombination was reduced and p-hydroxybenzoic acid degradation increased from 33.1% to 77.3%, by comparison of undoped and P-doped nanobelt catalysts. Once more, as the CB is less negative than $E^0(\cdot O_2^-/O_2)$, −0.33 eV, the superoxide may not be generated, and the presence of ozone would be beneficial as it possesses higher E^0 and may act as the main electron acceptor.

Liang et al. [118] suggested that, when a silicon-doped catalyst was synthesized, N-Si bonds may act as electron trap centers, as it forms impurity levels close to the conduction band of g-C_3N_4, facilitating the interaction between e^- and respective acceptors, oxygen or ozone.

Table 6. Application of non-metal-doped g-C3N4 for photocatalytic removal of contaminants.

Catalyst	Precursor		Method	E_{bg} (eV)	Radiation Source	Contaminant	Removal (Time)	Ref.
	C_3N_4	Dopant						
B-P-g-C_3N_4	Melamine	Boric and Phosphoric Acid	Thermal Polymerization and Hydrothermal	2.66	Visible	Diclofenac, 10 mg L^{-1}	100% (90 min)	[119]
Cl-g-C_3N_4	Dicyandiamide	Cyanuric Chloride	Solvothermal	1.78	Visible	Rhodamine B, 10 mg L^{-1}	99.6% (125 min)	[120]
	Melamine	Ammonium Chloride	Thermal Polymerization	2.71	Visible	Rhodamine B, 10 mg L^{-1}	57.8% (30 min)	[121]

Table 6. *Cont.*

Catalyst	Precursor		Method	E_{bg} (eV)	Radiation Source	Contaminant	Removal (Time)	Ref.
	C_3N_4	Dopant						
O-g-C_3N_4	Semicarbazide Hydrochloride		Molten-salt	2.25	Visible	Naproxen, 100 mg L^{-1}	100% (5 h)	[122]
	Urea	Hydrogen Peroxide	Thermal Polymerization and Solvothermal	2.94	Solar	Rhodamine B, 10 mg L^{-1}	95% (20 min)	[114]
						Methyl Orange, 10 mg L^{-1}	70% (4 h)	
	Urea	Cyanuric Acid	Thermal Polymerization	2.62	Visible	Rhodamine B, 10 mg L^{-1}	40% (120 min)	[115]
						Ofloxacin, 20 mg L^{-1}	97% (120 min)	
P-g-C_3N_4	Melamine	Phosphoric Acid	Thermal Polymerization	-	Visible	Dinotefuran, 2 mg L^{-1}	40.6% (5 h)	[123]
	Melamine	Phosphoric Acid	Solvothermal	1.66	Visible	p-Hydroxybenzoic Acid, 1 mg L^{-1}	77.3% (120 min)	[117]
S-g-C_3N_4	Melamine	Thiourea	Thermal Polymerization	2.51	Visible	Rhodamine B, 10 mg L^{-1}	29.7% (30 min)	[121]
S-Cl-g-C_3N_4	Melamine	Thiourea and Ammonium Chloride	Thermal Polymerization	2.55	Visible	Rhodamine B, 10 mg L^{-1}	100% (30 min)	[121]
Si-g-C_3N_4	Urea	Ammonium Fluorosilicate	Thermal Polymerization	2.75	Visible	Rhodamine B, 10 mg L^{-1}	75% (50 min)	[118]

Halogen doping of g-C_3N_4 has also been reported in the literature and shows interesting results. In the case of chlorine doping of g-C_3N_4, interesting aspects regarding the roles of interstitial and substitutional incorporation have been studied [120]. Substitutional chlorine doping may correspond to a higher specific surface area and oxidation performance, while the interstitial presence of chlorine contributes to a higher separation of photogenerated species. Thus, as both can occur simultaneously, their balanced doping may correspond to an optimization of the catalyst activity. The synthesis method applied and other parameters, such as the agitation and dopant ratio, can be fundamental to allow better control of the catalyst's characteristics. Moreover, as indicated previously during the discussion of other catalysts, halogen doping studies are very scarce, possibly due to their more difficult impregnation, being highly electronegative elements, and producing toxic by-products.

3.3.4. Co-Doping

The use of multiple elements simultaneously for the doping of g-C_3N_4 has already been explored, albeit with fewer studies, and has shown a great boost in the material photoactivity. The most significant feature of g-C_3N_4 co-doping is appointed to be an improvement in electron mobility [119]. Moreover, additional alterations also promote the increase in the material photocatalytic performance, such as modifications of its morphology. The use of different amounts of dopants may dictate the final structure of the catalyst, as found by Ganganboina et al. [119], where an increasing amount of phosphoric acid mL in the B-P-g-C_3N_4 synthesis transformed a solid into a sponge-like, and finally, a hexagonal tubular structure, which possessed a higher photoactivity (Figure 7). The hydrolysis of melamine, the g-C_3N_4 precursor, led to the production of cyanuric acid, which may form a supramolecular precursor complex involving both chemicals, and finally aggregate to form the catalyst structure. The dopants introduced may then adsorb on these supramolecular complexes and alter their configuration.

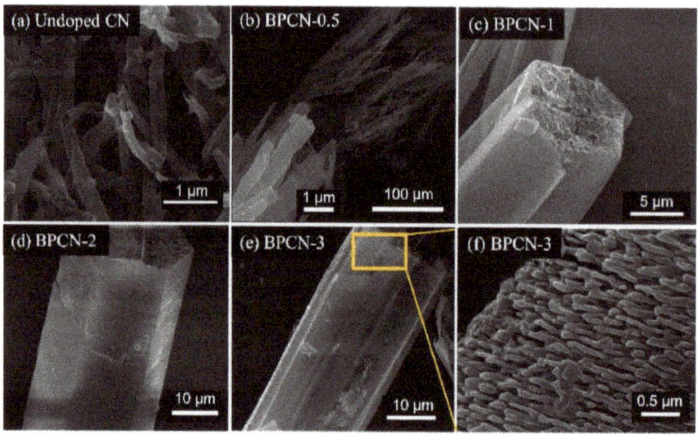

Figure 7. SEM images of (**a**) undoped g-C_3N_4, (**b**) 0.5 B-P-g-C_3N_4, (**c**) 1 B-P-g-C_3N_4, (**d**) 2 B-P-g-C_3N_4, (**e**) 3 B-P-g-C_3N_4, and (**f**) enlarged inside structure of 3 B-P-g-C_3N_4. (Retrieved with permission from Ref. [119] Copyright (2021) Elsevier).

3.4. Overall Considerations

The application of catalyst doping is a well-researched subject, noticeably producing positive effects regarding the improvement of photocatalytic activity. However, even with promising results, some considerations must be taken.

The similarities between the structures of WO_3 and TiO_2 as metal oxides allow analogous modification mechanisms to occur, providing great support as TiO_2 is already the subject of a great amount of research and information as it has been intensively optimized.

Rare-earth and noble metals provide excellent alterations, especially on the crystalline and electronic structure, due to the existence of more electron-rich orbitals, as well as their optical properties, e.g., through surface plasmon resonance. However, the sometimes complex and difficult impregnation of such elements, and their typically higher cost, need to be evaluated, although their use in small quantities can already provide significant results.

Other metallic elements can be an easier alternative, with comparatively lower costs regarding materials and impregnation techniques. In the case of g-C_3N_4, the exploration of transition metals doping is more widely explored, possibly due to the more distinct nature between the foreign and existent elements, which can provide more evident effects. These elements may lead to the easier formation of new energy sublevels and shift the light absorption to higher wavelengths. Such electronic interaction, especially as it occurs mainly with the conduction band of the catalysts, is also a great improvement for the less negative conduction band of WO_3, which difficult its reduction potential. The ozonide radical formation can also be enhanced by the modification of the conduction band, in the case of photocatalytic ozonation, facilitating the capture of electrons by ozone.

Nonetheless, transition metals doping has been reported to occasionally cause the lixiviation of such elements, e.g., iron and some heavy metals, presenting dangerous toxic effects and strict limits. Some progress in this matter has been made, with encapsulation techniques and methods that result in a stronger fixation, although it may increase the complexity of the process.

As the most abundant elements in nature, non-metals are the most broadly used elements in catalyst doping. Another great advantage is that a one-step method can usually be applied, simplifying the material's synthesis. Especially for g-C_3N_4, already being usually synthesized through a simple thermal polymerization, some compounds possess structures containing carbon and nitrogen together with other elements, which can be singularly applied, such as thiourea. Their electronic alteration may result in modification regarding the valence band, contributing, for example, to the less positive VB of g-C_3N_4 and improving hydroxyl radical production. Nonetheless, interaction with both conduction and valence bands can occur, even simultaneously, e.g., in sulfur doping.

In such a context, doping techniques represent a simple approach to overcome typical catalyst disadvantages and improve their performance. The difference in overall studies can challenge the conceptualization of standard knowledge, as multiple features can be altered and conflicting data exist. Thus, a more precise study may be necessary to evaluate each case.

4. Composite Catalysts

With the increasing conceptualization of different catalysts, the combination of different materials emerges to present a wide number of possibilities to improve the overall catalytic properties [124]. By mixing materials with different characteristics, their individual disadvantages may be overcome, and a final and more robust photocatalyst can be obtained. These composites are then a combination of materials with distinct natures, such as mixed metal oxides, carbon-semiconductors, polymeric structures, porous materials, and many others. In this section, an overview of the principal applications of the previously discussed photocatalysts in composite materials will be discussed [125]. The use of WO_3, g-C_3N_4, and TiO_2 combined in different composite catalysts is evaluated in Table 7.

The mixing of WO_3 and TiO_2 is a clear example of mixed metal oxides, a well-known group of photocatalysts defined by the combination of different semiconductors. The coupling of these photocatalysts aims to improve the photocatalytic activity of the resultant material by mostly increasing the charges separation efficiency and visible-light sensitization for the complete system [125]. Mugunthan et al. [126] proved the higher efficiency of the coupled TiO_2/WO_3 catalyst for diclofenac elimination under visible radiation. The group synthesized the photocatalyst through a hydrothermal method and studied the variation of TiO_2:WO_3 molar ratios. The composites presented higher surface areas and smaller particles compared to bare TiO_2, with the increasing amount of TiO_2 leading to larger S_{BET}, bus also particle sizes, and E_{bg}. Thus, an optimum TiO_2:WO_3 molar ratio (10:1), presented more balanced properties and higher removal of diclofenac (~90%). An excess amount of WO_3 is also appointed to possibly act as recombination centers for the photoinduced charges, reducing the process efficiency.

The surplus addition of one of the applied semiconductors can also promote the agglomeration of photocatalyst particles, which may hinder the absorption of light and CECs photodegradation, as it was suggested by El-Yazeed and Ahmed [127] during the impregnation of WO_3 particles onto TiO_2, which had a detrimental effect over a concentration of 10 wt%. This was also found by Wang et al. [128] for hollow spherical TiO_2/WO_3 composite, with uniform spheres being formed with a 5 wt% incorporation of WO_3, but with the further increase in concentration and particle agglomeration causing changes in the catalyst morphology and leading to flat and unequal forms. However, in this study, the catalyst containing 10 wt% WO_3 had a better performance in methylene blue and metoprolol degradation, due to the photoelectronic and surface area improvements.

The TiO_2-WO_3 heterojunction can greatly improve the photo-mechanisms involved in the photocatalytic process. Considering their bandgap positions, the charge separation can follow two mechanisms: type-II heterojunctions or Z-scheme mechanisms (Figure 8). For the type-II mechanism, the excited electrons from the CB of TiO_2, which has a more negative potential, will migrate to the CB of WO_3. Meanwhile, the opposite occurs in the positive holes, which tend to be transferred to the TiO_2 VB [129]. Nonetheless, the Z-scheme mechanism is appointed to be more favorable due to the larger redox potential, as the e^- of the CB of WO_3 and h^+ of the VB of TiO_2 tend to recombine fast, leading to the accumulation of e^- on the CB of TiO_2 and h^+ on the VB of WO_3, potentializing its respective reductive and oxidative potential [130].

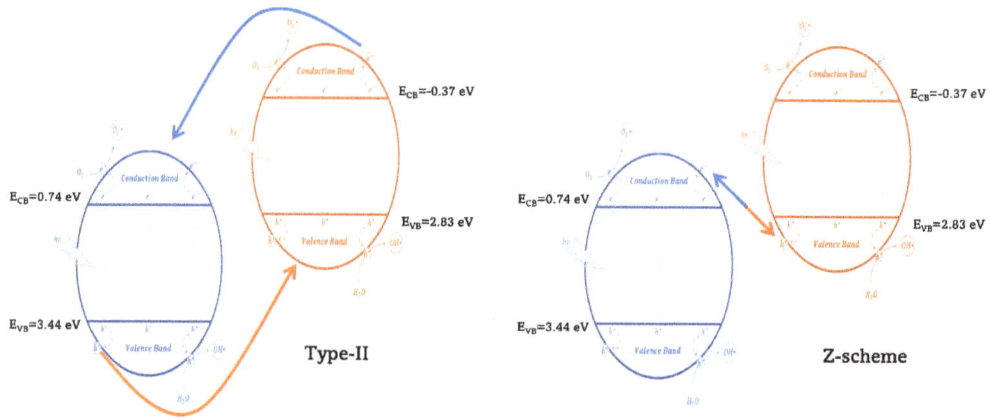

Figure 8. Representation of the mechanisms of Z-scheme and type-II heterojunctions [130–132].

Moreover, Wei et al. [131] indicated that, during the degradation of malachite green under visible light using TiO_2-WO_3, the superoxide radical ($\cdot O_2^-$) was one of the main mechanisms of the contaminant removal, even with the characteristic disadvantage of the more positive CB potential of WO_3 (0.74 eV vs. NHE) comparatively to the redox potential of $\cdot O_2^-/O_2$ (−0.33 eV vs. NHE). Thus, the composite material may also take advantage of alterations of the carriers' transference through interactions with TiO_2, allowing the adsorption and reduction of oxygen.

The use of doped catalysts in the coupled systems can further improve the overall efficiency through the synergic effect between all species. For instance, the use of noble metals, which promote the surface plasmon resonance effect, or non-metals that can increase visible light absorption, alongside the lower e^-/h^+ recombination and electronic bands interaction of the heterojunction can potentialize contaminants removal [133,134].

More complex composites may be synthesized by the continuous arrangement of materials with different natures. Chen et al. [135] combined carbon-based and ceramic materials with metal oxides, producing silica-TiO_2 microspheres coupled with graphene oxide (GO) and WO_3 quantum dots. Each material has a principal role, and the arrangement is conducted to overcome their individual drawbacks. GO possesses large surface areas and good electrical properties and may enhance the photocatalytic activity of semiconductors as it does not present one on its own. Meanwhile, silica is a low-cost and non-toxic template, and WO_3 and TiO_2 may harness the beneficial effects of their coupled mechanisms. Finally, the complex composite presented enhanced optoelectrical properties and higher photocatalytic activity under natural sunlight, removing 98% of RhB under 60 min.

The polymeric characteristics of g-C_3N_4 can also be greatly advantageous when coupled with other semiconductors and materials. The metal-free catalyst exhibits narrower bandgap energy, making the photoexcitation of the electrons in the VB under visible light possible. Nonetheless, one of its main drawbacks is the low VB potential, which typically lies below the redox potential for the reaction between positive holes and water to occur, producing hydroxyl radicals.

Following computational density functional theory (DFT) calculations, Deng et al. [136] investigated the g-C_3N_4/TiO_2 heterostructure. The study proposed that the composite may present a lower E_{bg} than the single catalysts. Moreover, it was found that an accumulation of photogenerated electrons over the CB of g-C_3N_4 occurred, while positive holes tended to accumulate on the VB of TiO_2, meaning that reduction reactions will favorably take place on g-C_3N_4, and oxidation will occur on TiO_2. These mechanisms may also reduce the recombination of the photogenerated pairs and lead to higher radical production, especially $\cdot OH$, due to the well-constructed heterojunction [137].

Table 7. Application of WO_3, g-C_3N_4, and TiO_2 mixed composites for photocatalytic removal of contaminants.

Composite	S_{BET} ($m^2\ g^{-1}$)	E_{bg} (eV)	Radiation Source	Contaminant	Removal (Time)	Ref.
g-C_3N_4/TiO_2	200.0	2.70	UV	Formic Acid, 50 mg L^{-1}	90.0% (5.5 h)	[138]
	40.2	2.81	Solar	Methylene Blue, 10 mg L^{-1}	94.9% (80 min)	[137]
				Rhodamine B, 15 mg L^{-1}	93.1% (80 min)	
	-	-	Visible	*E. coli*, 10^7 CFU mL^{-1}	100% (180 min)	[139]
	-	2.58	Solar	*E. coli*, 10^3 CFU mL^{-1}	96.8% (30 min)	[140]
WO_3/TiO_2	88.4	3.06	Visible	Malachite Green, 50 mg L^{-1}	99.0% (60 min	[131]
	11.7	2.40	Visible	Methylene Blue, 10 mg L^{-1}	87.8% (150 min)	[128]
				Metoprolol, 2 mg L^{-1}	67.1% (150 min	
	103.9	2.95	Visible	Diclofenac, 10 mg L^{-1}	100% (150 min)	[126]
WO_3/g-C_3N_4	-	-	UV-A + Visible + NIR	Ciprofloxacin, 10 mg L^{-1}	98.6% (180 min)	[141]
				Tetracycline, 10 mg L^{-1}	98.5% (180 min)	
	28.6	2.53	UV + Visible	Tartrazine, 25 mg L^{-1}	95.0% (20 min)	[142]
S-doped g-C_3N_4/TiO_2	-	3.00	Visible	Tetracycline, 10 mg L^{-1}	98.1% (60 min)	[143]
N-doped CHS/g-C_3N_4/TiO_2	78.0	-	Visible	Tetracycline, 20 mg L^{-1}	85.0% (120 min)	[144]
Ag_3PO_4/ g-C_3N_4/TiO_2	-	2.07	Visible	Metronidazole, 8.2 mg L^{-1}	97.2% (60 min)	[145]
C-doped WO_3/TiO_2	93.0	2.98	Solar	Diclofenac, 10 mg L^{-1}	100% (250 min)	[133]
Ag-doped WO_3/TiO_2	-	3.07	Visible	Methylene Blue, 3.2 g L^{-1}	72% (60 min)	[134]
GO/WO_3/TiO_2	-	2.89	Solar	Rhodamine B, 143.7 mg L^{-1}	98.2% (5 h)	[146]
	-	-	Solar	*E. coli*, 2×10^3 CFU mL^{-1}	97.3% (80 min)	[147]
CQDs/WO_3/TiO_2	96.2	2.61	Solar	Cephalexin, 10 mg L^{-1}	100% (90 min)	[148]
GO/CQDs/WO_3/TiO_2/SiO_2	202.6	-	Solar	Rhodamine B, 14.4 mg L^{-1}	98% (60 min)	[135]
Cd-doped WO_3/g-C_3N_4	8.5	1.53	Visible	Methylene Blue, 10 mg L^{-1}	96.0 (80 min)	[93]

The WO_3/g-C_3N_4 can thus be an interesting coupled system, as both catalysts can benefit from the synergic effect regarding their energy band potential and radical production. Huang et al. [141] produced g-C_3N_4 nanosheets coupled with WO_3 quantum dots (QDs), and they indicated that the combination of a strong reducing capacity of the CB of g-C_3N_4 to react with electron acceptors, such as O_2 and O_3, combined with the oxidation ability of the VB of WO_3 to form ·OH, promoted a greatly improved photocatalytic activity towards tetracycline and ciprofloxacin, with removals higher than 98%. Furthermore, the WO_3 QDs induced a local surface plasmon resonance, enhancing the charge carrier's separation efficiency.

Besides the reduction of E_{bg} and the synergic effect over photogenerated carriers, morphology modification may also be obtained to suppress the usual low specific surface area of g-C_3N_4. The coupling of g-C_3N_4 with TiO_2, WO_3, or other materials may lead to the formation of different porous structures and the existence of multiple functional groups that may interact with a wider range of organic pollutants [93,149].

In the following sections, the use of TiO_2, g-C_3N_4, and WO_3 in different composite materials as well as their main and most recently found advantages and mechanisms will be assessed.

4.1. TiO_2

Over the years, TiO_2 was vastly employed in the production of different composite catalysts, allowing a better understanding of their overall mechanisms and advantages.

The coupling of TiO_2 with other photocatalytic materials appeared as a vast family of mixed catalysts, with combined improved properties. Due to the existence of multiple photocatalysts, numerous combination possibilities are the aim of a rising number of investigations. Some studies involving TiO_2 composite catalysts are present in Table 8.

Zinc Oxide (ZnO) is, jointly with TiO_2, one of the most investigated semiconductors for contaminants removal, sharing their good photochemical characteristics but with relatively better electrical properties. Nonetheless, their heterostructures have been explored in different forms and systems for pollutant elimination. The increasing addition of ZnO in ZnO/TiO_2 fibers has been shown to also promote changes in the physical properties of the material, creating a rougher surface with a higher specific surface area and faster interaction with contaminants [150]. Changes in the morphology of the material have been also attested. Das et al. [151], indicated the transformation of uniform rod-like structures of ZnO into more of a flower-like form by the addition of TiO_2, with the enhancement of the BET surface until an optimal amount. A lower recombination rate of the electron-hole pairs was also demonstrated, increasing the lifespan of the photoinduced charge carriers.

The combination of TiO_2 with bismuth-based materials is a topic that gained much attention, as this family of compounds presents a series of attractive features, such as their low toxicity, easy functionalization, cost-effectiveness, and ability to absorb in near infra-red regions. Some examples of the most studied Bi-based photocatalysts are bismuth oxide (Bi_2O_3), bismuth vanadate ($BiVO_4$), and bismuth oxyhalide (BiOX), involving combination with halogen elements (X) [152–154].

The $BiVO_4/N$-TiO_2 heterostructure synthesized by Cipagauta-Díaz et al. [152] presented increased specific surface areas and absorption of visible light with the increasing $BiVO_4$ content. The catalyst also presented zones of heterogeneity that allow contact between the semiconductors and improve the separation of photogenerated charges and the lifetime of charge carriers. Ultimately, the best composite catalyst resulted in 98% removal of ofloxacin under 90 min. Even with the morphology and light absorption properties improvement, an excessive amount of $BiVO_4$ (>5 wt%) was demonstrated to be prejudicial to the photocatalytic performance, possibly due to the formation of $BiVO_4$ agglomerates in the catalyst surface, which hamper the homogeneous light absorption.

The electron-accepting nature of Bi_2O_3 was also proven to enhance the lifetime of the photocarriers and considerably increase the visible light absorption, even if the E_{bg} of the resulting materials did not suffer great alteration. Sharma et al. [153] indicated that Bi_2O_3 creates heterojunctions that facilitate the transport of photogenerated carriers and improve the visible light-capturing capacity in a $CuO/Bi_2O_3/TiO_2$ ternary composite catalyst. However, in high concentrations, Bi_2O_3 can reduce the catalyst efficiency, as pure Bi_2O_3 possesses less activity and can act as a recombination center.

Table 8. Application of TiO_2 composites for photocatalytic removal of contaminants.

Composite	S_{BET} (m^2 g^{-1})	E_{bg} (eV)	Radiation Source	Contaminant	Removal (Time)	Ref.
ZnO/TiO_2	-	3.15	UV	Eriochrome Black T, 6.4×10^3 mg L^{-1}	82% (6 h)	[155]
	84.7	3.15	Solar	Methylene Blue, 6.4 mg L^{-1}	95% (60 min)	[151]
				Methyl Orange, 6.5 mg L^{-1}	99% (60 min)	
$BiVO_4$/N-TiO_2	92.0	2.56	Visible	Ofloxacin, 20 mg L^{-1}	98% (90 min)	[152]
				Rhodamine B, 20 mg L^{-1}	92% (90 min)	
Bi_2O_3/TiO_2	102.9	-	Visible	Rhodamine B, 10 mg L^{-1}	100% (100 min)	[153]
$CuO/Bi_2O_3/TiO_2$	83.6	-	Visible	Rhodamine B, 10 mg L^{-1}	100% (60 min)	

Table 8. *Cont.*

Composite	S_{BET} ($m^2 g^{-1}$)	E_{bg} (eV)	Radiation Source	Contaminant	Removal (Time)	Ref.
Fe_2O_3/TiO_2	-	2.49	Solar	Methylene Blue, 10 mg L^{-1}	92% (180 min)	[156]
	-			Phenol, 10 mg L^{-1}	52% (180 min)	
	56.9	3.08	Visible	Naproxen and Ibuprofen, 10 mg L^{-1} each	100% NPX (15 min) and 91% IBF (240 min)	[157]
$Au/Fe_2O_3/TiO_2$	-	1.55	Visible	2,4 Dichlorophenol, 10 mg L^{-1}	94% (90 min)	[158]
	-	1.55	Visible	4-Bromophenol, 10 mg L^{-1}	97% (60 min)	
SiO_2/Fe_3O_4/Sn-TiO_2	-	1.32	UV	Tetracycline, 10 mg L^{-1}	98.2% (40 min)	[159]
P/Ag/$Ag_2O/Ag_3PO_4/TiO_2$	307.2	2.98	Visible	*E.coli*, 10^7 CFU mL^{-1}	100% (20 min)	[160]
				Salmonella, 10^7 CFU mL^{-1}	100% (30 min)	
				Enterococcus sp., 10^7 CFU mL^{-1}	100% (6 h)	
				S. aureus, 10^7 CFU mL^{-1}	100% (3 h)	
CNT/Au-TiO_2	-	1.95	Solar	Methylene Blue, 3 mg L^{-1}	80% (30 min)	[161]
GO/TiO_2	-	3.02	Solar	*E. coli*, 10^7 CFU mL^{-1}	99.9% (30 min)	[162]
Chitosan/N-TiO_2	52.0	2.82	UV	Patulin, 500 μg kg^{-1}	100% (35 min)	[163]
Perlite/F-Ce-TiO_2	14.8	2.96	Visible	*Microcystis aeruginosa*, 2.7 × 10^6 cell mL^{-1}	98.1% (9 h)	[164]

Iron oxides (e.g., Fe_2O_3, Fe_3O_4) have been successfully used in photocatalytic systems, as highly chemically stable and non-toxic materials with narrow bandgaps and interesting magnetic properties [124]. α-Fe_2O_3 possesses a broad absorption over the entire UV-Vis region. Thus, their use in composite photocatalytic can significantly extend the material's optical absorption, allowing it to preserve a good photocatalytic activity in both UV and visible range, and fully harness solar radiation (Figure 9) [165]. Bouziani et al. [156] demonstrated the expansion of absorption range in sol-gel synthesized α-Fe_2O_3/TiO_2, as single TiO_2 presented minimal absorption over 400 nm while the composite catalyst showed good absorption over all UV-Vis spectra. Furthermore, the 10 wt% α-Fe_2O_3/TiO_2 had better crystallization and a greater number of active sites, providing a faster degradation of methylene blue and phenol, i.e., 96% and 52%, within 120 min of sunlight irradiation.

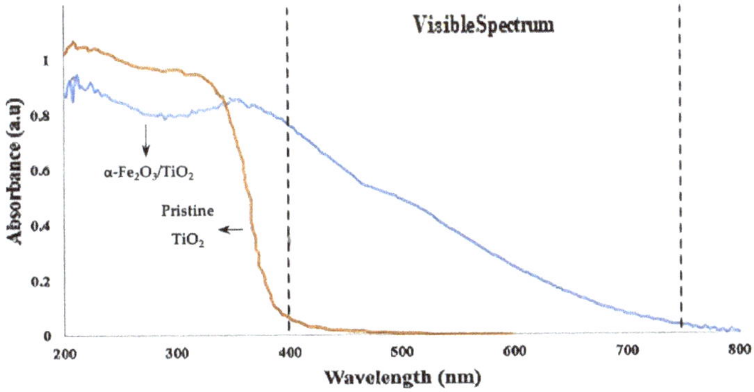

Figure 9. UV-vis spectra of Fe_2O_3/TiO_2 composite and commercial TiO_2 (Adapted and retrieved with permission from Ref. [162] Copyright (2020) Elsevier).

The synthesis of α-Fe_2O_3/TiO_2 composites may also occur using metal-organic frameworks (MOFs) as sacrificial templates. These materials are highly porous and lead to the formation of homogeneous structures with excellent electronic mobility and active sites. Li et al. [158] confirmed the potential of these catalysts through the study of MOF-derived Au-doped α-Fe_2O_3/TiO_2 and its reaction with 2,4 dichlorophenol and 4-bromophenol. The ternary system obtained degradation rates higher than 95% in 90 min, accounted for the high surface area and the number of reactive sites, and the movement of excited electrons between the more negatively charged CB of TiO_2 to the CB of Fe_2O_3 in the formed heterojunction. Peña-Velasco et al. [157] also concluded that, when synthesizing a Fe_2O_3/TiO_2 composite derived from a Fe-based MOF (MIL-101), Fe^{3+} may partially substitute Ti^{4+} in the TiO_2 structure, which may introduce lattice defects and vacancies that boost the photocatalytic activity, achieving 91% and 100% removal of ibuprofen and naproxen.

Asgari et al. [53] reported the efficient use of magnetic Fe_3O_4/TiO_2 composite catalyst in photocatalytic ozonation reactions. The presence of the photocatalyst enhanced ozone decomposition in the medium and the generation of oxidative radicals, boosting the degradation of ceftazide, eliminated under 15 min with a reaction rate constant of 0.1682 min^{-1}, almost two times faster than single ozonation. The magnetic property of the composite may also facilitate the posterior removal of the catalyst from the medium, a key parameter for the scale-up of the process. Remarkably, Lu et al. [159] appointed that interference may occur in the Fe_3O_4/TiO_2 catalyst as a result of Fe ion diffusion to the surrounding TiO_2, causing excessive doping. The authors utilized a SiO_2 coating, as a ceramic material able to shield the Fe_3O_4, but also reduce particle agglomeration, forming a ternary composite.

Carbonaceous materials, such as carbon nanotubes (CNTs), carbon quantum dots (CQDs), activated carbon (AC), and graphene oxide (GO), can provide support for the photocatalyst, but also great mechanical and electrical properties that may enhance its photoactivity. CNTs possess high mechanical strength and conductivity, but also present an important charge storage capacity and can act as a receiver of the photogenerated electrons from the catalyst CB. Mohammed [161] indicated an incredibly higher specific surface area of CNT/Au-TiO_2 hybrid catalysts, in simultaneous to higher absorption of visible light, effective transport of charges between CNTs and TiO_2, reduced electron-hole pair recombination, and better adsorption of molecules. CNTs may also promote a higher ozone decomposition in photocatalytic ozonation reactions, as ozone present a high affinity towards basic carbons due to its high density of π electrons on the basal planes. Orge et al. [166] verified the higher efficiency of CNT/TiO_2 in the photocatalytic ozonation of oxamic acid, which was completely removed under 60 min. The group also pointed out that ozone decomposition may be favored by CNTs with low acidic characteristics.

Graphene oxide (GO) also shows appealing hydrophilic features, tunable attributes, and particular optical properties that may be applied to composite materials. Different GO-hybrid catalysts have shown better exploitation of visible and solar irradiation, fast electron transference, and suppressed agglomeration. Thomas et al. [162] exploited the antibacterial properties of GO in GO/$BiVO_4$ and GO/TiO_2 composites, with a higher production of ROS and the characteristic morphology, small size, and sharp edges. Both binary composites presented promising results regarding their application in disinfection treatment. However, GO/TiO_2 had a better performance with a 3-log reduction of *E. coli k12*. It was indicated that Bi ions interactions with the surface of bacteria cells are lengthier than with Ti ions, which can bind to the glycoproteins and penetrate the cell.

In ozone-assisted photocatalysis, Chávez et al. [167] combined the magnetic properties of Fe_3O_4 and the high surface area of activated carbon (AC) with a TiO_2 catalyst for the treatment of a real secondary effluent spiked with different CECs. The study proved the higher efficiency of photocatalytic ozonation, as single ozonation presented low mineralization levels and photocatalytic oxidation did not achieve the full elimination of CECs. Thus, the combined process showed faster removal, lower presence of intermediates, and better usage of ozone. Nonetheless, it was appointed that interactions between ozone and AC may alter the latter surface chemical properties, as a pH_{ZPC} decrease was found because of acidic oxygen groups on the surface of the carbon structure. This modification may affect the adsorptive behavior of the material. Other studies involving the application of TiO_2 composites in photocatalytic ozonation treatments are present in Table 9.

Table 9. Application of TiO_2 composites in photocatalytic and ozone-based processes for contaminants removal.

Composite	Radiation Source	Contaminant	Oxidant	Results	Ref.
CNT/TiO_2	UV	Oxamic acid, 89 mg L^{-1}	O_2	70% removal in 60 min	[166]
			O_3	100% removal in 60 min	
-				24% removal in 60 min	
Fe_3O_4/TiO_2	UV-A	Ceftazide, 10 mg L^{-1}	O_2	34.6% removal in 15 min	[168]
			O_3	100% removal in 15 min	
-				86.7% removal in 15 min	
AC/Fe_3O_4/ TiO_2	Solar	Metoprolol, Ibuprofen, Clofibric acid and DEET, 2 mg L^{-1} each	O_2	56% and 45% removals of contaminants mixture in 120 min in, respectively, synthetic and real secondary effluent, and up to 40% of DOC removal	[167]
			O_3	100% removal of contaminants mixture in 15 min and up to 70% of DOC removal	
-				100% removal of contaminants mixture in 30 min and up to 25% of DOC removal	
GO/Fe_3O_4/TiO_2	Solar	Cotinine, Caffeine, Ciproflaxin, Metoprolol, Sulfamethoxazole, Bezafibrate, Tritosulfuron, Ibuprofen, Clofibric acid, and DEET, 0.5 mg L^{-1} each	O_3	70% of TOC removal in 120 min at pH = 4 in urban wastewater	[169]
-				63% of TOC removal in 120 min at pH = 4 in urban wastewater	

4.2. WO_3

To overcome WO_3 drawbacks and obtain a more robust photocatalytic material, various hybrid structures have been explored. The coupling with other metal oxides and semiconductors is more substantially pursued, possibly as a more facile approach to surpass its high electron-hole recombination rate, and the less negative conduction band and compromised reductive potential, as it can benefit from the electron-hole interaction of the formed heterojunctions. The use of these WO_3 composites in the degradation of different contaminants is demonstrated in Table 10.

In recent years, phosphate-based photocatalysts, especially Ag_3PO_4, have been the focus of numerous studies due to their superior quantum efficiency under visible light irradiation, 90% [22]. However, it still faces relatively large particle sizes (0.5–2 μm), instability, and photo-corrosion, hindering a highly efficient photoactivity and its recyclability. Thus, Ag_3PO_4 and WO_3 composites may benefit from their heterostructures and be presented as more robust photocatalysts. Liu et al. [170] demonstrated that Ag_3PO_4/WO_3 had relatively stronger absorption in the long-wavelength range (>500 nm), while single WO_3 presented no activity in the same range. Regarding the ratio between the semiconductors, when Ag_3PO_4 exceeded 60%, a decrease in the photocatalytic activity towards RhB degradation, related to an excessive number of Ag_3PO_4 particles deposited at the WO_3 surface that may block light penetration. However, other studies showed that the formation of metallic Ag during the composite synthesis may occur and possess a negative effect, inhibiting the transport of positive holes from the Ag_3PO_4 VB to its surface, and the deposition of inactive substances over the composite surface, blocking active sites [171].

The instability of Ag_3PO_4 can be further increased by adding other components with better electronic properties. Graphene, which can be doped to boost its characteristics, is an excellent electron mediator and has been demonstrated to significantly increase the recyclability of Ag_3PO_4/WO_3 composites, providing higher chemical stability, surface area, and electron mobility by the organic material on the ternary composite [172].

Other semiconductors have been combined with WO_3, as mentioned, effectively obtaining improved features. Sodium niobate ($NaNbO_3$) is a semiconductor with a perovskite structure, low density, and high crystallinity, but with a broad bandgap energy (3.0–4.7 eV). The respective $NaNbO_3/WO_3$ has been reported to possess great photocatalytic activity, with a higher separation of photogenerated charges and reductive potential, benefiting from the considerably more negative conduction band of $NaNbO_3$ [173]. $ZnWO_4$ is another photocatalyst that presents higher bandgap energy (3.5–5.7 eV), but its formed composites with WO_3 can broaden its visible light absorption and form built-in electric fields as a result of a fast electron transference between both materials [174].

Table 10. Application of WO_3 composites for photocatalytic removal of contaminants.

Composite	S_{BET} ($m^2\ g^{-1}$)	E_{bg} (eV)	Radiation Source	Contaminant	Removal (Time)	Ref.
Ag_3PO_4/WO_3	23.9	-	Visible	Rhodamine B, 20 mg L^{-1}	98% (90 min)	[170]
$Ag_3PO_4/WO_3 \cdot H_2O$	-	2.43	Visible	Methylene Blue, 10 mg L^{-1} and Tetracycline 20 mg L^{-1}	98.9% MB and 70.4% (35 min)	[171]
Ag_3PO_4/NG/WO_3	-	2.36	Visible	Indomethacin, 5 mg L^{-1}	99.3% (50 min)	[172]
$NaNbO_3/WO_3$	7.2	2.60	Visible	2,4-Dichlorophenoxyacetic acid, 10 mg L^{-1}	60% (210 min)	[173]
Bi_2S_3/WO_3	53.8	1.90	Visible	Rhodamine B, 10 mg L^{-1}	90.7% (100 min)	[175]
Ag/$ZnWO_4/WO_3$	-	-	UV-Visible	Methylene Blue, 200 mg L^{-1}	94% (80 min)	[174]
GO/WO_3	18.7	2.32	Visible	Rhodamine B, 20 mg L^{-1}	96% (120 min)	[176]
				Ciprofloxacin, 20 mg L^{-1}	90% (120 min)	
Ag/GO/Chitosan/WO_3	26.4	2.40	Visible	Methylene Blue, 10 mg L^{-1}	99% (8 min)	[177]
SBA-15/Ag-WO_3	208	1.70	Visible	Atrazine, 20 mg L^{-1}	68% (18 min)	[178]

Gondal et al. [178] reported efficient herbicide removal using SBA-15, a highly porous siliceous material, as a template for a g-C_3N_4 composite catalyst, to obtain a structure with a high surface area and facilitate the reutilization of the material. The composite also benefits from the presence of the noble metal, with the surface plasmon resonance effect, which can increase light absorption, but it can also act as an electron trap center, as its Fermi level is lower than the semiconductor and electrons tend to accumulate on the Ag particles. Finally, the SBA-15/Ag-WO_3 presented a ~70% removal of atrazine, with a reaction rate constant of 0.065 min^{-1}.

4.3. *g-C_3N_4*

The characteristics of graphitic carbon nitride that hinder its broader utilization, such as its low specific surface area, obstructed active sites, and visible light utilization, can be strongly surpassed by the construction of composite materials. Its polymeric structure may be linked to other materials, with their own photocatalytic properties or characteristics that can boost the g-C_3N_4 performance. Their application in photocatalytic studies is shown in Table 11.

Table 11. Application of g-C_3N_4 composites for photocatalytic removal of contaminants.

Composite	S_{BET} ($m^2\ g^{-1}$)	E_{bg} (eV)	Radiation Source	Contaminant	Removal (Time)	Ref.
Bi_4O_7/g-C_3N_4	109	-	Visible	*Aspergillus fumigatus*, 10^6 CFU mL^{-1}	81% (6 h)	[179]
$BiVO_4$/g-C_3N_4	7	2.43	Visible	Nonylphenol Ethoxylate, 50 ppm	100% (120 min)	[180]
$BiVO_4/Bi_2O_6$/g-C_3N_4	95.9	-	Visible	Rhodamine B, 20 mg L^{-1}	100% (60 min)	[181]
				Tetracycline, 20 mg L^{-1}	100% (60 min)	
$LaVO_4$/g-C_3N_4	-	-	Visible	Tetracycline, 20 mg L^{-1}	83.4% (30 min)	[182]
				Naproxen, 20 mg L^{-1}	80% (120 min)	
Ag-ZnO/S-g-C_3N_4	57.2	2.51	Solar	Methylene Blue, 10 mg L^{-1}	97% (40 min)	[183]

Table 11. *Cont.*

Composite	S_{BET} (m^2 g^{-1})	E_{bg} (eV)	Radiation Source	Contaminant	Removal (Time)	Ref.
HTCC/O-g-C_3N_4	-	0.95	Visible	Human Adenovirus Type 2, 10^5 MPN mL^{-1}	100% (120 min)	[184]
PAN/g-C_3N_4	13.3	-	UV	Oilfield Produced Water	96.6% (8 h)	[185]
			Visible		85.4% (8 h)	
PEI/g-C_3N_4	70.2	1.74	Visible	Tetracycline, 40 mg L^{-1}	80% (120 min)	[186]
PEI/g-C_3N_4	-	-	Solar	*E. coli*, 2 × 10^6 CFU mL^{-1}	100% (45 min)	[187]
				Enterococcus faecalis, 2 × 10^6 CFU mL^{-1}	67.7% (60 min)	
Au-SiO_2/g-C_3N_4	365	-	Visible	Rhodamine B, 10 mg L^{-1}	99.8% (90 min)	[188]
β-SiAlON/g-C_3N_4	-	-	Visible	Murexide, 250 mg L^{-1}	90% (8 h)	[189]

There is an expanding number of studies regarding the combination of g-C_3N_4 with other polymeric materials. Their typically low-cost production, large surface areas, and presence of different functional groups and electrostatic charges can enhance pollutants interaction. Polyethyleneimine (PEI), for example, is a cationic polymer containing many amino groups that have been used in combination with catalyst and carbon-based materials to enhance the electrochemical properties and overall activity of composites. Yan et al. [186] synthesized in one step a PEI/g-C_3N_4 composite through the thermal copolymerization of urea mixed with PEI, obtaining a tremella-like structure containing -NH_X groups, that may be beneficial for water dispersion and photon absorption, and an increased BET surface area, up to 250%. The best composite catalyst also presented 80% removal of tetracycline with a reaction rate constant of 0.0226 min^{-1}, 3.2 times higher than g-C_3N_4. The efficiency of PEI/g-C_3N_4 has also been attested for disinfection purposes, by Zeng et al. [187], resulting in 6.2 and 4.2 log reductions of *E. coli* and *E. faecalis* in 45 and 60 min, respectively. PEI can increase O_2 reduction and alter the surface charges of the catalyst, promoting the adhesion of bacteria through electrostatic attraction.

Polyacrylonitrile (PAN) is also used for the synthesis of nanowires and other hybrid materials, with excellent flexibility and adsorption capacities. Alias et al. [185] applied electrospun PAN/g-C_3N_4 nanofibers for oilfield-produced water (OPW) treatment. The PAN matrix can adsorb organic compounds such as oil and accumulate them near g-C_3N_4, which allows the treatment of even more diluted solutions. The proposed composite obtained 96.6% and 85.4% degradation of OPW under UV and visible irradiation, respectively.

g-C_3N_4 has been used in combination with other metal oxides and semiconductor materials to merge its advantages and obtain heterojunctions with synergic effects. $LaVO_4$ and other vanadate complex oxides such as $BiVO_4$ present interesting features, especially their surface catalytic and optical properties, with good visible light absorption due to the V 3d orbital state. Jing et al. [182] indicated a high matching degree of $LaVO_4$ and g-C_3N_4 band structures in their respective hybrid catalyst, promoting an efficient migration of photocarriers and improving the photocatalytic activity. $LaVO_4$ also presents a more negative CB, while g-C_3N_4 has a more positive VB, potentializing the redox reactions in the respective bands of the heterostructure.

Dong et al. [180] demonstrated the advantages of $BiVO_4$ in combination with g-C_3N_4, which resulted in a material with a larger specific surface area, from 1.6 to 16.3 m^2 g^{-1}, and better visible light absorption. The presence of bismuth precursors may cause the compression of g-C_3N_4, which facilitates electron transference to the catalyst surface and the photogenerated pair recombination [181].

The matching bands of Bi-based oxides and g-C_3N_4 catalysts, and their better reductive/oxidative potential and ROS production have been reported as efficient pathogens disinfection treatment. Zhang et al. [179] demonstrated the capacity of Bi_4O_7/g-C_3N_4 to eliminate germinated fungi spores of *Aspergillus fumigatus* under visible light, attaining

81% disinfection of an initial 1×10^4 CFU mL^{-1}. The boosted production of $\cdot O_2^-$ and $\cdot OH$ by the composite were fundamental for better performance, as they were appointed to be the main active species on the decomposition of the fungus cell wall and cytoplasm.

The significant improvement of treatment performance in photocatalytic ozonation using g-C_3N_4 composites has been also attested, and some studies are present in Table 12. Jourshabani et al. [190] investigated the Fe_2O_3 and S-g-C_3N_4 combination in a composite catalyst and obtained substantially higher mineralization of Bisphenol A (BPA) under visible radiation, 97.8% TOC removal in 3 h, compared with the photocatalytic oxidation and photolytic ozonation treatments, 41.1%, and 40.1% respectively. The presence of the Fe-based material enhances the visible light absorption of g-C_3N_4 and the adsorption of both contaminants and ozone, which leads to faster elimination, higher production of ROS, and lower dosages of ozone, consequently reducing the process cost.

Zhang et al. [184] reported the use of a metal-free heterostructure consisting of oxygen-doped g-C_3N_4 microspheres integrated with hydrothermal carbonation carbon (HTCC) for human adenovirus disinfection. HTCC is an interesting carbon-based material recently found to act as a visible-light semiconductor, with a lower cost production compared to other carbonaceous compounds, such as graphene, although it suffers from a low photocatalytic activity due to fast recombination of electron-hole pairs [191]. The group applied a facile solvo-hydrothermal method, and the resulting composite presented a robust photocatalytic performance and biocompatibility towards human cells, achieving 5-log inactivation of the virus within 120 min.

Table 12. Application of g-C_3N_4 composites in photocatalytic and ozone-based processes for contaminants removal.

Composite	Radiation Source	Contaminant	Oxidant	Removal (Time)	References
GO/g-C_3N_4	Solar	Oxalic Acid, 10 mg L^{-1}	O_2	49.5% (40 min)	[192]
			O_3	93.2% (40 min)	
-				82% (40 min)	
SBA-15/Ag-g-C_3N_4	Solar	Oxalic Acid, 10 mg L^{-1}	O_2	16.8% (11 min)	[193]
			O_3	100% (11 min)	
-				4% (11 min)	
Fe_2O_3/S-g-C_3N_4	Visible	Bisphenol A, 50 mg L^{-1}	O_2	41.1% TOC (3 h)	[190]
			O_3	97.8% TOC (3 h)	
				40.1% TOC (3 h)	

Ceramic materials may be employed in composite catalysts as highly porous materials with excellent mechanical properties. These materials also act as great templates for photocatalytic species, producing high surface areas and recycling properties. Akulinkin et al. [189] used β-SiAlON, an aluminosilicate porous material, as a base for a g-C_3N_4 composite. The semiconductor was loaded in the pores of the ceramic substrate and the catalyst was obtained in plate forms. The composite presented great removal of murexide dye, as a consequence of its extended surface area and stability, maintaining its photocatalytic activity after seven cycles.

5. Future Perspectives

TiO_2 is by far the most used photocatalyst and possesses its basic properties already optimized, with a proper position of its CB and VB, which allow good redox potentials, high crystallinity, and overall photocatalytic performance. Although some disadvantages still prevent its full application, such as fast electron-hole pair recombination and, more especially, its broad bandgap energy, which implies the use of UV irradiation.

Thus, the development of new visible light active photocatalysts has been extremely encouraged and studied by the scientific community. Among them, g-C_3N_4 and WO_3 appear as promising candidates, presenting lower bandgap energies, higher absorption of visible light, and still maintaining the required properties common to TiO_2, such as good chemical stability and facile and low-cost production. Other alternative catalysts that have gotten more attention in recent years include the Bi-based (BiO_3, $BiVO_4$, BiOX) and phosphate (Ag_3PO_4, $CePO_4$) catalysts, which have interesting visible-light and overall catalytic activities.

Even with a better performance in the visible region, these selected catalysts still suffer from low specific surface areas, high electron-hole pair recombination, and unfavorable band potential, as g-C_3N_4 typically presents low VB and WO_3 less negative CB potentials, that diminish, respectively, ·OH production from water and O_2 reduction into $\cdot O_2^-$.

To overcome these obstacles, modifications of the photocatalysts can be explored, such as the introduction of multiple elements in catalyst doping, and the combination with materials of different natures to obtain composite photocatalysts. The incorporation of foreign elements in the structure of the catalyst may take place in different positions, with all possessing respective advantages. By the application of characterization methods, a specific analysis of each modification needs to be conducted to better understand its mechanisms. Generally, doping can alter the bandgap properties and may lead to shifts in the optical response of the catalysts, allowing it to present better responses towards higher wavelengths. These techniques may also reduce the photogenerated pairs recombination, modification of surface properties, and functional groups, which highly influence the interaction with CECs.

Composite materials encompass the use of mixed metallic oxides and other semiconductors, ceramics, polymers, and other structures to obtain a combination of their individual strengths and advantages. Outstanding properties may be achieved, such as high specific surface areas, electron mobility, different morphologies, easier recyclability, posterior separation, and heterojunctions which may increase the reduction and oxidation reactions, an important aspect for the commented alternative catalysts.

Even so, each alteration may also have side effects. Possible poisoning, lixiviation, active site blockage, and loss of photoactivity may occur. Some compounds, such as noble metal or rare-earth elements, and more elaborated materials can also represent a high cost involved and short availability. The precise evaluation of the concentration of elemental dopants and added materials are crucial, with excessive doping resulting in the opposite effect, acting as recombination centers, or the extreme agglomeration of the structures in hybrid materials. The combination of different materials in composite catalysts may also lead to a faster deactivation of the final product, shortening their lifetime and increasing the process cost. Thus, a case-by-case analysis is fundamental to better understand these side effects and obtain a more tailored material, which can be an exhaustive task. Additionally, the catalysts' reuse and possible post-treatments for their reactivation need to be more profoundly explored.

The proper application and reactor to be used are also fundamental in the conceptualization of the photocatalytic treatment apparatus and need to be evaluated alongside the production of the catalytic material. This task is one of the greatest challenges, as it encompasses the optimization of multiple parameters, to ensure efficient light penetration, interactions between contaminants and the catalyst, proper residence time, etc. Even being a cost-free radiation source, sunlight photocatalyst activation may require special reactor designs, presenting additional costs that need to be evaluated case-by-case. Nonetheless, great advancements have been obtained in compound parabolic collector (CPC) reactor developments, allowing proper light penetration within the equipment and better exploitation of the photon source. A facile retrieval of the catalyst after the reaction is also a key factor and dependent on the selected reaction equipment. Some achievements have been obtained in this post-recovery phase, such as the exploitation of magnetic properties of some materials, which can assist in the separation and reutilization of the photocatalyst.

The number of studies involving the application of such materials in real effluents is considerably limited. There are different common parameters and compounds present in these complex water matrices that compromise and interact with the species participants in the photocatalytic ozonation reaction. Broader investigations are thus vital to understand these effects and construct a more robust system.

6. Conclusions

The application of semiconductors in photocatalytic based for CECs abatement has shown remarkable potential as a water treatment technology. The production of highly oxidative radicals may eliminate a variety of these chemical and biological compounds that represent a danger to human and environmental health. These treatments have been pointed out to be effective even in more complex matrices, which proves their capacity to be applied to real effluents under different conditions. The use of additional oxidants such as ozone can also boost the overall process efficiency, increasing the production of radicals and electron retrieval.

The investigations regarding new visible light active photocatalysts show promising results, but more complete studies still need to be conducted to collect more information. The applications of doping and composite materials open a great variety of possibilities for materials with more robust and feasible characteristics to be obtained. Thus, more elaborative comparisons between the new and standard semiconductors and their adaptations need to be performed, to better understand the advantages of further exploration of the already well-founded TiO_2 based materials and the development of alternative materials. The addition of ozone in the photocatalytic process has been proven to enhance the overall efficiency, and more studies of its application with alternative and adapted catalysts will be valuable. Nonetheless, the photocatalytic-based process is a favorable route for the degradation of pathogens and CECs and future large-scale water reclamation technologies.

Author Contributions: Conceptualization, E.F., J.G. and R.C.M.; methodology, E.F., J.G. and R.C.M.; writing—original draft, E.F.; writing—review and editing, J.G. and R.C.M.; supervision, J.G. and R.C.M.; funding acquisition, E.F., J.G. and R.C.M. All authors have read and agreed to the published version of the manuscript.

Funding: This work was financed by European Structural and Investment Funds through Portugal2020 by the project PhotoSupCatal—Development of supported catalytic systems for wastewater treatment by photo-assisted processes (POCI-01-0247-FEDER-047545). The authors gratefully acknowledge FCT (Fundação para a Ciência e Tecnologia, Portugal) for the PhD Grant (2020.06130.BD) and the financial support (CEECIND/01207/2018). Thanks are due to FCT/MCTES for the financial support to CIEPQPF (UIDB/00102/2020).

Data Availability Statement: Not applicable.

Conflicts of Interest: The authors declare no conflict of interest.

References

1. Gittins, J.R.; Hemingway, J.R.; Dajka, J.C. How a Water-Resources Crisis Highlights Social-Ecological Disconnects. *Water Res.* **2021**, *194*, 116937. [CrossRef]
2. Gleick, P.H. *The World's Water*; Island Press: Washington, DC, USA, 2015; Volume 8.
3. McCance, W.; Jones, O.A.H.; Cendón, D.I.; Edwards, M.; Surapaneni, A.; Chadalavada, S.; Wang, S.; Currell, M. Combining Environmental Isotopes with Contaminants of Emerging Concern (CECs) to Characterise Wastewater Derived Impacts on Groundwater Quality. *Water Res.* **2020**, *182*, 1–15. [CrossRef]
4. Rizzo, L.; Malato, S.; Antakyali, D.; Beretsou, V.G.; Đolić, M.B.; Gernjak, W.; Heath, E.; Ivancev-Tumbas, I.; Karaolia, P.; Lado Ribeiro, A.R.; et al. Consolidated vs New Advanced Treatment Methods for the Removal of Contaminants of Emerging Concern from Urban Wastewater. *Sci. Total Environ.* **2019**, *655*, 986–1008. [CrossRef]
5. Valbonesi, P.; Pro, M.; Vasumini, I.; Fabbri, E. Contaminants of Emerging Concern in Drinking Water: Quality Assessment by Combining Chemical and Biological Analysis. *Sci. Total Environ.* **2021**, *758*, 143624. [CrossRef]
6. Olalla, A.; Moreno, L.; Valcárcel, Y. Prioritisation of Emerging Contaminants in the Northern Antarctic Peninsula Based on Their Environmental Risk. *Sci. Total Environ.* **2020**, *742*, 140417. [CrossRef]

7. Lencioni, V.; Bellamoli, F.; Paoli, F. Multi-Level Effects of Emerging Contaminants on Macroinvertebrates in Alpine Streams: From DNA to the Ecosystem. *Ecol. Indic.* **2020**, *117*, 106660. [CrossRef]
8. Llamas-dios, M.I.; Vadillo, I.; Jiménez-gavilán, P.; Candela, L.; Corada-fernández, C. Assessment of a Wide Array of Contaminants of Emerging Concern in a Mediterranean Water Basin (Guadalhorce River, Spain): Motivations for an Improvement of Water Management and Pollutants Surveillance. *Sci. Total Environ.* **2021**, *788*, 147822. [CrossRef]
9. Tian, Z.; Wark, D.A.; Bogue, K.; James, C.A. Suspect and Non-Target Screening of Contaminants of Emerging Concern in Streams in Agricultural Watersheds. *Sci. Total Environ.* **2021**, *795*, 148826. [CrossRef]
10. Shah, A.I.; Din Dar, M.U.; Bhat, R.A.; Singh, J.P.; Singh, K.; Bhat, S.A. Prospectives and Challenges of Wastewater Treatment Technologies to Combat Contaminants of Emerging Concerns. *Ecol. Eng.* **2020**, *152*, 105882. [CrossRef]
11. Liu, Z.; Demeestere, K.; Hulle, S. Van Comparison and Performance Assessment of Ozone-Based AOPs in View of Trace Organic Contaminants Abatement in Water and Wastewater: A Review. *J. Environ. Chem. Eng.* **2021**, *9*, 105599. [CrossRef]
12. Giwa, A.; Yusuf, A.; Balogun, H.A.; Sambudi, N.S.; Bilad, M.R.; Adeyemi, I.; Chakraborty, S.; Curcio, S. Recent Advances in Advanced Oxidation Processes for Removal of Contaminants from Water: A Comprehensive Review. *Process Saf. Environ. Prot.* **2021**, *146*, 220–256. [CrossRef]
13. Sgroi, M.; Snyder, S.; Roccaro, P. Comparison of AOPs at Pilot Scale: Energy Costs for Micro-Pollutants Oxidation, Disinfection by-Products Formation and Pathogens Inactivation. *Chemosphere* **2021**, *273*, 128527. [CrossRef]
14. Khan, J.A.; Sayed, M.; Khan, S.; Shah, N.S.; Dionysiou, D.D.; Boczkaj, G. Advanced Oxidation Processes for the Treatment of Contaminants of Emerging Concern. In *Contaminants of Emerging Concern in Water and Wastewater: Advanced Treatment Processes*; Elsevier Inc.: Cambridge, MA, USA, 2019; pp. 299–365, ISBN 9780128135617.
15. Itzel, F.; Gehrmann, L.; Bielak, H.; Ebersbach, P.; Boergers, A.; Herbst, H.; Maus, C.; Simon, A.; Dopp, E.; Hammers-Wirtz, M.; et al. Investigation of Full-Scale Ozonation at a Municipal Wastewater Treatment Plant Using a Toxicity-Based Evaluation Concept. *J. Toxicol. Environ. Health—Part A Curr. Issues* **2017**, *80*, 1242–1258. [CrossRef] [PubMed]
16. Irani, R.; Khoshfetrat, A.B.; Forouzesh, M. Real Municipal Wastewater Treatment Using Simultaneous Pre and Post-Ozonation Combined Biological Attached Growth Reactor: Energy Consumption Assessment. *J. Environ. Chem. Eng.* **2021**, *9*, 104595. [CrossRef]
17. Gulde, R.; Rutsch, M.; Clerc, B.; Schollée, J.E.; von Gunten, U.; McArdell, C.S. Formation of Transformation Products during Ozonation of Secondary Wastewater Effluent and Their Fate in Post-Treatment: From Laboratory- to Full-Scale. *Water Res.* **2021**, *200*, 117200. [CrossRef]
18. Fernandes, E.; Contreras, S.; Medina, F.; Martins, R.C.; Gomes, J. N-Doped Titanium Dioxide for Mixture of Parabens Degradation Based on Ozone Action and Toxicity Evaluation: Precursor of Nitrogen and Titanium Effect. *Process Saf. Environ. Prot.* **2020**, *138*, 80–89. [CrossRef]
19. Sgroi, M.; Anumol, T.; Vagliasindi, F.G.A.; Snyder, S.A.; Roccaro, P. Comparison of the New $Cl_2/O_3/UV$ Process with Different Ozone- and UV-Based AOPs for Wastewater Treatment at Pilot Scale: Removal of Pharmaceuticals and Changes in Fluorescing Organic Matter. *Sci. Total Environ.* **2021**, *765*, 142720. [CrossRef] [PubMed]
20. Cai, Q.Q.; Jothinathan, L.; Deng, S.H.; Ong, S.L.; Ng, H.Y.; Hu, J.Y. *Fenton- and Ozone-Based AOP Processes for Industrial Effluent Treatment*; Elsevier: Amsterdam, The Netherlands, 2021; ISBN 9780128210116.
21. Fernandes, E.; Martins, R.C.; Gomes, J. Photocatalytic Ozonation of Parabens Mixture Using 10% N-TiO_2 and the Effect of Water Matrix. *Sci. Total Environ.* **2020**, *718*, 137321. [CrossRef] [PubMed]
22. Dong, P.; Xi, X.; Hou, G. Typical Non–TiO_2-Based Visible-Light Photocatalysts. In *Semiconductor Photocatalysis—Materials, Mechanisms and Applications*; BoD–Books on Demand: Norderstedt, Germany, 2016; pp. 209–229.
23. Barakat, M.A.; Kumar, R. *Photocatalytic Activity Enhancement of Titanium Dioxide Nanoparticles*; Springer: Cham, Switzerland, 2016; ISBN 9783319242699.
24. Gomes, J.; Roccamante, M.; Contreras, S.; Medina, F.; Oller, I.; Martins, R.C. Scale-up Impact over Solar Photocatalytic Ozonation with Benchmark-P25 and N-TiO_2 for Insecticides Abatement in Water. *J. Environ. Chem. Eng.* **2021**, *9*, 104915. [CrossRef]
25. Huang, F.; Yan, A.; Zhao, H. Influences of Doping on Photocatalytic Properties of TiO_2 Photocatalyst. In *Semiconductor Photocatalysis: Materials, Mechanisms and Applications*; InTech: Osaka, Japan, 2016; pp. 31–80.
26. García-López, E.I.; Palmisano, L. (Eds.) *Material Science in Photocatalysis*; Elsevier: Amsterdam, The Netherlands, 2021; ISBN 9781119130536.
27. Nguyen, T.K.A.; Pham, T.T.; Nguyen-Phu, H.; Shin, E.W. The Effect of Graphitic Carbon Nitride Precursors on the Photocatalytic Dye Degradation of Water-Dispersible Graphitic Carbon Nitride Photocatalysts. *Appl. Surf. Sci.* **2021**, *537*, 148027. [CrossRef]
28. Fernandes, R.A.; Sampaio, M.J.; Dražić, G.; Faria, J.L.; Silva, C.G. Efficient Removal of Parabens from Real Water Matrices by a Metal-Free Carbon Nitride Photocatalyst. *Sci. Total Environ.* **2020**, *716*, 135346. [CrossRef] [PubMed]
29. Orge, C.A.; Sampaio, M.J.; Faria, J.L.; Pereira, M.F.R.; Silva, C.G. Efficiency and Stability of Metal-Free Carbon Nitride in the Photocatalytic Ozonation of Oxamic Acid under Visible Light. *J. Environ. Chem. Eng.* **2020**, *8*, 104172. [CrossRef]
30. Liu, H.; Ma, S.; Shao, L.; Liu, H.; Gao, Q.; Li, B.; Fu, H.; Fu, S.; Ye, H.; Zhao, F.; et al. Defective Engineering in Graphitic Carbon Nitride Nanosheet for Efficient Photocatalytic Pathogenic Bacteria Disinfection. *Appl. Catal. B Environ.* **2020**, *261*, 118201. [CrossRef]
31. Li, Y.; Zhang, C.; Shuai, D.; Naraginti, S.; Wang, D.; Zhang, W. Visible-Light-Driven Photocatalytic Inactivation of MS2 by Metal-Free g-C_3N_4: Virucidal Performance and Mechanism. *Water Res.* **2016**, *106*, 249–258. [CrossRef] [PubMed]

32. Nagy, D.; Nagy, D.; Szilágyi, I.M.; Fan, X. Effect of the Morphology and Phases of WO_3 Nanocrystals on Their Photocatalytic Efficiency. *RSC Adv.* **2016**, *6*, 33743–33754. [CrossRef]
33. Palanisamy, G.; Bhuvaneswari, K.; Pazhanivel, T.; Bharathi, G. Enriched Photocatalytic Activity of Rhodamine B Dye from Aqueous Solution Using Hollow Sphere Tungsten Trioxide Nanoparticles. *Optik* **2020**, *204*, 164171. [CrossRef]
34. Mena, E.; Rey, A.; Contreras, S.; Beltrán, F.J. Visible Light Photocatalytic Ozonation of DEET in the Presence of Different Forms of WO_3. *Catal. Today* **2015**, *252*, 100–106. [CrossRef]
35. Razali, N.A.M.; Salleh, W.N.W.; Rosman, N.; Ismail, N.H.; Ahmad, S.Z.N.; Aziz, F.; Jye, L.W.; Ismail, A.F. Palm Oil Mill Effluent Treatment Using Tungsten Trioxide: Adsorption and Photocatalytic Degradation. *Mater. Today Proc.* **2021**, *42*, 22–27. [CrossRef]
36. Basavarajappa, P.S.; Patil, S.B.; Ganganagappa, N.; Reddy, K.R.; Raghu, A.V.; Reddy, C.V. Recent Progress in Metal-Doped TiO_2, Non-Metal Doped/Codoped TiO_2 and TiO_2 Nanostructured Hybrids for Enhanced Photocatalysis. *Int. J. Hydrogen Energy* **2020**, *45*, 7764–7778. [CrossRef]
37. Varma, K.S.; Tayade, R.J.; Shah, K.J.; Joshi, P.A.; Shukla, A.D.; Gandhi, V.G. Photocatalytic Degradation of Pharmaceutical and Pesticide Compounds (PPCs) Using Doped TiO_2 Nanomaterials: A Review. *Water-Energy Nexus* **2020**, *3*, 46–61. [CrossRef]
38. Karuppasamy, P.; Ramzan Nilofar Nisha, N.; Pugazhendhi, A.; Kandasamy, S.; Pitchaimuthu, S. An Investigation of Transition Metal Doped TiO_2 photocatalysts for the Enhanced Photocatalytic Decoloration of Methylene Blue Dye under Visible Light Irradiation. *J. Environ. Chem. Eng.* **2021**, *9*, 105254. [CrossRef]
39. Lee, H.; Jang, H.S.; Kim, N.Y.; Joo, J.B. Cu-Doped TiO_2 Hollow Nanostructures for the Enhanced Photocatalysis under Visible Light Conditions. *J. Ind. Eng. Chem.* **2021**, *99*, 352–363. [CrossRef]
40. Moreira, A.J.; Malafatti, J.O.D.; Giraldi, T.R.; Paris, E.C.; Pereira, E.C.; de Mendonça, V.R.; Mastelaro, V.R.; Freschi, G.P.G. Prozac® Photodegradation Mediated by Mn-Doped TiO_2 Nanoparticles: Evaluation of by-Products and Mechanisms Proposal. *J. Environ. Chem. Eng.* **2020**, *8*, 104543. [CrossRef]
41. Ellouzi, I.; Bouddouch, A.; Bakiz, B.; Benlhachemi, A.; Abou Oualid, H. Glucose-Assisted Ball Milling Preparation of Silver-Doped Biphasic TiO_2 for Efficient Photodegradation of Rhodamine B: Effect of Silver-Dopant Loading. *Chem. Phys. Lett.* **2021**, *770*, 138456. [CrossRef]
42. Wu, M.C.; Lin, T.H.; Hsu, K.H.; Hsu, J.F. Photo-Induced Disinfection Property and Photocatalytic Activity Based on the Synergistic Catalytic Technique of Ag Doped TiO_2 Nanofibers. *Appl. Surf. Sci.* **2019**, *484*, 326–334. [CrossRef]
43. Saber, N.B.; Mezni, A.; Alrooqi, A.; Altalhi, T. Fabrication of Efficient Au@TiO_2/RGO Heterojunction Nanocomposite: Boosted Photocatalytic Activity under Ultraviolet and Visible Light Irradiation. *J. Mater. Res. Technol.* **2021**, *12*, 2238–2246. [CrossRef]
44. Manasa, M.; Chandewar, P.R.; Mahalingam, H. Photocatalytic Degradation of Ciprofloxacin & Norfloxacin and Disinfection Studies under Solar Light Using Boron & Cerium Doped TiO_2 Catalysts Synthesized by Green EDTA-Citrate Method. *Catal. Today* **2021**, *375*, 522–536. [CrossRef]
45. Komaraiah, D.; Radha, E.; Sivakumar, J.; Reddy, M.V.R.; Sayanna, R. Influence of Fe^{3+} Ion Doping on the Luminescence Emission Behavior and Photocatalytic Activity of Fe^{3+}, Eu^{3+}-Codoped TiO_2 Thin Films. *J. Alloys Compd.* **2021**, *868*, 159109. [CrossRef]
46. Mancuso, A.; Sacco, O.; Vaiano, V.; Sannino, D.; Pragliola, S.; Venditto, V.; Morante, N. Visible Light Active Fe-Pr Co-Doped TiO_2 for Water Pollutants Degradation. *Catal. Today* **2021**. [CrossRef]
47. Crişan, M.; Răileanu, M.; Drăgan, N.; Crişan, D.; Ianculescu, A.; Niţoi, I.; Oancea, P.; Şomăcescu, S.; Stănică, N.; Vasile, B.; et al. Sol-Gel Iron-Doped TiO_2 Nanopowders with Photocatalytic Activity. *Appl. Catal. A Gen.* **2015**, *504*, 130–142. [CrossRef]
48. Pascariu, P.; Cojocaru, C.; Homocianu, M.; Samoila, P.; Dascalu, A.; Suchea, M. New La^{3+} Doped TiO_2 Nanofibers for Photocatalytic Degradation of Organic Pollutants: Effects of Thermal Treatment and Doping Loadings. *Ceram. Int.* **2022**, *48*, 4953–4964. [CrossRef]
49. Hernández-Ramírez, A.; Medina-Ramírez, I.; Bustos, E.; Manríquez, J.; Peralta-Hernández, J.M. *Photocatalytic Semiconductors*; Springer: Berlin/Heidelberg, Germany, 2015; ISBN 9783319109985.
50. Coronado, J.M.; Fresno, F.; Hernández-Alonso, M.D.; Portela, R. *Design of Advanced Photocatalytic Materials for Energy and Environmental Applications*; Springer: London, UK, 2013; Volume 71, ISBN 9781447150602.
51. Kayani, Z.N.; Intizar, T.; Riaz, S.; Naseem, S. Antibacterial, Magnetic and Dielectric Properties of Nano-Structured V Doped TiO_2 Thin Films Deposited by Dip Coating Technique. *Mater. Chem. Phys.* **2021**, *267*, 124659. [CrossRef]
52. Lashuk, B.; Yargeau, V. A Review of Ecotoxicity Reduction in Contaminated Waters by Heterogeneous Photocatalytic Ozonation. *Sci. Total Environ.* **2021**, *787*, 147645. [CrossRef]
53. Asgari, E.; Sheikhmohammadi, A.; Nourmoradi, H.; Nazari, S.; Aghanaghad, M. Degradation of Ciprofloxacin by Photocatalytic Ozonation Process under Irradiation with UVA: Comparative Study, Performance and Mechanism. *Process Saf. Environ. Prot.* **2021**, *147*, 356–366. [CrossRef]
54. Mecha, A.C.; Onyango, M.S.; Ochieng, A.; Momba, M.N.B. Ultraviolet and Solar Photocatalytic Ozonation of Municipal Wastewater: Catalyst Reuse, Energy Requirements and Toxicity Assessment. *Chemosphere* **2017**, *186*, 669–676. [CrossRef] [PubMed]
55. Ribeiro, A.R.; Umbuzeiro, G. de A. Effects of a Textile Azo Dye on Mortality, Regeneration, and Reproductive Performance of the Planarian, Girardia Tigrina. *Environ. Sci. Eur.* **2014**, *26*, 1–8. [CrossRef]
56. Gomes, J.F.; Leal, I.; Bednarczyk, K.; Gmurek, M.; Stelmachowski, M.; Diak, M.; Emília Quinta-Ferreira, M.; Costa, R.; Quinta-Ferreira, R.M.; Martins, R.C. Photocatalytic Ozonation Using Doped TiO_2 Catalysts for the Removal of Parabens in Water. *Sci. Total Environ.* **2017**, *609*, 329–340. [CrossRef] [PubMed]

57. Prakash, J.; Samriti; Kumar, A.; Dai, H.; Janegitz, B.C.; Krishnan, V.; Swart, H.C.; Sun, S. Novel Rare Earth Metal–Doped One-Dimensional TiO_2 Nanostructures: Fundamentals and Multifunctional Applications. *Mater. Today Sustain.* **2021**, *13*, 100066. [CrossRef]
58. Cerrato, E.; Gaggero, E.; Calza, P.; Paganini, M.C. The Role of Cerium, Europium and Erbium Doped TiO_2 Photocatalysts in Water Treatment: A Mini-Review. *Chem. Eng. J. Adv.* **2022**, *10*, 100268. [CrossRef]
59. Mancuso, A.; Sacco, O.; Sannino, D.; Pragliola, S.; Vaiano, V. Enhanced Visible-Light-Driven Photodegradation of Acid Orange 7 Azo Dye in Aqueous Solution Using Fe-N Co-Doped TiO_2. *Arab. J. Chem.* **2020**, *13*, 8347–8360. [CrossRef]
60. Lin, L.; Wang, H.; Luo, H.; Xu, P. Enhanced Photocatalysis Using Side-Glowing Optical Fibers Coated with Fe-Doped TiO_2 Nanocomposite Thin Films. *J. Photochem. Photobiol. A Chem.* **2015**, *307–308*, 88–98. [CrossRef]
61. Gusain, R.; Kumar, N.; Ray, S.S. Factors Influencing the Photocatalytic Activity of Photocatalysts in Wastewater Treatment. In *Photocatalysts in Advanced Oxidation Processes for Wastewater Treatment*; John Wiley & Sons: Hoboken, NJ, USA, 2020; pp. 229–270.
62. Mirzaei, A.; Eddah, M.; Roualdes, S.; Ma, D.; Chaker, M. Multiple-Homojunction Gradient Nitrogen Doped TiO_2 for Photocatalytic Degradation of Sulfamethoxazole, Degradation Mechanism, and Toxicity Assessment. *Chem. Eng. J.* **2021**, *422*, 130507. [CrossRef]
63. Ramezanisani, S.; Rajabi, M.; Mohseni, F. Influence of Nitrogen Doping on Visible Light Photocatalytic Activity of TiO_2 Nanowires with Anatase-Rutile Junction. *Chem. Phys. Lett.* **2020**, *744*, 137217. [CrossRef]
64. Suwannaruang, T.; Kamonsuangkasem, K.; Kidkhunthod, P.; Chirawatkul, P.; Saiyasombat, C.; Chanlek, N.; Wantala, K. Influence of Nitrogen Content Levels on Structural Properties and Photocatalytic Activities of Nanorice-like N-Doped TiO_2 with Various Calcination Temperatures. *Mater. Res. Bull.* **2018**, *105*, 265–276. [CrossRef]
65. Erdogan, N.; Bouziani, A.; Park, J.; Micusik, M.; Kim, S.Y.; Majkova, E.; Omastova, M.; Ozturk, A. Synthesis and Enhanced Photocatalytic Activity of Nitrogen-Doped Triphasic TiO_2 Nanoparticles. *J. Photochem. Photobiol. A Chem.* **2019**, *377*, 92–100. [CrossRef]
66. Toe, E.D.; Kurniawan, W.; Mariquit, E.G.; Hinode, H. Synthesis of N-Doped Mesoporous TiO_2 by Facile One-Step Solvothermal Process for Visible Light Photocatalytic Degradation of Organic Pollutant. *J. Environ. Chem. Eng.* **2018**, *6*, 5125–5134. [CrossRef]
67. Yadav, V.; Saini, V.K.; Sharma, H. How Different Dopants Leads to Difference in Photocatalytic Activity in Doped TiO_2? *Ceram. Int.* **2020**, *46*, 27308–27317. [CrossRef]
68. Bakre, P.V.; Tilve, S.G.; Shirsat, R.N. Influence of N Sources on the Photocatalytic Activity of N-Doped TiO_2. *Arab. J. Chem.* **2020**, *13*, 7637–7651. [CrossRef]
69. Zhao, W.; Liu, S.; Zhang, S.; Wang, R.; Wang, K. Preparation and Visible-Light Photocatalytic Activity of N-Doped TiO_2 by Plasma-Assisted Sol-Gel Method. *Catal. Today* **2019**, *337*, 37–43. [CrossRef]
70. Gao, H.; Zhou, J.; Dai, D.; Qu, Y. Photocatalytic Activity and Electronic Structure Analysis of N-Doped Anatase TiO_2: A Combined Experimental and Theoretical Study. *Chem. Eng. Technol.* **2009**, *32*, 867–872. [CrossRef]
71. Assayehegn, E.; Solaiappan, A.; Chebude, Y.; Alemayehu, E. Fabrication of Tunable Anatase/Rutile Heterojunction N/TiO_2 Nanophotocatalyst for Enhanced Visible Light Degradation Activity. *Appl. Surf. Sci.* **2020**, *515*, 145966. [CrossRef]
72. Chaturvedi, R.; Singh, P.K. Materials Today: Proceedings Synthesis and Characterization of Nano Crystalline Nitrogen Doped Titanium Dioxide. *Mater. Today Proc.* **2021**, *45*, 3666–3669. [CrossRef]
73. Wafi, A.; Szabó-Bárdos, E.; Horváth, O.; Makó, É.; Jakab, M.; Zsirka, B. Coumarin-Based Quantification of Hydroxyl Radicals and Other Reactive Species Generated on Excited Nitrogen-Doped TiO_2. *J. Photochem. Photobiol. A Chem.* **2021**, *404*, 112913. [CrossRef]
74. Huang, J.; Dou, L.; Li, J.; Zhong, J.; Li, M.; Wang, T. Excellent Visible Light Responsive Photocatalytic Behavior of N-Doped TiO_2 toward Decontamination of Organic Pollutants. *J. Hazard. Mater.* **2021**, *403*, 123857. [CrossRef]
75. Zhang, J.; Xie, M.; Zhao, H.; Zhang, L.r.; Wei, G.; Zhao, G. Preferential and Efficient Degradation of Phenolic Pollutants with Cooperative Hydrogen-Bond Interactions in Photocatalytic Process. *Chemosphere* **2021**, *269*, 129404. [CrossRef]
76. Ravidhas, C.; Anitha, B.; Venkatesh, R.; Monica, S.E.S.; Gopalakrishna, D.; Raj, A.M.E.; Ravichandran, K. Role of Fluorine Doping on Luminescence Centers and Enhanced Photocatalytic Performance of Nebulizer Sprayed TiO_2 Films under Visible Light. *J. Lumin.* **2018**, *198*, 272–283. [CrossRef]
77. Hwang, J.; Kalanur, S.S.; Seo, H. Identification of Visible Photocatalytic and Photoelectrochemical Properties of I-TiO_2 via Electronic Band Structure. *Electrochim. Acta* **2017**, *252*, 482–489. [CrossRef]
78. Zhou, L.; Cai, M.; Zhang, X.; Cui, N.; Chen, G.; Zou, G. yan In-Situ Nitrogen-Doped Black TiO_2 with Enhanced Visible-Light-Driven Photocatalytic Inactivation of Microcystis Aeruginosa Cells: Synthesization, Performance and Mechanism. *Appl. Catal. B Environ.* **2020**, *272*, 119019. [CrossRef]
79. Tzeng, J.H.; Weng, C.H.; Yen, L.T.; Gaybullaev, G.; Chang, C.J.; de Luna, M.D.G.; Lin, Y.T. Inactivation of Pathogens by Visible Light Photocatalysis with Nitrogen-Doped TiO_2 and Tourmaline-Nitrogen Co-Doped TiO_2. *Sep. Purif. Technol.* **2021**, *274*, 118979. [CrossRef]
80. Boningari, T.; Inturi, S.N.R.; Suidan, M.; Smirniotis, P.G. Novel One-Step Synthesis of Sulfur Doped-TiO_2 by Flame Spray Pyrolysis for Visible Light Photocatalytic Degradation of Acetaldehyde. *Chem. Eng. J.* **2018**, *339*, 249–258. [CrossRef]
81. Bento, R.T.; Correa, O.V.; Pillis, M.F. Photocatalytic Activity of Undoped and Sulfur-Doped TiO_2 Films Grown by MOCVD for Water Treatment under Visible Light. *J. Eur. Ceram. Soc.* **2019**, *39*, 3498–3504. [CrossRef]
82. Portela, R. Non-Metal Doping for Band-Gap Engineering Raquel. In *Design of Advanced Photocatalytic Materials for Energy and Environmental Applications. In Green Energy and Technology*; Coronado, J.M., Fresno, F., Hernández-Alonso, M.D., Portela, R., Eds.; Springer: London, UK, 2013; Volume 71, pp. 287–309, ISBN 9781447150602.

83. He, J.; Kumar, A.; Khan, M.; Lo, I.M.C. Critical Review of Photocatalytic Disinfection of Bacteria: From Noble Metals- and Carbon Nanomaterials-TiO_2 Composites to Challenges of Water Characteristics and Strategic Solutions. *Sci. Total Environ.* **2021**, *758*, 143953. [CrossRef] [PubMed]
84. Wang, Y.; Wu, Y.; Yang, H.; Xue, X.; Liu, Z. Doping TiO_2 with Boron or/and Cerium Elements: Effects on Photocatalytic Antimicrobial Activity. *Vacuum* **2016**, *131*, 58–64. [CrossRef]
85. Jahdi, M.; Mishra, S.B.; Nxumalo, E.N.; Mhlanga, S.D.; Mishra, A.K. Smart Pathways for the Photocatalytic Degradation of Sulfamethoxazole Drug Using F-Pd Co-Doped TiO_2 Nanocomposites. *Appl. Catal. B Environ.* **2020**, *267*, 118716. [CrossRef]
86. Khedr, T.M.; El-Sheikh, S.M.; Hakki, A.; Ismail, A.A.; Badawy, W.A.; Bahnemann, D.W. Highly Active Non-Metals Doped Mixed-Phase TiO_2 for Photocatalytic Oxidation of Ibuprofen under Visible Light. *J. Photochem. Photobiol. A Chem.* **2017**, *346*, 530–540. [CrossRef]
87. Venkatesan, A.; Al-onazi, W.A.; Elshikh, M.S.; Pham, T.H.; Suganya, S.; Boobas, S.; Priyadharsan, A. Study of Synergistic Effect of Cobalt and Carbon Codoped with TiO_2 Photocatalyst for Visible Light Induced Degradation of Phenol. *Chemosphere* **2022**, *305*, 135333. [CrossRef]
88. Ariza-Tarazona, M.C.; Villarreal-Chiu, J.F.; Hernández-López, J.M.; Rivera De la Rosa, J.; Barbieri, V.; Siligardi, C.; Cedillo-González, E.I. Microplastic Pollution Reduction by a Carbon and Nitrogen-Doped TiO_2: Effect of PH and Temperature in the Photocatalytic Degradation Process. *J. Hazard. Mater.* **2020**, *395*, 122632. [CrossRef]
89. Thongpool, V.; Phunpueok, A.; Jaiyen, S.; Sornkwan, T. Results in Physics Synthesis and Photocatalytic Activity of Copper and Nitrogen Co-Doped Titanium Dioxide Nanoparticles. *Results Phys.* **2020**, *16*, 102948. [CrossRef]
90. Wu, Y.; Dong, Y.; Xia, X.; Liu, X.; Li, H. Facile Synthesis of N-F Codoped and Molecularly Imprinted TiO_2 for Enhancing Photocatalytic Degradation of Target Contaminants. *Appl. Surf. Sci.* **2016**, *364*, 829–836. [CrossRef]
91. Razali, N.A.M.; Salleh, W.N.W.; Aziz, F.; Jye, L.W.; Yusof, N.; Ismail, A.F. Review on Tungsten Trioxide as a Photocatalysts for Degradation of Recalcitrant Pollutants. *J. Clean. Prod.* **2021**, *309*, 127438. [CrossRef]
92. Palharim, P.H.; Fusari, B.L.D.d.R.; Ramos, B.; Otubo, L.; Teixeira, A.C.S.C. Effect of HCl and HNO_3 on the Synthesis of Pure and Silver-Based WO_3 for Improved Photocatalytic Activity under Sunlight. *J. Photochem. Photobiol. A Chem.* **2022**, *422*, 113550. [CrossRef]
93. Hanif, M.A.; Akter, J.; Islam, M.A.; Sapkota, K.P.; Hahn, J.R. Visible-Light-Driven Enhanced Photocatalytic Performance Using Cadmium-Doping of Tungsten (VI) Oxide and Nanocomposite Formation with Graphitic Carbon Nitride Disks. *Appl. Surf. Sci.* **2021**, *565*, 150541. [CrossRef]
94. Mehmood, F.; Iqbal, J.; Jan, T.; Gul, A.; Mansoor, Q.; Faryal, R. Structural, Photoluminescence, Electrical, Anti Cancer and Visible Light Driven Photocatalytic Characteristics of Co Doped WO_3 Nanoplates. *Vib. Spectrosc.* **2017**, *93*, 78–89. [CrossRef]
95. Quyen, V.T.; Kim, J.T.; Park, P.M.; Huong, P.T.; Viet, N.M.; Thang, P.Q. Enhanced the Visible Light Photocatalytic Decomposition of Antibiotic Pollutant in Wastewater by Using Cu Doped WO_3. *J. Environ. Chem. Eng.* **2021**, *9*, 104737. [CrossRef]
96. Luxmi, V.; Kumar, A. Enhanced Photocatalytic Performance of m-WO_3 and m-Fe-Doped WO_3 Cuboids Synthesized via Sol-Gel Approach Using Egg Albumen as a Solvent. *Mater. Sci. Semicond. Process.* **2019**, *104*, 104690. [CrossRef]
97. Govindaraj, T.; Mahendran, C.; Marnadu, R.; Shkir, M.; Manikandan, V.S. The Remarkably Enhanced Visible-Light-Photocatalytic Activity of Hydrothermally Synthesized WO_3 Nanorods: An Effect of Gd Doping. *Ceram. Int.* **2021**, *47*, 4267–4278. [CrossRef]
98. Yazdanbakhsh, A.R.; Eslami, A.; Massoudinejad, M.; Avazpour, M. Enhanced Degradation of Sulfamethoxazole Antibiotic from Aqueous Solution Using Mn-WO_3/LED Photocatalytic Process: Kinetic, Mechanism, Degradation Pathway and Toxicity Reduction. *Chem. Eng. J.* **2020**, *380*, 122497. [CrossRef]
99. Zheng, Y.; Chen, G.; Yu, Y.; Zhou, Y.; He, F. Synthesis of Carbon Doped $WO_3 \cdot 0.33H_2O$ Hierarchical Photocatalyst with Improved Photocatalytic Activity. *Appl. Surf. Sci.* **2016**, *362*, 182–190. [CrossRef]
100. Tijani, J.O.; Ugochukwu, O.; Fadipe, L.A.; Bankole, M.T.; Abdulkareem, A.S.; Roos, W.D. Photocatalytic Degradation of Local Dyeing Wastewater by Iodine-Phosphorus Co-Doped Tungsten Trioxide Nanocomposites under Natural Sunlight Irradiation. *J. Environ. Manag.* **2019**, *236*, 519–533. [CrossRef]
101. Han, F.; Li, H.; Fu, L.; Yang, J.; Liu, Z. Synthesis of S-Doped WO_3 Nanowires with Enhanced Photocatalytic Performance towards Dye Degradation. *Chem. Phys. Lett.* **2016**, *651*, 183–187. [CrossRef]
102. Tahir, M.B.; Sagir, M. Carbon Nanodots and Rare Metals (RM = La, Gd, Er) Doped Tungsten Oxide Nanostructures for Photocatalytic Dyes Degradation and Hydrogen Production. *Sep. Purif. Technol.* **2019**, *209*, 94–102. [CrossRef]
103. Ma, T.; Shen, Q.; Xue, B.Z.J.; Guan, R.; Liu, X.; Jia, H.; Xu, B. Facile Synthesis of Fe-Doped g-C_3N_4 for Enhanced Visible-Light Photocatalytic Activity. *Inorg. Chem. Commun.* **2019**, *107*, 107451. [CrossRef]
104. Pham, T.H.; Park, J.W.; Kim, T.Y. Enhanced Photodegradation of Paracetamol from Water by Cobalt Doped Graphitic Carbon Nitride. *Sol. Energy* **2021**, *215*, 151–156. [CrossRef]
105. Pham, T.H.; Jung, S.H.; Kim, T.Y. Enhanced Photodegradation of Toxic Volatile Organic Pollutants Using Ni-Doped Graphitic Carbon Nitride under Natural Solar Light. *Sol. Energy* **2021**, *224*, 18–26. [CrossRef]
106. Tripathi, A.; Narayanan, S. Potassium Doped Graphitic Carbon Nitride with Extended Optical Absorbance for Solar Light Driven Photocatalysis. *Appl. Surf. Sci.* **2019**, *479*, 1–11. [CrossRef]
107. Tri, N.L.M.; Kim, J.; Giang, B.L.; Tahtamouni, T.M.A.; Huong, P.T.; Lee, C.; Viet, N.M.; Trung, D.Q. Ag-Doped Graphitic Carbon Nitride Photocatalyst with Remarkably Enhanced Photocatalytic Activity towards Antibiotic in Hospital Wastewater under Solar Light. *J. Ind. Eng. Chem.* **2019**, *80*, 597–605. [CrossRef]

108. Bawazeer, T.M.; Al-Shehri, B.M.; Alsoufi, M.S.; Shkir, M.; Hamdy, M.S. Solvent-Free Facile Fabrication of Gold Nanoparticles Loaded Carbon Nitride and Their Photocatalytic Performance under Visible Light Illumination. *Optik* **2021**, *241*, 167205. [CrossRef]
109. Li, G.; Wang, B.; Zhang, J.; Wang, R.; Liu, H. Er-Doped g-C3N4 for Photodegradation of Tetracycline and Tylosin: High Photocatalytic Activity and Low Leaching Toxicity. *Chem. Eng. J.* **2020**, *391*, 123500. [CrossRef]
110. Viet, N.M.; Trung, D.Q.; Giang, B.L.; Tri, N.L.M.; Thao, P.; Pham, T.H.; Kamand, F.Z.; Al Tahtamouni, T.M. Noble Metal -Doped Graphitic Carbon Nitride Photocatalyst for Enhancement Photocatalytic Decomposition of Antibiotic Pollutant in Wastewater under Visible Light. *J. Water Process Eng.* **2019**, *32*, 100954. [CrossRef]
111. Xu, J.; Brenner, T.J.K.; Chen, Z.; Neher, D.; Antonietti, M.; Shalom, M. Upconversion-Agent Induced Improvement of g-C_3N_4 Photocatalyst under Visible Light. *ACS Appl. Mater. Interfaces* **2014**, *6*, 16481–16486. [CrossRef]
112. Xiong, T.; Cen, W.; Zhang, Y.; Dong, F. Bridging the g-C_3N_4 Interlayers for Enhanced Photocatalysis. *ACS Catal.* **2016**, *6*, 2462–2472. [CrossRef]
113. Liu, J.; Liang, H.; Li, C.; Bai, J. Construction of V-Doped Graphitic Carbon Nitride with Nanotube Structure for Sustainable Photodegradation of Tetracycline. *Vacuum* **2022**, *204*, 111342. [CrossRef]
114. Zhang, B.; Li, X.; Zhao, Y.; Song, H.; Wang, H. Facile Synthesis of Oxygen Doped Mesoporous Graphitic Carbon Nitride with High Photocatalytic Degradation Efficiency under Simulated Solar Irradiation. *Colloids Surfaces A Physicochem. Eng. Asp.* **2019**, *580*, 123736. [CrossRef]
115. Praus, P.; Smýkalová, A.; Foniok, K.; Novák, V.; Hrbáč, J. Doping of Graphitic Carbon Nitride with Oxygen by Means of Cyanuric Acid: Properties and Photocatalytic Applications. *J. Environ. Chem. Eng.* **2021**, *9*, 105498. [CrossRef]
116. Krumova, K.; Cosa, G.; Aubry, J.; Kanofsky, J.R. Overview of Reactive Oxygen Species Katerina. In *Singlet Oxygen: Applications in Biosciences and Nanosciences*; The Royal Society of Chemistry: Cambridge, UK, 2016; pp. 3–18, ISBN 9781412916837.
117. Wang, S.; He, F.; Zhao, X.; Zhang, J.; Ao, Z.; Wu, H.; Yin, Y.; Shi, L.; Xu, X.; Zhao, C.; et al. Phosphorous Doped Carbon Nitride Nanobelts for Photodegradation of Emerging Contaminants and Hydrogen Evolution. *Appl. Catal. B Environ.* **2019**, *257*, 117931. [CrossRef]
118. Liang, Z.; Ba, G.; Li, H.; Du, N.; Hou, W. Facile Synthesis of Silicon-Doped Polymeric Carbon Nitride with Enhanced Photocatalytic Performance. *J. Alloys Compd.* **2020**, *815*, 152488. [CrossRef]
119. Ganganboina, A.B.; Nguyen, M.D.; Nguyen, T.H.L.; Kuncoro, E.P.; Doong, R.-A. Boron and Phosphorus Co-Doped One-Dimensional Graphitic Carbon Nitride for Enhanced Visible-Light-Driven Photodegradation of Diclofenac. *Chem. Eng. J.* **2021**, *425*, 131520. [CrossRef]
120. Cao, M.; Wang, K.; Tudela, I.; Fan, X. Improve Photocatalytic Performance of g-C_3N_4 through Balancing the Interstitial and Substitutional Chlorine Doping. *Appl. Surf. Sci.* **2021**, *536*, 147784. [CrossRef]
121. Yi, F.; Gan, H.; Jin, H.; Zhao, W.; Zhang, K.; Jin, H.; Zhang, H.; Qian, Y.; Ma, J. Sulfur- and Chlorine-Co-Doped g-C_3N_4 Nanosheets with Enhanced Active Species Generation for Boosting Visible-Light Photodegradation Activity. *Sep. Purif. Technol.* **2020**, *233*, 115997. [CrossRef]
122. Wang, Y.; Zhou, J.; Wang, F.; Xie, Y.; Liu, S.; Ao, Z.; Li, C. Hydrogen Generation from Photocatalytic Treatment of Wastewater Containing Pharmaceuticals and Personal Care Products by Oxygen-Doped Crystalline Carbon Nitride. *Sep. Purif. Technol.* **2022**, *296*, 121425. [CrossRef]
123. Liu, X.; Zhou, Y.; Ma, Y.; Fang, S.; Kong, F.; Pang, X. Photocatalytic Degradation of Dinotefuran by Layered Phosphorus-Doped Carbon Nitride and Its Mechanism. *J. Photochem. Photobiol. A Chem.* **2021**, *414*, 113287. [CrossRef]
124. Pandit, B.B.; Sonawane, S.; Aniruddha, V.P. (Eds.) *Handbook of Nanomaterials for Wastewater Treatment: Fundamental and Scale Up Issues*, 1st ed.; Elsevier: Amsterdam, The Netherlands, 2021; ISBN 9789896540821.
125. Khan, M.M.; Pradhan, D.; Sohn, Y. (Eds.) *Nanocomposites for Visible Light-Induced Photocatalysis*; Springer: Berlin/Heidelberg, Germany, 2017; ISBN 9783319624457.
126. Mugunthan, E.; Saidutta, M.B.; Jagadeeshbabu, P.E. Visible Light Assisted Photocatalytic Degradation of Diclofenac Using TiO_2-WO_3 Mixed Oxide Catalysts. *Environ. Nanotechnol. Monit. Manag.* **2018**, *10*, 322–330. [CrossRef]
127. El-Yazeed, W.S.A.; Ahmed, A.I. Photocatalytic Activity of Mesoporous WO_3/TiO_2 Nanocomposites for the Photodegradation of Methylene Blue. *Inorg. Chem. Commun.* **2019**, *105*, 102–111. [CrossRef]
128. Wang, Q.; Zhang, W.; Hu, X.; Xu, L.; Chen, G.; Li, X. Hollow Spherical WO_3/TiO_2 Heterojunction for Enhancing Photocatalytic Performance in Visible-Light. *J. Water Process Eng.* **2021**, *40*, 101943. [CrossRef]
129. Su, X.; Liu, C.j.; Liu, Y.; Yang, Y.; Liu, X.; Chen, S. Construction of $BiVO_4$ Nanosheets@WO_3 Arrays Heterojunction Photoanodes by Versatile Phase Transformation Strategy. *Trans. Nonferrous Met. Soc. China* **2021**, *31*, 533–544. [CrossRef]
130. Meng, S.; Sun, W.; Zhang, S.; Zheng, X.; Fu, X.; Chen, S. Insight into the Transfer Mechanism of Photogenerated Carriers for WO_3/TiO_2 Heterojunction Photocatalysts: Is It the Transfer of Band-Band or Z-Scheme? Why? *J. Phys. Chem. C* **2018**, *122*, 26326–26336. [CrossRef]
131. Wei, Y.; Huang, Y.; Fang, Y.; Zhao, Y.; Luo, D.; Guo, Q.; Fan, L.; Wu, J. Hollow Mesoporous TiO_2/WO_3 Sphere Heterojunction with High Visible-Light-Driven Photocatalytic Activity. *Mater. Res. Bull.* **2019**, *119*, 110571. [CrossRef]
132. Zhou, Z.; Niu, X.; Zhang, Y.; Wang, J. Janus MoSSe/WSeTe Heterostructures: A Direct Z-Scheme Photocatalyst for Hydrogen Evolution. *J. Mater. Chem. A* **2019**, *7*, 21835–21842. [CrossRef]
133. Cordero-García, A.; Guzmán-Mar, J.L.; Hinojosa-Reyes, L.; Ruiz-Ruiz, E.; Hernández-Ramírez, A. Effect of Carbon Doping on WO_3/TiO_2 Coupled Oxide and Its Photocatalytic Activity on Diclofenac Degradation. *Ceram. Int.* **2016**, *42*, 9796–9803. [CrossRef]

134. Basumatary, B.; Basumatary, R.; Ramchiary, A.; Konwar, D. Evaluation of Ag@TiO_2/WO_3 Heterojunction Photocatalyst for Enhanced Photocatalytic Activity towards Methylene Blue Degradation. *Chemosphere* **2022**, *286*, 131848. [CrossRef] [PubMed]
135. Chen, C.; Liu, X.; Long, H.; Ding, F.; Liu, Q.; Chen, X. Preparation and Photocatalytic Performance of Graphene Oxide/WO_3 Quantum Dots/TiO_2@SiO_2 Microspheres. *Vacuum* **2019**, *164*, 66–71. [CrossRef]
136. Deng, M.; Cao, X.; Li, Z.; Tang, B. Hybrid Triazine-Based g-C_3N_4(001)/ Anatase TiO_2(001) Heterojunction: Insights into Enhanced Photocatalytic Mechanisms Via DFT Calculation. *J. Photochem. Photobiol. A Chem.* **2022**, *423*, 113577. [CrossRef]
137. Barzegar, M.H.; Sabzehmeidani, M.M.; Ghaedi, M.; Avargani, V.M.; Moradi, Z.; Roy, V.A.L.; Heidari, H. S-Scheme Heterojunction g-C_3N_4/TiO_2 with Enhanced Photocatalytic Activity for Degradation of a Binary Mixture of Cationic Dyes Using Solar Parabolic Trough Reactor. *Chem. Eng. Res. Des.* **2021**, *174*, 307–318. [CrossRef]
138. Gahlot, S.; Dappozze, F.; Mishra, S.; Guillard, C. High Surface Area g-C_3N_4 and g-C_3N_4-TiO_2 Photocatalytic Activity under UV and Visible Light: Impact of Individual Component. *J. Environ. Chem. Eng.* **2021**, *9*, 105587. [CrossRef]
139. Li, G.; Nie, X.; Chen, J.; Jiang, Q.; An, T.; Wong, P.K.; Zhang, H.; Zhao, H.; Yamashita, H. Enhanced Visible-Light-Driven Photocatalytic Inactivation of Escherichia Coli Using g-C_3N_4/TiO_2 Hybrid Photocatalyst Synthesized Using a Hydrothermal-Calcination Approach. *Water Res.* **2015**, *86*, 17–24. [CrossRef]
140. Liu, Y.; Zeng, X.; Hu, X.; Hu, J.; Wang, Z.; Yin, Y.; Sun, C.; Zhang, X. Two-Dimensional g-C_3N_4/TiO_2 Nanocomposites as Vertical Z-Scheme Heterojunction for Improved Photocatalytic Water Disinfection. *Catal. Today* **2019**, *335*, 243–251. [CrossRef]
141. Huang, J.; Wang, B.; Hao, Z.; Zhou, Z.; Qu, Y. Boosting Charge Separation and Broadening NIR Light Response over Defected WO_3 Quantum Dots Coupled g-C_3N_4 Nanosheets for Photocatalytic Degrading Antibiotics. *Chem. Eng. J.* **2021**, *416*, 129109. [CrossRef]
142. Cadan, F.M.; Ribeiro, C.; Azevedo, E.B. Improving g-C_3N_4:WO_3 Z-Scheme Photocatalytic Performance under Visible Light by Multivariate Optimization of g-C_3N_4 Synthesis. *Appl. Surf. Sci.* **2021**, *537*, 147904. [CrossRef]
143. Divakaran, K.; Baishnisha, A.; Balakumar, V.; Perumal, K.N.; Meenakshi, C.; Kannan, R.S. Photocatalytic Degradation of Tetracycline under Visible Light Using TiO_2@sulfur Doped Carbon Nitride Nanocomposite Synthesized via In-Situ Method. *J. Environ. Chem. Eng.* **2021**, *9*, 105560. [CrossRef]
144. Dehkordi, A.B.; Badiei, A. Insight into the Activity of TiO_2@nitrogen-Doped Hollow Carbon Spheres Supported on g-C_3N_4 for Robust Photocatalytic Performance. *Chemosphere* **2022**, *288*, 132392. [CrossRef] [PubMed]
145. Abbasi-Asl, H.; Sabzehmeidani, M.M.; Ghaedi, M. Efficient Degradation of Metronidazole Antibiotic by TiO_2/Ag_3PO_4/g–C_3N_4 Ternary Composite Photocatalyst in a Continuous Flow-Loop Photoreactor. *J. Environ. Chem. Eng.* **2021**, *9*, 105963. [CrossRef]
146. Wang, C.; Luo, S.; Liu, C.; Chen, C. WO_3 Quantum Dots Enhanced the Photocatalytic Performances of Graphene Oxide/TiO_2 Films under Flowing Dye Solution. *Inorg. Chem. Commun.* **2020**, *115*, 107875. [CrossRef]
147. Zeng, X.; Wang, Z.; Wang, G.; Gengenbach, T.R.; McCarthy, D.T.; Deletic, A.; Yu, J.; Zhang, X. Highly Dispersed TiO_2 Nanocrystals and WO_3 Nanorods on Reduced Graphene Oxide: Z-Scheme Photocatalysis System for Accelerated Photocatalytic Water Disinfection. *Appl. Catal. B Environ.* **2017**, *218*, 163–173. [CrossRef]
148. Sun, X.; He, W.; Yang, T.; Ji, H.; Liu, W.; Lei, J.; Liu, Y.; Cai, Z. Ternary TiO_2/WO_3/CQDs Nanocomposites for Enhanced Photocatalytic Mineralization of Aqueous Cephalexin: Degradation Mechanism and Toxicity Evaluation. *Chem. Eng. J.* **2021**, *412*, 128679. [CrossRef]
149. Wang, G.; Zhou, F.; Yuan, B.; Xiao, S.; Kuang, A.; Zhong, M.; Dang, S.; Long, X.; Zhang, W. Strain-Tunable Visible-Light-Responsive Photocatalytic Properties of Two-Dimensional Cds/g-C_3N_4: A Hybrid Density Functional Study. *Nanomaterials* **2019**, *9*, 244. [CrossRef] [PubMed]
150. Someswararao, M.V.; Dubey, R.S.; Subbarao, P.S.V. Electrospun Composite Nanofibers Prepared by Varying Concentrations of TiO_2/ZnO Solutions for Photocatalytic Applications. *J. Photochem. Photobiol.* **2021**, *6*, 100016. [CrossRef]
151. Das, A.; Kumar, P.M.; Bhagavathiachari, M.; Nair, R.G. Hierarchical ZnO-TiO_2 Nanoheterojunction: A Strategy Driven Approach to Boost the Photocatalytic Performance through the Synergy of Improved Surface Area and Interfacial Charge Transport. *Appl. Surf. Sci.* **2020**, *534*, 147321. [CrossRef]
152. Cipagauta-Díaz, S.; Estrella-González, A.; Navarrete-Magaña, M.; Gómez, R. N Doped-TiO_2 Coupled to $BiVO_4$ with High Performance in Photodegradation of Ofloxacin Antibiotic and Rhodamine B Dye under Visible Light. *Catal. Today* **2021**, *394*, 445–457. [CrossRef]
153. Sharma, S.; Kumar, N.; Mari, B.; Chauhan, N.S.; Mittal, A.; Maken, S.; Kumari, K. Solution Combustion Synthesized TiO_2/Bi_2O_3/CuO Nano-Composites and Their Photocatalytic Activity Using Visible LEDs Assisted Photoreactor. *Inorg. Chem. Commun.* **2021**, *125*, 108418. [CrossRef]
154. Chawla, H.; Chandra, A.; Ingole, P.P.; Garg, S. Recent Advancements in Enhancement of Photocatalytic Activity Using Bismuth-Based Metal Oxides Bi_2MO_6 (M = W, Mo, Cr) for Environmental Remediation and Clean Energy Production. *J. Ind. Eng. Chem.* **2021**, *95*, 1–15. [CrossRef]
155. Munguti, L.; Dejene, F. Influence of Annealing Temperature on Structural, Optical and Photocatalytic Properties of ZnO–TiO_2 Composites for Application in Dye Removal in Water. *Nano-Struct. Nano-Objects* **2020**, *24*, 100594. [CrossRef]
156. Bouziani, A.; Park, J.; Ozturk, A. Synthesis of α-Fe_2O_3/TiO_2 Heterogeneous Composites by the Sol-Gel Process and Their Photocatalytic Activity. *J. Photochem. Photobiol. A Chem.* **2020**, *400*, 112718. [CrossRef]

157. Peña-Velasco, G.; Hinojosa-Reyes, L.; Morán-Quintanilla, G.A.; Hernández-Ramírez, A.; Villanueva-Rodríguez, M.; Guzmán-Mar, J.L. Synthesis of Heterostructured Catalyst Coupling MOF Derived Fe_2O_3 with TiO_2 for Enhanced Photocatalytic Activity in Anti-Inflammatory Drugs Mixture Degradation. *Ceram. Int.* **2021**, *47*, 24632–24640. [CrossRef]
158. Li, Y.; Yang, B.; Liu, B. MOF Assisted Synthesis of $TiO_2/Au/Fe_2O_3$ Hybrids with Enhanced Photocatalytic Hydrogen Production and Simultaneous Removal of Toxic Phenolic Compounds. *J. Mol. Liq.* **2021**, *322*, 114815. [CrossRef]
159. Lu, F.; Chen, K.; Feng, Q.; Cai, H.; Ma, D.; Wang, D.; Li, X.; Zuo, C.; Wang, S. Insight into the Enhanced Magnetic Separation and Photocatalytic Activity of Sn-Doped TiO_2 Core-Shell Photocatalyst. *J. Environ. Chem. Eng.* **2021**, *9*, 105840. [CrossRef]
160. Liu, N.; Ming, J.; Sharma, A.; Sun, X.; Kawazoe, N.; Chen, G.; Yang, Y. Sustainable Photocatalytic Disinfection of Four Representative Pathogenic Bacteria Isolated from Real Water Environment by Immobilized TiO_2-Based Composite and Its Mechanism. *Chem. Eng. J.* **2021**, *426*, 131217. [CrossRef]
161. Mohammed, M.K.A. Sol-Gel Synthesis of Au-Doped TiO_2 Supported SWCNT Nanohybrid with Visible-Light-Driven Photocatalytic for High Degradation Performance toward Methylene Blue Dye. *Optik* **2020**, *223*, 165607. [CrossRef]
162. Thomas, C.-T.; Manisekaran, R.; Santoyo-Salazar, J.; Schoefs, B.; Velumani, S.; Castaneda, H.; Jantrania, A. Graphene Oxide Decorated TiO_2 and $BiVO_4$ Nanocatalysts for Enhanced Visible-Light-Driven Photocatalytic Bacterial Inactivation. *J. Photochem. Photobiol. A Chem.* **2021**, *418*, 113374. [CrossRef]
163. Huang, C.; Peng, B. Photocatalytic Degradation of Patulin in Apple Juice Based on Nitrogen-Doped Chitosan-TiO_2 Nanocomposite Prepared by a New Approach. *Lwt* **2021**, *140*, 110726. [CrossRef]
164. Wang, X.; Wang, X.; Zhao, J.; Song, J.; Wang, J.; Ma, R.; Ma, J. Solar Light-Driven Photocatalytic Destruction of Cyanobacteria by F-Ce-TiO_2/Expanded Perlite Floating Composites. *Chem. Eng. J.* **2017**, *320*, 253–263. [CrossRef]
165. Asadi, A.M.S.; Malakootian, M.; Kowsari, E.; Alidadi, H. Ionic Liquid-Assisted Sol-Gel Synthesis of Fe_2O_3-TiO_2 for Enhanced Photocatalytic Degradation of Bisphenol a under UV Illumination: Modeling and Optimization Using Response Surface Methodology. *Optik* **2020**, *204*, 164229. [CrossRef]
166. Orge, C.A.; Soares, O.S.G.P.; Faria, J.L.; Pereira, M.F.R. Synthesis of TiO_2-Carbon Nanotubes through Ball-Milling Method for Mineralization of Oxamic Acid (OMA) by Photocatalytic Ozonation. *J. Environ. Chem. Eng.* **2017**, *5*, 5599–5607. [CrossRef]
167. Chávez, A.M.; Quiñones, D.H.; Rey, A.; Beltrán, F.J.; Álvarez, P.M. Simulated Solar Photocatalytic Ozonation of Contaminants of Emerging Concern and Effluent Organic Matter in Secondary Effluents by a Reusable Magnetic Catalyst. *Chem. Eng. J.* **2020**, *398*, 125642. [CrossRef]
168. Asgari, E.; Farzadkia, M.; Esrafili, A.; Badi, M.Y.; Jokandan, S.F.; Sobhi, H.R. Application of a Photocatalytic Ozonation Process Using TiO_2 Magnetic Nanoparticles for the Removal of Ceftazide from Aqueous Solutions: Evaluation of Performance, Comparative Study and Mechanism. *Optik* **2020**, *212*, 164667. [CrossRef]
169. Chávez, A.M.; Solís, R.R.; Beltrán, F.J. Magnetic Graphene TiO_2-Based Photocatalyst for the Removal of Pollutants of Emerging Concern in Water by Simulated Sunlight Aided Photocatalytic Ozonation. *Appl. Catal. B Environ.* **2020**, *262*, 118275. [CrossRef]
170. Liu, X.; Xu, J.; Ni, Z.; Wang, R.; You, J.; Guo, R. Adsorption and Visible-Light-Driven Photocatalytic Properties of Ag_3PO_4/WO_3 Composites: A Discussion of the Mechanism. *Chem. Eng. J.* **2019**, *356*, 22–33. [CrossRef]
171. Tang, J.; Meng, R.; Xue, Y.; Zhang, S.; Li, Q. Fabrication of a Novel $Ag_3PO_4/WO_3{\cdot}H_2O$ Composite with Enhanced Visible Light Photocatalytic Performance for the Degradation of Methylene Blue and Oxytetracycline. *Inorg. Chem. Commun.* **2021**, *132*, 108792. [CrossRef]
172. Li, H.; Zhang, Y.; Zhang, Q.; Wang, Y.; Fan, Y.; Gao, X.; Niu, J. Boosting Visible-Light Photocatalytic Degradation of Indomethacin by an Efficient and Photostable $Ag_3PO_4/NG/WO_3$ Composites. *Appl. Surf. Sci.* **2019**, *490*, 481–491. [CrossRef]
173. Hernández-Moreno, E.J.; Martínez de la Cruz, A.; Hinojosa-Reyes, L.; Guzmán-Mar, J.; Gracia-Pinilla, M.A.; Hernández-Ramírez, A. Synthesis, Characterization, and Visible Light–Induced Photocatalytic Evaluation of $WO_3/NaNbO_3$ Composites for the Degradation of 2,4-D Herbicide. *Mater. Today Chem.* **2021**, *19*, 100406. [CrossRef]
174. Wei, L.; Zhang, H.; Cao, J. Electrospinning of $Ag/ZnWO_4/WO_3$ Composite Nanofibers with High Visible Light Photocatalytic Activity. *Mater. Lett.* **2019**, *236*, 171–174. [CrossRef]
175. Rong, J.; Zhang, T.; Qiu, F.; Rong, X.; Zhu, X.; Zhang, X. Preparation of Hierarchical Micro/Nanostructured Bi_2S_3-WO_3 composites for Enhanced Photocatalytic Performance. *J. Alloys Compd.* **2016**, *685*, 812–819. [CrossRef]
176. Govindaraj, T.; Mahendran, C.; Manikandan, V.S.; Archana, J.; Shkir, M.; Chandrasekaran, J. Fabrication of WO_3 Nanorods/RGO Hybrid Nanostructures for Enhanced Visible-Light-Driven Photocatalytic Degradation of Ciprofloxacin and Rhodamine B in an Ecosystem. *J. Alloys Compd.* **2021**, *868*, 159091. [CrossRef]
177. Elhakim, A.A.; El-Kemary, M.; Ibrahim, M.M.; El-Mehasseb, I.M.; El-Sheshtawy, H.S. Direct Z-Scheme of WO_3/GO Decorated with Silver Nanoparticles for Synergetic Adsorption and Photocatalytic Activity for Organic and Inorganic Water Pollutants Removal. *Appl. Surf. Sci.* **2021**, *564*, 150410. [CrossRef]
178. Gondal, M.A.; Suliman, M.A.; Dastageer, M.A.; Chuah, G.K.; Basheer, C.; Yang, D.; Suwaiyan, A. Visible Light Photocatalytic Degradation of Herbicide (Atrazine) Using Surface Plasmon Resonance Induced in Mesoporous Ag-WO_3/SBA-15 Composite. *J. Mol. Catal. A Chem.* **2016**, *425*, 208–216. [CrossRef]
179. Yan, L.; Hou, J.; Li, T.; Wang, Y.; Liu, C.; Zhou, T.; Jiang, W.; Wang, D.; Che, G. Tremella-like Integrated Carbon Nitride with Polyvinylimine-Doped for Enhancing Photocatalytic Degradation and Hydrogen Evolution Performances. *Sep. Purif. Technol.* **2021**, *279*, 119766. [CrossRef]

180. Zeng, X.; Liu, Y.; Xia, Y.; Uddin, M.H.; Xia, D.; McCarthy, D.T.; Deletic, A.; Yu, J.; Zhang, X. Cooperatively Modulating Reactive Oxygen Species Generation and Bacteria-Photocatalyst Contact over Graphitic Carbon Nitride by Polyethylenimine for Rapid Water Disinfection. *Appl. Catal. B Environ.* **2020**, *274*, 119095. [CrossRef]
181. Alias, N.H.; Jaafar, J.; Samitsu, S.; Yusof, N.; Othman, M.H.D.; Rahman, M.A.; Ismail, A.F.; Aziz, F.; Salleh, W.N.W.; Othman, N.H. Photocatalytic Degradation of Oilfield Produced Water Using Graphitic Carbon Nitride Embedded in Electrospun Polyacrylonitrile Nanofibers. *Chemosphere* **2018**, *204*, 79–86. [CrossRef]
182. Jing, L.; Xu, Y.; Liu, J.; Zhou, M.; Xu, H.; Xie, M.; Li, H.; Xie, J. Direct Z-Scheme Red Carbon Nitride/Rod-like Lanthanum Vanadate Composites with Enhanced Photodegradation of Antibiotic Contaminants. *Appl. Catal. B Environ.* **2020**, *277*, 119245. [CrossRef]
183. Dong, S.; Lee, G.J.; Zhou, R.; Wu, J.J. Synthesis of g-C_3N_4/$BiVO_4$ Heterojunction Composites for Photocatalytic Degradation of Nonylphenol Ethoxylate. *Sep. Purif. Technol.* **2020**, *250*, 117202. [CrossRef]
184. Hu, H.; Kong, W.; Wang, J.; Liu, C.; Cai, Q.; Kong, Y.; Zhou, S.; Yang, Z. Engineering 2D Compressed Layered G-C3N4 Nanosheets by the Intercalation of $BiVO_4$-Bi_2WO_6 Composites for Boosting Photocatalytic Activities. *Appl. Surf. Sci.* **2021**, *557*, 149796. [CrossRef]
185. Zhang, H.; Wei, W.; Hou, F.; Guo, W.; Zhang, Q.; Wang, T.; Wei, A. Visible-Light-Driven Photocatalytic Disinfection of Aspergillus Fumigatus under Germinative Biomorph: Efficiency and Mechanism. *Sep. Purif. Technol.* **2021**, *265*, 1–10. [CrossRef]
186. Iqbal, S.; Bahadur, A.; Javed, M.; Hakami, O.; Irfan, R.M.; Ahmad, Z.; AlObaid, A.; Al-Anazy, M.M.; Baghdadi, H.B.; Abd-Rabboh, H.S.M.; et al. Design Ag-Doped ZnO Heterostructure Photocatalyst with Sulfurized Graphitic C_3N_4 Showing Enhanced Photocatalytic Activity. *Mater. Sci. Eng. B Solid-State Mater. Adv. Technol.* **2021**, *272*, 115320. [CrossRef]
187. Zhang, C.; Zhang, M.; Li, Y.; Shuai, D. Visible-Light-Driven Photocatalytic Disinfection of Human Adenovirus by a Novel Heterostructure of Oxygen-Doped Graphitic Carbon Nitride and Hydrothermal Carbonation Carbon. *Appl. Catal. B Environ.* **2019**, *248*, 11–21. [CrossRef]
188. Zhou, X.; Zhang, G.; Shao, C.; Li, X.; Jiang, X.; Liu, Y. Fabrication of g-C_3N_4/SiO2-Au Composite Nanofibers with Enhanced Visible Photocatalytic Activity. *Ceram. Int.* **2017**, *43*, 15699–15707. [CrossRef]
189. Akulinkin, A.; Bolgaru, K.; Reger, A. Facile Synthesis of Porous g-C_3N_4/β-SiAlON Material with Visible Light Photocatalytic Activity. *Mater. Lett.* **2021**, *305*, 130788. [CrossRef]
190. Jourshabani, M.; Dominic, J.A.; Achari, G.; Shariatinia, Z. Synergetic Photocatalytic Ozonation Using Modified Graphitic Carbon Nitride for Treatment of Emerging Contaminants under UVC, UVA and Visible Irradiation. *Chem. Eng. Sci.* **2019**, *209*, 115181. [CrossRef]
191. Wang, T.; Sun, M.; Sun, H.; Shang, J.; Wong, P.K. Efficient Z-Scheme Visible-Light-Driven Photocatalytic Bacterial Inactivation by Hierarchical MoS_2-Encapsulated Hydrothermal Carbonation Carbon Core-Shell Nanospheres. *Appl. Surf. Sci.* **2019**, *464*, 43–52. [CrossRef]
192. Yin, J.; Liao, G.; Zhu, D.; Lu, P.; Li, L. Photocatalytic Ozonation of Oxalic Acid by g-C_3N_4/Graphene Composites under Simulated Solar Irradiation. *J. Photochem. Photobiol. A Chem.* **2016**, *315*, 138–144. [CrossRef]
193. Ling, Y.; Liao, G.; Feng, W.; Liu, Y.; Li, L. Excellent Performance of Ordered Ag-g-C_3N_4/SBA-15 for Photocatalytic Ozonation of Oxalic Acid under Simulated Solar Light Irradiation. *J. Photochem. Photobiol. A Chem.* **2017**, *349*, 108–114. [CrossRef]

catalysts

MDPI

Review

Photochemical Water Splitting via Transition Metal Oxides

Fateh Mikaeili, Tessa Gilmore * and Pelagia-Iren Gouma

Department of Material Science and Engineering, The Ohio State University, Columbus, OH 43210, USA
* Correspondence: gilmore.281@osu.edu

Abstract: Rapid population growth and ever-increasing energy consumption have resulted in increased environmental pollution and energy demands in recent years. Accordingly, studies and research on innovative and efficient ways of wastewater clean-up and exploiting eco-friendly and renewable energy sources such as sunlight have become a necessity. This review focuses on recent progress with photocatalysis for water splitting capabilities. It introduces photocatalysis and hydrogen as a fuel source, before moving on to explain water splitting. Then, the criteria for ideal photocatalytic materials are discussed along with current material systems and their limitations. Finally, it concludes on the TiO_2 systems and their potential in future photocatalysis research.

Keywords: photocatalysis; photochemical; water splitting; hydrogen; TiO_2

1. Introduction to Photocatalysis

Within one hour, the sun provides more energy to the surface of the earth than the human population needs for one year. Considering the ever-increasing energy demands and concerning environmental issues, the pursuit of such an enormous source of eco-friendly and renewable energy is more important now than ever [1,2]. Therefore, a significant number of current projects have been dedicated to tackling the challenges of making chemical fuel from sunlight.

Solar-to-chemical energy conversion provides a beneficial way to store solar energy that is both sustainable and efficient. The question then arises as to how we split water molecules in order to generate hydrogen gas for fuel. The first instance of water splitting through hydrolysis was reported around 1800, and since then numerous studies of alternative ways to achieve this reaction were reported. Unfortunately, the capital cost of mining hydrogen from water remained higher than the present-day industrial process of methane steam reforming [3]. However, with the increased pressure from government and independent groups for renewable and efficient energy, hydrogen generation through water splitting has been given more focus.

When water comes into contact with sunlight within a photocatalyst platform, it dissociates to produce a clean energy source of hydrogen without further energy requirements from the photocatalytic process [4,5]. Researchers have suggested and investigated various systems for this photocatalytic water splitting process [6], some of which will be discussed in this review. The fundamental elements related to efficient hydrogen generation through one-step photochemical water splitting are first examined. Then, the essential steps of a photochemical reaction, along with chemical, electronic, and physical criteria for the selection of photocatalytic material, and an overview of the current material systems is reviewed. Finally, this review will discuss the TiO_2 systems and their potential in further photocatalytic research.

2. Hydrogen as Fuel and Its Production

The smallest and lightest element in the universe is Hydrogen. Its unique chemical and physical properties offer both advantages and challenges in developing it as a prevalent fuel source. The uniqueness of hydrogen compared to other chemical fuels comes from its high diffusivity, inimitability, combustibility, very low viscosity, and electrochemical properties [7]. All of these characteristics together make hydrogen different or more favorable than other gaseous fuels. Moreover, it has been reported that hydrogen engines perform more efficiently than gasoline engines. This is due to the low autoignition temperature of hydrogen as well as the higher octane rating compared to conventional gasoline. Hydrogen also has the highest heating value, which is 52,000 Btu/lb, among all the available fuels [8].

Fossil fuels are limited in the ways they can be mined or extracted. The advantage of hydrogen is its abundant different sources such as oil, coal, natural gas (NG), biomass, water, etc. Currently, the primary source of hydrogen is NG due to the low production cost and high efficiency. Figure 1 below illustrates the current most notable sources of hydrogen for industrial use. Hydrogen is largely generated from NG (48%), raw petroleum products (30%), coal (18%), and electrolysis of water (4%) [9].

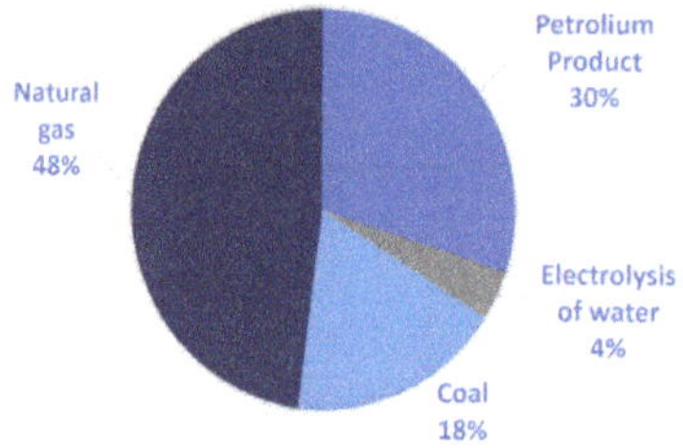

Figure 1. Percentage of production of hydrogen from petroleum products, natural gases, coal, and electrolysis of water. Values taken from [9].

Almost 96% of today's industrial hydrogen production is sourced from fossil fuels. In truth, the current hydrogen fuel is simply relying on another byproduct of non-renewable fuel sources and would ultimately cost more than any of the initial forms of the fossil fuels themselves. However, petroleum supplies will become limited in the near future [10]; therefore, it is critical for renewable hydrogen sources to gain attention and support.

3. Water Splitting

As mentioned in the previous section, water could be used as the main source of hydrogen production. By definition, the dissociation of the water molecules into their constituents (hydrogen and oxygen) is known as water splitting (Reaction 1). This phenomenon was first observed around the 1860s through water electrolysis, where an electric current was passed through water to complete the water-splitting reaction. Production of hydrogen from water is highly energy-demanding. There are many different proposed systems to use other renewable energy sources, such as hydropower, wind turbines, photovoltaic cells, and so on, to provide the electrical energy required for water hydrolysis. However, the electricity that is consumed is more valuable than the hydrogen that is produced. To fully understand the energy requirements for the water-splitting reaction, the thermodynamics and electrochemistry of the process will be discussed [11].

$$2H_2O \rightarrow 2H_2 + O_2 \quad (1)$$

3.1. *Thermodynamics and Electrochemistry of Water Splitting*

As previously mentioned, dissociation of a water molecule is a reaction that is energetically uphill. Typically, and without considering the overpotential, it requires 1.23 V per one mole of water for the complete splitting of water into oxygen and hydrogen. Due to the very low ionization power ($Kw = 1.0 \times 10^{-14}$), water splitting becomes thermodynamically unfavorable (Gibbs free energy LG0 = 237 kJ/mol, 2.46 eV per molecule) at standard temperature and pressure [9]. LG is calculated at 25 °C using the thermodynamic parameters (ΔH, ΔT, and ΔS) required for the water-splitting process:

$$\Delta G = \Delta H - T\Delta S = 285.83\ kJ - 48.7\ kJ = 237.13\ kJ \quad (2)$$

With the obtained value of the Gibbs free energy and using the help of the Nernst Equation, the standard cell potential E° of the reaction could be obtained (Equation (3)).

$$E^{\circ} = -\frac{\Delta G^{\circ}}{zF}. \quad (3)$$

In the above equation, z is the number of electrons transferred in the reaction (in this case, two), and F is a proportionality constant and is in the Faraday units (96,485 C/mol). Using the Nernst equation mentioned above, the standard potential of the water electrolysis can be calculated as 1.229 V at 25 °C. This cell potential belongs to the difference in potentials of the two half-cell reactions occurring at the cathode (reduction; hydrogen evolution reaction (HER)) and anode (oxidation; oxygen evolution reaction (OER)). The Nernst equations for the half-cell reactions of water splitting are mentioned below:

$$\text{Anode (oxidation): } 4OH^- + 4h^+ \rightarrow O_2 + 2H_2O \quad E = 1.23\ V \text{ versus NHE} \quad (4)$$

$$\text{Cathode (reduction): } 2H_2O + 2e^- \rightarrow H_2 + 2OH^- \quad E = 0.00\ V \text{ versus NHE.} \quad (5)$$

Leading to an overall reaction of the following:

$$2H_2O(l) + 4e^- + 4h^+ \rightarrow O_2(g) + 2H_2(g) \quad E = 1.23\ V \quad (6)$$

where NHE is the normal hydrogen electrode.

3.2. Solar Water Splitting

The above calculations illustrate how thermodynamically unfavorable this reaction is; however, nature shows us that it has been done for millions of years in plant leaves through photosynthesis—a process in which the plant splits absorbed water using sunlight and transforms it into fuels in the form of hydrocarbons. Therefore, a reasonable platform would be the combination of solar energy with the plentiful water resources available to us. This platform is called solar water splitting and is generally completed by photobiological, thermochemical, or photocatalytic water splitting. Table 1 below contains a summary of each method.

Table 1. Summary of solar water splitting methods.

Solar Water Splitting Methods	Description	Comments
Thermochemical	Uses high temperature—from concentrated solar power and chemical reactions.	The simplest method. The requirements for large solar concentrators make this method very expensive [12].
Photobiological	A process in which light is used in biological systems to dissociate water into molecular oxygen and hydrogen.	Low yields of hydrogen production, toxic effects of enzymes, limitations on scaling up [13].
Photocatalytic	Theoretically, only light energy, water, and catalyst are needed.	Low cost, relatively higher solar H_2 efficiency, the capability of separating H_2 and O_2, and flexible reactor size.

4. Photocatalytic Water Splitting

Photocatalytic water splitting uses sunlight, water, and a semiconducting photocatalyst to dissociate water molecules through the two redox reactions mentioned above. The breakthrough study was started in 1972 by Fujishima and Honda [14] in a photoelectrochemical cell using TiO_2 as their photocatalyst. Afterwards, photocatalytic water splitting received an enormous amount of attention due to its potential. During the past 40 years, various photocatalyst materials and systems were used to split water under ultraviolet light or visible light. Photocatalytic water splitting could be categorized into either photochemical water splitting or photoelectrochemical (PEC) water splitting [15].

The schematics of the two different systems are summarized in Figures 2 and 3 below. Both types include three basic steps: a semiconductor photocatalyst absorbs more photon energy than the band gap energy of the photocatalyst and excites the electron–hole pairs; the photogenerated charge carriers are then separated out and move toward different sites of the photocatalyst's surface; finally, at these sites water reacts with the charge carriers in two separate redox reactions and therefore is reduced by electrons to produce H_2 in the same time its oxidized by holes to produce O_2 [16].

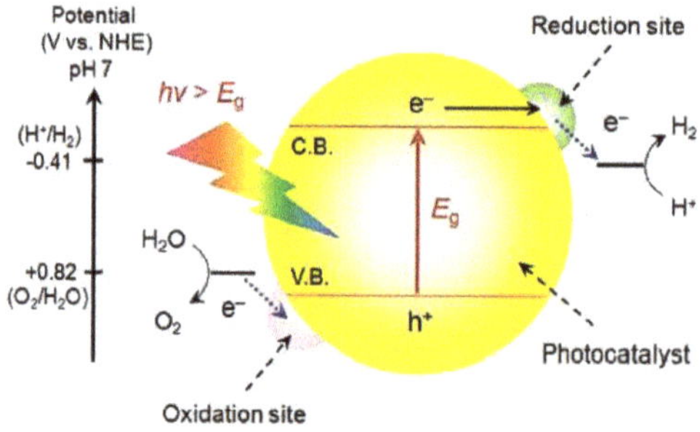

Figure 2. Conventional one-step system (photochemical water splitting). Reprinted with permission from [16]. Copyright 2022 American Chemical Society.

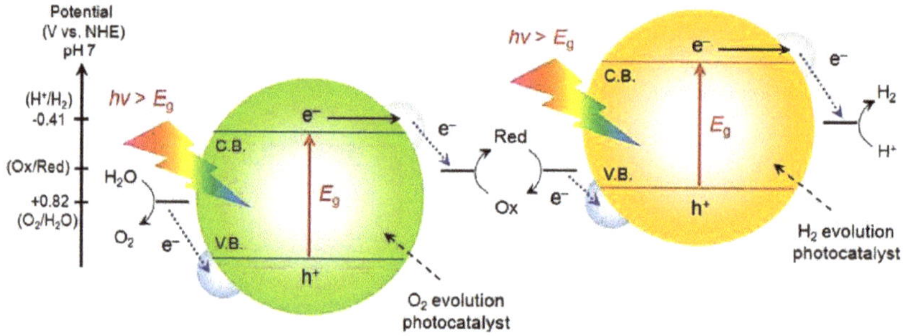

Figure 3. Two-step photoexcitation system with two photocatalysts in tandem (PEC water splitting). Reprinted with permission from [16]. Copyright 2022 American Chemical Society.

Even though the general concept of photochemical and PEC systems is the same, the setup is different. In photochemical reactions, the water-splitting reaction takes place at the semiconductor–electrolyte junction, whereas in a PEC setup the reaction takes place at two different sites. In this method, illuminating the cathode or anode would provide the required potential [15].

In order to understand the difference between these two setups, an important characteristic of the semiconductor should be taken into consideration. This characteristic is the band edge position of the semiconductor [17]. A suitable semiconductor for water splitting has a valance band position that is more positive than the O_2/H_2O energy level

(1.23–0.059 pH, V versus NHE) and a conduction band position that is more negative than the H^+/H_2 energy level (0–0.059 pH, V versus NHE). In other words, in the ideal case, a single semiconductor material should have a band gap that is large enough to split water, so the conduction band energy and valance band energy should straddle the electrochemical potentials E0 (H^+/H_2) and E0 (O_2/H_2O) [18]. However, in the case of a single semiconducting material, the second requirement is not satisfied in most of the material systems (as will be discussed in depth in further sections). Scaife [19] mentioned in 1980 that it is exceptionally difficult to find a single semiconductor photocatalyst with both characteristics. This difficulty is why many studies focus on two semiconductor photocatalytic systems (PEC water splitting). By using two different materials, each one will act as either a photoanode or photocathode, which when used in tandem, satisfies the band gab requirement.

As seen in Figure 3, a two-step PEC system involves water splitting in two parts: one for the hydrogen evolution using a semiconductor that satisfies the conduction band position for that reaction, and the other for the O_2 evolution. In this method, a semiconductor that only partially satisfies the band edge position for the redox reaction could still be used in conjunction with another semiconductor to facilitate a water-splitting reaction [20]. However, there are many drawbacks to and critics of this method.

First, it requires the number of photons to be double that for the one-step system to achieve overall water splitting. The number of photons required in two-step photocatalytic water splitting is eight, whereas in one-step overall water splitting it is four. This difference causes the amount of hydrogen and oxygen produced in a two-step process to be half that of the one-step process at light absorption values and apparent quantum yield of unity [21].

Second, there are still some drawbacks that involve promoting electron transfer between two semiconductors and opposing and suppressing the possible backward reactions that involve shuttle redox mediators [22]. Therefore, since the number of backward electron-transfer routes increases, which is the result of an increase in the number of elementary steps, this route is kinetically unfavorable. Overall, two different studies summarized the techno-economic analyses which determined that a high capital cost prevents PEC devices from being implemented into solar hydrogen production. For these reasons, we will focus on one-step photochemical water-splitting systems.

5. Photochemical Water Splitting

In order to understand the photocatalytic process in photochemical water splitting (Figure 4), which is a quite complicated process, a simple step-by-step description of the process will be discussed in this section. Overall, the following steps occur:

1. Photon absorption: Photocatalysts absorb photons and generate electrons and holes at the surface. When the material absorbs the photons with an energy that is equal to or more than the band gap energy of the semiconductor, an electron jumps from the valance band to the conduction band, leaving a hole in the valance band. Electrons and holes release energy (heat) and move the conduction and valance bands to the minimum and the maximum positions, respectively [23].
2. Charge transport: After the charge carriers have been excited, there are different scenarios that could occur. The first scenario, which is highly unfavorable, is that excited-state conduction band electrons and valance band holes recombine. In the case where there is not a suitable force to separate these charged carriers, the energy stored in them will dissipate in a very short time (typically a couple of nanoseconds) in the form of heat. In the event that there is a defect state, a trap on the surface of the material, or a suitable scavenger, the recombination is potentially avoided. Another scenario is that the excited electrons and holes will move to the respective reaction sites [24]. In bulk, if the carriers do not recombine, it is only possible for either an electron or hole to be accumulated at the anode or cathode, whereas in a nanostructured semiconductor both of the photogenerated charge carriers could be present at the same surface. Low dimensionality, few numbers of defects, and a high

surface area in nanostructured materials result in key differences in electron transport compared to their bulk counterparts [25,26].

3. After the charges have moved to the reaction sites in the material/water interface, they can participate in the surface chemical reactions between these carriers and the compounds (e.g., water) [15].

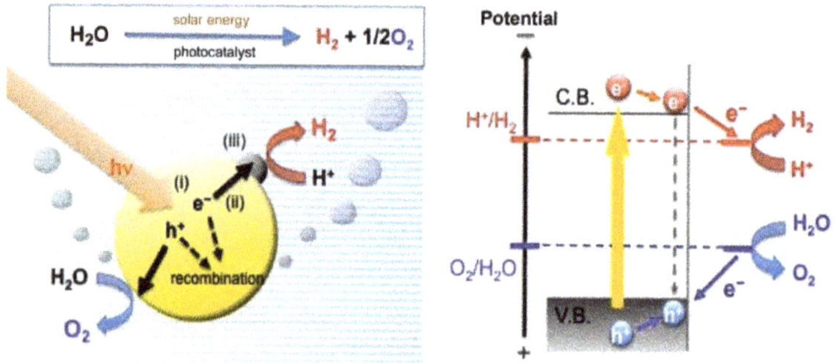

Figure 4. Schematics of photochemical water splitting [27]. Figure reused with permission from https://pubs.acs.org/doi/10.1021/cr1002326 (accessed on 1 September 2022); further permission related to this article should be directed to ACS.

6. Criteria for Selection and Synthesis of an Ideal Photocatalytic Material

As mentioned in the previous section, there are many different factors and steps in the photochemical water-splitting process. Therefore, the development of photocatalytic materials that can maximize the conversion of solar energy to solar fuel demands significant consideration. At the molecular level, several physicochemical functions need to be integrated into one stable chemical system that can set the criteria, which must be satisfied simultaneously.

6.1. Chemical and Electronic Properties of the Materials

1. The band gap of the semiconducting material should lie between 1.6 eV (1.23 eV + overpotential) and 2.7 eV (larger than 2.43 eV). It is known that only 4% of the sunlight on earth is UV. Thus, in order to achieve maximum efficiency, the material should be able to harvest within the visible light spectrum [28].
2. Band edge positions mean that band edges must straddle between the redox potentials of H_2O (0.00 eV and 1.23 eV), as illustrated in Figure 4. Semiconductor materials must satisfy the minimum band gap requirement (~1.4 eV) [29]. Although recently a couple of different ways of determining the valance band edge position of new material systems have been proposed, the task of predicting the exact band position of a new photocatalyst is arduous. For this reason, the applicability of a specific photocatalyst material for overall water splitting is unknown in advance [30,31].
3. Charge transfer is necessary at the photocatalytic surface and must be fast enough to prevent photo-corrosion and shifting of the band edges. This prevention avoids the recombination of the charge carriers and can further provide efficient oxidation and reduction sites on the surface of the material. A prevalent approach to hinder the recombination process is the use of a cocatalyst. Cocatalysts have been used in many investigations on hydrogen generation; however, some cocatalysts are highly active and induce a reverse reaction, i.e., the generation of water from molecular oxygen and hydrogen, which must be reduced [17].
4. The stability of the material in an aqueous medium is an essential requirement (at least for 20 years). Many investigated material systems are susceptible to photo-corrosion such as CdS. Although high efficiencies are reported in the early stages of experiments,

due to oxidation of the material the process deteriorates over time before eventually stopping [32].

6.2. Physical and Crystal Structure Properties of the Material

Aside from the electronic and chemical properties that were mentioned above, there are structural factors of semiconductors that play a significant role in the final efficiency of the photochemical water-splitting process. Recent capabilities in advanced characterization enabled material scientists to investigate these factors, some of which are briefly discussed below:

1. Crystalline phases: Semiconductors with different polymorphs have been shown to have different photocatalytic overall water-splitting features. One famous example of this is TiO_2, which has the three main polymorphs, anatase, rutile, and brookite [33,34]. Initially, it was proposed that overall water splitting could only occur using the rutile phase [35,36]. However, it was later discovered that the reaction is also feasible when anatase or brookite are treated with continued UV irradiation [37]. Early studies using infrared absorption–excitation energy scanning spectroscopy revealed that there are numerous trapped states close to the valance bands of anatase and brookite but not rutile. Later, it was concluded that the elongated emission of UV on these two phases promotes the phase transformation to a quasi-rutile structure that elevates water-splitting reactions. On the other hand, a recent experiment by the Akira group in 2017 [38] illustrated the opposite conclusion, that brookite should have better photocatalytic activities, by studying the depth of electron traps in the three different polymorphs. The group concluded that brookite has moderate trap levels that both preserves the reactivity of electrons and hinders the recombination process of the charge carriers. These inconsistent results and conclusions show that there is a lack of fundamental understanding of the different behavior of the photocatalytic activity of a semiconductor with different polymorphs.
2. Crystallinity: A semiconductor with a high degree of crystallization has fewer structural imperfections such as vacancies and dislocations. These defects are known to be recombination sites for photogenerated charge carriers. Therefore, it is commonly suggested that a highly crystalline semiconductor has a lower electron–hole pair recombination rate [39]. However, in two distinct studies in 2019 published in the *Nature Materials* journal, two different conclusions have been derived. Wang et al. [40] studied the overall water splitting of single-crystal Ta_3N_5 nanorods with no surface defects and grain boundaries and compared the water-splitting efficiency of the same material system with different nanorods with relatively more defects. Their result demonstrates that the hydrogen production rate is almost three times higher in the case of nanowires with no defects. In another study by Li et al. in 2019 [41], N-doped titania samples with differing amounts of oxygen vacancies on the surface were studied. From this experiment, it was demonstrated that by increasing the oxygen vacancies on the surface by close to 13 times, these oxygen vacancies acted as electron traps and increased the final efficiency of hydrogen evolution by nearly two times. These two different conclusions show that, although it is known that defects act as recombination sites, it is possible that different types of defects could have different effects in the lifetime of photogenerated carriers that needs further investigation.
3. Particle size and morphology: As one might expect, the particle size and morphology of the photocatalyst can have profound effects on the performance of the material [42]. In the case of particle size, the literature suggests that there is a degree of compromise that occurs [43]. That is, small particle sizes facilitate the diffusion of electron and hole pairs to the surface of the semiconductor, which results in a lower recombination rate. However, when the particle is too small, insufficient absorption of light can occur due to a lower surface area [44]. Therefore, for any particular material system, it is critical to optimize the effect of particle size. Morphology has a significant effect on determining the anisotropic distribution of charge within a semiconductor and

the overall efficiency of the water-splitting process. For example, La-doped $NaTaO_3$ demonstrated the presence of different spatially resolved reaction sites where the edges and grooves acted as reduction and oxidation sites, respectively [45]. Unfortunately, there still exists a number of studies which do not specifically address the effect of morphology on the performance, such as of ultra-aspect ratio nanofibers and nanoribbons.

4. Heterostructures and cocatalysts: As mentioned previously, the cocatalyst is another important component that is widely used in most of the literature on overall water splitting in a single system. The literature states that the main purpose of the cocatalyst is to provide redox reaction sites. Theoretically, a cocatalyst should allow the photoexcited electrons and holes to have a smooth migration towards the surface and the reaction sites. This theory comes from the model of semiconductor and conductor interfaces that generate a built-in electric field, which should promote interfacial charge transfer [10,46,47]. However, this understanding and model is not experimentally supported, and the results do not align with the current understanding of the system. Even so, since an increased efficiency of overall water splitting in most of the material system has been reported, it is now used in ongoing research in this field [17]. On the other hand, these active sites could also promote the reverse reaction of the hydrogen evolution reaction and oxygen evolution reaction, and these reverse reactions are usually preferred to the forward reaction on a thermodynamic basis. Therefore, the use of cocatalysts in studies that require a fundamental understanding of this research area should be carried out with care. Ultimately, due to the nature of the nanostructures, uniform deposition of the cocatalysts on the materials is not usually achieved, which makes systematic studies difficult.

7. Current Material Systems and Limitations

As McKone and Lewis point out in their review of the ideal photocatalysts [48], regardless of the chosen solar fuel systems, certain constraints are present for their further development. These constraints arise from the three key system requirements: efficiency, stability, and scalability, which are summarized in Figure 5. Hundreds of metal oxides, nitride, and sulfide, with the electronic configuration of d0 and d10 metal cations, were tested for water splitting. Other than these materials, perovskite materials that are formed by group I and group II metals, along with some lanthanides, can also be used to catalyze photochemical water splitting.

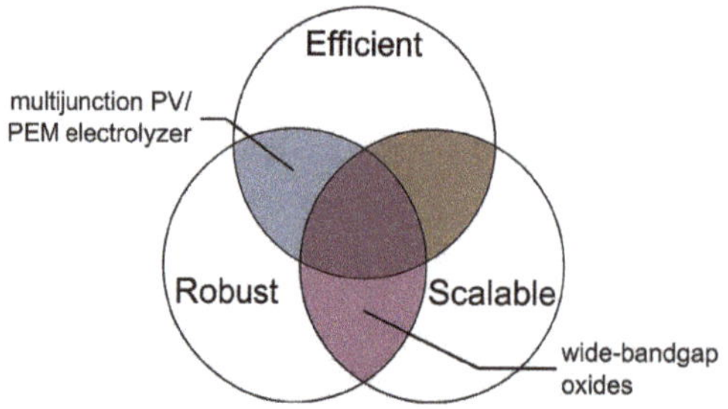

Figure 5. Venn diagram depicting the three requirements for an ideal solar water-splitting system [48]. Figure reused with permission from https://pubs.acs.org/doi/10.1021/cm4021518 (accessed on 5 September 2022); further permission related to this article should be directed to ACS.

As mentioned before, the key requirement for a one-step water-splitting catalyst is how the CB edge and VB edge should straddle the H^+/H_2 and O_2/H_2O redox potentials. Figure 6 represents different semiconductors with respect to the relative position of their band edges. Furthermore, in order to be in the center of the Venn diagram above, the following necessities must be met: the catalyst system should be made from one of the earth-abundant materials, a scalable synthesis method should be present, the system should be robust and susceptible to photo-corrosion in water, and the system should have sufficient carrier mobility. Then, in order to increase efficiency, different strategies can be implemented, such as band gap engineering to reduce the band gap and make the photocatalyst activated by visible light; morphological advancement at the surface level that improves the shape and size of the particle; and different methods of synthesis.

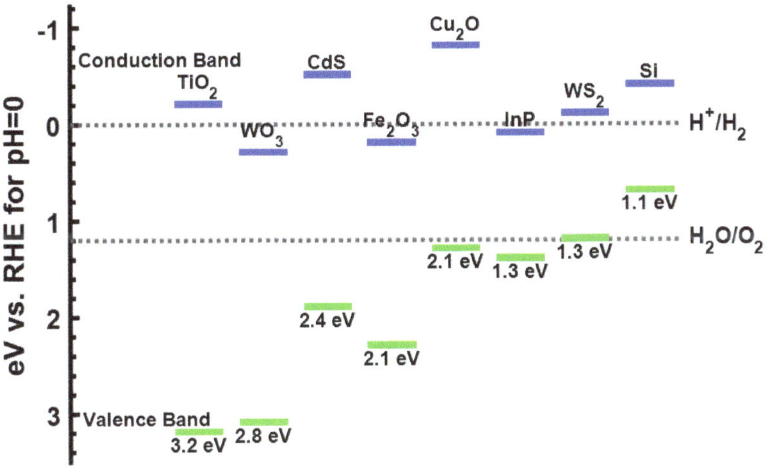

Figure 6. Band edge position for different material systems proposed for water splitting, where RHE is reversible hydrogen electrode, and band gap potentials are shown. Values from [49].

Based on Figure 6 and Table 2 below, most of the options do not meet the primary requirements of the band edge positions for single-step photocatalytic water splitting. A couple of remainder candidates, such as CdS, that depict high efficiencies in many studies suffer from instability in water and oxidizing shortly after catalyzing the water-splitting reaction.

Table 2. List of recent one-step photochemical water-splitting materials.

Semiconductor (Available Wavelength)	Cocatalyst		Light Source	Reactant Solution	Efficiency	Refs
			Ultraviolet Light			
TiO_2 (<385 nm)	Pt, RuO_2	-	450 W Xe lamp (>300 nm)	H_2O with pH adjustment	AQY [1]: 30% at 360 nm	[50]
$NaTaO_3$:La (<300 nm)	0.2 wt% NiO	-	400 W Hg lamp	H_2O	AQY: 56% at 270 nm	[45]
			Visible Light			
TaON (<495 nm)	3 wt% RuOx/2.5 wt% Cr_2O_3-4 wt% IrO_2	ZrO_2	450 W Hg lamp (>400 nm)	H_2O	AQY: 0.1% at 420 nm	[51]
Cu_2O 2.0 (<620 nm)	3 wt% IrO_2	-	300 W Xe lamp (>440 nm)	H_2O	AQY: 0.3% at 550 nm	[52]
CoO (<515 nm)	-	-	AMI.5G solar simulator	H_2O	STH: 5%	[53]
C_3N_4 (<442 nm)	3 wt% Pt-1 wt% CoOx	-	300 W Xe lamp (>420 nm)	H_2O	AQY: 0.3% at 550 nm	[54]

[1] Apparent quantum yield.

The rest of the recently discovered/studied photochemical water-splitting examples mentioned in Table 2 are all suffering from low efficiencies under visible light. Among all the materials mentioned above, TiO_2 is still studied due to its favorable conditions. However, due to its large band gap, the pure form of titania only activates in the presence of

UV light. Although many studies illustrated high efficiencies when TiO_2 is used under UV light, the attempts to achieve similar results when the material is transformed into visible-light-active material through band gap engineering remained inefficient. The second barrier that is widely mentioned is titania's low carrier mobility that leads to a fast recombination rate of the charged carriers.

8. TiO_2 Systems

The crystallographic structure of TiO_2 is shown in Figure 7 and listed in Table 3. Well-known phases of TiO_2 are rutile, anatase, and brookite. Rutile is a tetragonal (a = 4.594 Å, c = 2.958 Å), anatase is also a tetragonal (a = 3.785 Å, c = 9.514 Å), and brookite is orthorhombic (a = 9.184 Å, b = 5.447 Å, c = 5.145 Å). The c coordinate of anatase is higher than other phases, and the c/a ratio of its unit cell is greater than the rutile and brookite phases. Among all three phases, anatase is the most active allotropic in terms of photocatalytic activity when compared to rutile, brookite, and TiO_2-B (artificial phase).

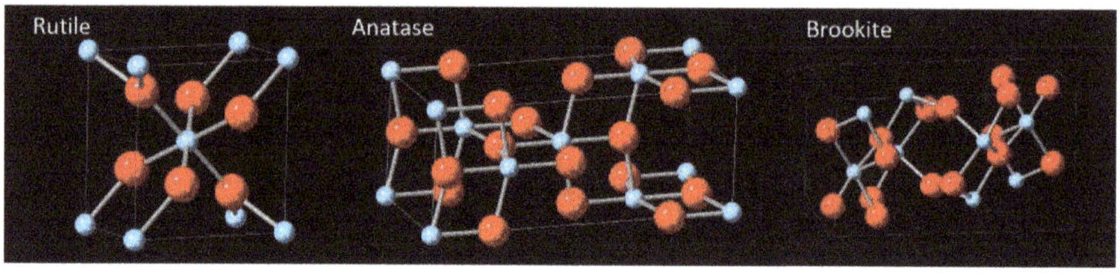

Figure 7. Crystallographic structures of TiO_2 phases.

Table 3. Physical properties of TiO_2 phases.

Structure	Lattice Constant (Å)			System
	a	b	c	
Rutile	4.594	4.594	2.958	Tetragonal
Anatase	3.785	3.785	9.514	Tetragonal
Brookite	9.184	5.447	5.145	Orthorhombic

Rutile is thermodynamically stable at ambient conditions. Anatase is kinetically stable and transforms to rutile at higher temperatures depending on the particle size, ambient pressure, and other parameters [55]. The brookite phase is also metastable and difficult to synthesize, and so it is seldom studied. Table 4 lists the oxide polymorphs of TiO_2 that have been synthesized at the nanoscale, along with the synthesis techniques, processing conditions, and relative particle size of the transforming material [56]. If the particle sizes of the three crystalline phases are equal, anatase is the most stable at sizes less than 11 nm thermodynamically, brookite is the most stable between 11–35 nm, and rutile is the most stable at sizes greater than 35 nm. These values are summarized in Table 4.

For bulk TiO_2, the rutile phase is thought to be a more stable phase than the anatase at room temperature. However, Zhang et al. [57] has suggested, and Gouma et al. has confirmed [58], that the total free energy of rutile is higher than anatase when the particle size is smaller than the critical size of 14 nm; therefore, anatase becomes a more stable phase. After that, some studies have claimed that rutile and anatase together have higher photocatalytic activity than pure anatase. For example, the commercial TiO_2, Degussa P25, is a mixture of anatase (~75%) and rutile (~25%) phases, and it is the most widely used photocatalyst for environmental applications.

Anatase can be synthesized at lower temperatures (400 °C) to nanoscale size; it shows better photocatalytic properties because nanoscale size increases the number of pores and enhances solid–solid interactions. Conversely, at higher temperatures (above 300 °C), nano-TiO_2 has worse photocatalytic activity.

The properties of brookite are poorly known due to the difficulties of obtaining it as a pure phase [59]. Qiuling et al. [60] also shows that the synthesized two phases (anatase/brookite) have higher photocatalytic activity due to the electron transfer from brookite CB to anatase CB, resulting in effective electron–hole separation.

Table 4. Different phases of TiO_2.

Phase	Conditions	Particle Size (nm)	Techniques	Refs
Brookite	-	11–35 nm	sol-gel	[61]
Anatase	room temperature	<11 nm	sol-gel	[62]
Rutile	>850 °C	>35 nm	sol-gel	[63]

Two principal factors mentioned before must be taken into account when designing systems for conducting photocatalytic reactions. The first is the efficiency of e^- and h^+ pair generation when absorbing solar irradiation. The second is the efficiency of the e^-/h^+ pair separation before they lose their redox ability due to recombination. In order to enhance the photocatalytic efficiency, the main issues that should be taken care of are the increase in the charge separation and recombination lifetimes of charge carriers, increase in solar spectrum response range, and changing of the selectivity or yielding of a particular product.

The efficiency of the photocatalytic process, the stability under light, the selectivity of the product, and the activation range of the wavelength are factors in determining the right aim of semiconductor photocatalysis systems. For the TiO_2 system, the primary barrier remains to be its activation by ultraviolet light (12′ 400 nm) due to the wide band gap of TiO_2 (~3.2 eV for anatase and ~3.0 eV for rutile), for which it only absorbs UV light. Another major drawback of TiO_2 is photogenerated electron–hole recombination, which deteriorates photocatalytic activity [64]. Therefore, an obvious goal is reducing the band gap of TiO_2 in order to shift the absorption band to the visible region and to enhance the electron–hole separation process. Modifications of TiO_2 have been accomplished by different strategies such as doping, cooping, sensitization, and coupling. The coupling of two semiconductors provides different energy levels, which can allow a more efficient charge separation in order to enhance the lifetime of charge carriers and also to increase interfacial charge transfer. In order to overcome the drawback of TiO_2, the rare earth (lanthanide) elements can also be used.

Doped TiO_2 Systems

Rare earth elements are ideal dopants for modifying the crystal structure, electronic structure, and optical properties of TiO_2 due to the 4f electronic configuration and spectroscopic properties [65,66]. Lanthanide ions could act as effective electron scavengers to trap the CB electrons from TiO_2 [67]. Xu et al. studied doping with rare earth ions including La^{3+}, Ce^{3+}, Er^{3+}, Pr^{3+}, Gd^{3+}, Nd^{3+}, and Sm^{3+} in TiO_2, which improved photocatalytic activity in the degradation of nitrite. Wang et al. [68] also illustrated doping with lanthanide ions (La^{3+}, Er^{3+}, Pr^{3+}, Nd^{3+}, and Sm^{3+}), which improved the photoelectrochemical properties and increased the photocurrent response and the photon current conversion efficiency in the range of 300–400 nm.

9. Conclusions

Rapid growth in population and energy usage have resulted in increased environmental pollution and energy demands in recent years. In order to meet this demand, research on innovative and efficient methods of using renewable energy sources such as sunlight have become a necessity. In this review, we have discussed hydrogen as a fuel, its current industrial sources from fossil fuels, and its generation from the electrolysis of water. We then illustrated the mechanics (thermos, chemical, and otherwise) that allow for water splitting, specifically solar water splitting. Afterwards, we moved on to reviewing photocatalytic water splitting and the difference between the one-step and two-step processes. From here, we concluded that one-step water splitting is more cost efficient and should be the focus of future research. We then discussed the criteria for ideal photocatalytic materials, including that the band positions should lie between 1.6 and 2.7 eV in order to use light in the visible spectrum, and the limitations with the current material systems. We finally focused on the TiO_2 systems and their potential and drawbacks as photocatalysts for water splitting.

It is suggested that the TiO_2 systems be further investigated as photocatalysts. Specifically, investigations including effectively decreasing the band gap of TiO_2 in order to absorb light in the visible spectrum. This will increase the efficiency of TiO_2 as a photocatalytic material and will further research in the areas of hydrogen fuel from a renewable energy source.

Author Contributions: Methodology, investigation, and writing—original draft preparation, F.M.; writing—review and editing, T.G.; conceptualization, supervision, and funding acquisition, P.-I.G. All authors have read and agreed to the published version of the manuscript.

Funding: This research was funded by the National Science Foundation, grant number CMMI-1833345.

Conflicts of Interest: The authors declare no conflict of interest.

References

1. Chen, X.; Shen, S.; Guo, L.; Mao, S.S. Semiconductor-based photocatalytic hydrogen generation. *Chem. Rev.* **2010**, *110*, 6503–6570. [CrossRef] [PubMed]
2. Lewis, N.S. Toward cost-effective solar energy use. *Science* **2007**, *315*, 798–801. [CrossRef] [PubMed]
3. Hisatomi, T.; Domen, K. Introductory lecture: Sunlight-driven water splitting and carbon dioxide reduction by heterogeneous semiconductor systems as key processes in artificial photosynthesis. *Faraday Discuss.* **2017**, *198*, 11–35. [CrossRef] [PubMed]
4. Midilli, A.; Ay, M.; Dincer, I.; Rosen, M. On hydrogen and hydrogen energy strategies: I: Current status and needs. *Renew. Sustain. Energy Rev.* **2005**, *9*, 255–271. [CrossRef]
5. Pinaud, B.A.; Benck, J.D.; Seitz, L.C.; Forman, A.J.; Chen, Z.; Deutsch, T.G.; James, B.D.; Baum, K.N.; Baum, G.N.; Ardo, S.; et al. Technical and economic feasibility of centralized facilities for solar hydrogen production via photocatalysis and photoelectrochemistry. *Energy Environ. Sci.* **2013**, *6*, 1983–2002. [CrossRef]
6. Wang, Q.; Hisatomi, T.; Jia, Q.; Tokudome, H.; Zhong, M.; Wang, C.; Pan, Z.; Takata, T.; Nakabayashi, M.; Shibata, N.; et al. Scalable water splitting on particulate photocatalyst sheets with a solar-to-hydrogen energy conversion efficiency exceeding 1%. *Nat. Mater.* **2016**, *15*, 611. [CrossRef]
7. Spellman, F.R. *Water & Wastewater Infrastructure: Energy Efficiency and Sustainability*; CRC Press: Boca Raton, FL, USA, 2013.
8. King, R.; Rand, M. The hydrogen engine. *Nature* **1954**, *174*, 975–976. [CrossRef]
9. Chouhan, N.; Liu, R.-S.; Zhang, J. *Photochemical Water Splitting: Materials and Applications*; CRC Press: Boca Raton, FL, USA, 2017.
10. Ma, Y.; Wang, X.; Jia, Y.; Chen, X.; Han, H.; Li, C. Titanium dioxide-based nanomaterials for photocatalytic fuel generations. *Chem. Rev.* **2014**, *114*, 9987–10043. [CrossRef]
11. Hashimoto, K.; Kawai, T.; Sakata, T. Photocatalytic reactions of hydrocarbons and fossil fuels with water. Hydrogen production and oxidation. *J. Phys. Chem.* **1984**, *88*, 4083–4088. [CrossRef]
12. Ferrandon, M.S.; Lewis, M.A.; Tatterson, D.F.; Gross, A.; Doizi, D.; Croize, L.; Dauvois, V.; Roujou, J.L.; Zanella, Y.; Carles, P. Hydrogen production by the Cu-Cl thermochemical cycle: Investigation of the key step of hydrolysing $CuCl_2$ to Cu_2OCl_2 and HCl using a spray reactor. *Int. J. Hydrog. Energy* **2010**, *35*, 992–1000. [CrossRef]
13. Maness, P.-C.; Yu, J.; Eckert, C.; Ghirardi, M.L. Photobiological hydrogen production-prospects and challenges. *Microbe* **2009**, *4*, 659–667. [CrossRef]
14. Fujishima, A.; Honda, K. Electrochemical photolysis of water at a semiconductor electrode. *Nature* **1972**, *238*, 37–38. [CrossRef] [PubMed]
15. Jafari, T.; Moharreri, E.; Amin, A.S.; Miao, R.; Song, W.; Suib, S.L. Photocatalytic water splitting-the untamed dream: A review of recent advances. *Molecules* **2016**, *21*, 900. [CrossRef]
16. Maeda, K.; Domen, K. Photocatalytic water splitting: Recent progress and future challenges. *J. Phys. Chem. Lett.* **2010**, *1*, 2655–2661. [CrossRef]
17. Chen, S.; Takata, T.; Domen, K. Particulate photocatalysts for overall water splitting. *Nat. Rev. Mater.* **2017**, *2*, 17050. [CrossRef]
18. Zou, Z.; Arakawa, H. Direct water splitting into H_2 and O_2 under visible light irradiation with a new series of mixed oxide semiconductor photocatalysts. *J. Photochem. Photobiol. A Chem.* **2003**, *158*, 145–162. [CrossRef]
19. Scaife, D. Oxide semiconductors in photoelectrochemical conversion of solar energy. *Sol. Energy* **1980**, *25*, 41–54. [CrossRef]
20. Tabata, M.; Maeda, K.; Higashi, M.; Lu, D.; Takata, T.; Abe, R.; Domen, K. Modified Ta_3N_5 powder as a photocatalyst for O_2 evolution in a two-step water splitting system with an iodate/iodide shuttle redox mediator under visible light. *Langmuir* **2010**, *26*, 9161–9165. [CrossRef]

21. Abe, R.; Shinmei, K.; Hara, K.; Ohtani, B. Robust dye-sensitized overall water splitting system with two-step photoexcitation of coumarin dyes and metal oxide semiconductors. *Chem. Commun.* **2009**, 3577–3579. [CrossRef]
22. Maeda, K.; Teramura, K.; Lu, D.; Takata, T.; Saito, N.; Inoue, Y.; Domen, K. Photocatalyst releasing hydrogen from water. *Nature* **2006**, *440*, 295. [CrossRef]
23. Guo, K.; Chen, X.; Han, J.; Liu, Z. Synthesis of ZnO/Cu_2S core/shell nanorods and their enhanced photoelectric performance. *J. Sol-Gel Sci. Technol.* **2014**, *72*, 92–99. [CrossRef]
24. Hussain, M.; Ceccarelli, R.; Marchisio, D.; Fino, D.; Russo, N.; Geobaldo, F. Synthesis, characterization, and photocatalytic application of novel TiO_2 nanoparticles. *Chem. Eng. J.* **2010**, *157*, 45–51. [CrossRef]
25. Li, R.; Zhang, F.; Wang, D.; Yang, J.; Li, M.; Zhu, J.; Zhou, X.; Han, H.; Li, C. Spatial separation of photogenerated electrons and holes among {010} and {110} crystal facets of $BiVO_4$. *Nat. Commun.* **2013**, *4*, 1432. [CrossRef] [PubMed]
26. Mu, L.; Zhao, Y.; Li, A.; Wang, S.; Wang, Z.; Yang, J.; Wang, Y.; Liu, T.; Chen, R.; Zhu, J.; et al. Enhancing charge separation on high symmetry $SrTiO_3$ exposed with anisotropic facets for photocatalytic water splitting. *Energy Environ. Sci.* **2016**, *9*, 2463–2469. [CrossRef]
27. Walter, M.G.; Warren, E.L.; McKone, J.R.; Boettcher, S.W.; Mi, Q.; Santori, E.A.; Lewis, N.S. Solar water splitting cells. *Chem. Rev.* **2010**, *110*, 6446–6473. [CrossRef]
28. Bolton, J.R.; Strickler, S.J.; Connolly, J.S. Limiting and realizable efficiencies of solar photolysis of water. *Nature* **1985**, *316*, 495. [CrossRef]
29. Ong, W.-J.; Tan, L.-L.; Ng, Y.H.; Yong, S.-T.; Chai, S.-P. Graphitic carbon nitride (g-C_3N_4)-based photocatalysts for artificial photosynthesis and environmental remediation: Are we a step closer to achieving sustainability? *Chem. Rev.* **2016**, *116*, 7159–7329. [CrossRef]
30. Kraut, E.; Grant, R.; Waldrop, J.; Kowalczyk, S. Precise determination of the valence-band edge in X-ray photoemission spectra: Application to measurement of semiconductor interface potentials. *Phys. Rev. Lett.* **1980**, *44*, 1620. [CrossRef]
31. Wu, Y.; Chan, M.; Ceder, G. Prediction of semiconductor band edge positions in aqueous environments from first principles. *Phys. Rev. B* **2011**, *83*, 235301. [CrossRef]
32. Zhang, H.; Huang, H.; Ming, H.; Li, H.; Zhang, L.; Liu, Y.; Kang, Z. Carbon quantum dots/Ag_3PO_4 complex photocatalysts with enhanced photocatalytic activity and stability under visible light. *J. Mater. Chem.* **2012**, *22*, 10501–10506. [CrossRef]
33. Chen, S.; Zhu, Y.; Li, W.; Li, L.; Yang, Z.; Liu, C.; Yao, W.; Lu, X.; Feng, X. Synthesis, features, and applications of mesoporous titania with TiO_2(B). *Chin. J. Catal.* **2010**, *31*, 605–614. [CrossRef]
34. Marchand, R.; Brohan, L.; Tournoux, M. TiO_2(B) a new form of titanium dioxide and the potassium octatitanate $K_2Ti_8O_{17}$. *Mater. Res. Bull.* **1980**, *15*, 1129–1133. [CrossRef]
35. Maeda, K. Direct splitting of pure water into hydrogen and oxygen using rutile titania powder as a photocatalyst. *Chem. Commun.* **2013**, *49*, 8404–8406. [CrossRef] [PubMed]
36. Li, R.; Weng, Y.; Zhou, X.; Wang, X.; Mi, Y.; Chong, R.; Han, H.; Li, C. Achieving overall water splitting using titanium dioxide-based photocatalysts of different phases. *Energy Environ. Sci.* **2015**, *8*, 2377–2382. [CrossRef]
37. Zhang, J.; Xu, Q.; Feng, Z.; Li, M.; Li, C. Importance of the relationship between surface phases and photocatalytic activity of TiO_2. *Angew. Chem. Int. Ed.* **2008**, *47*, 1766–1769. [CrossRef]
38. Yamakata, A.; Vequizo, J.J.M.; Matsunaga, H. Distinctive behavior of photogenerated electrons and holes in anatase and rutile TiO_2 powders. *J. Phys. Chem. C* **2015**, *119*, 24538–24545. [CrossRef]
39. Rajeshwar, K.; Thomas, A.; Janáky, C. Photocatylitic Activity of Inorganic Semiconductor Surfaces: Myths, Hype, and Realty. *J. Phys. Chem. Lett.* **2015**, *6*, 139–147. [CrossRef]
40. Wang, Z.; Inoue, Y.; Hisatomi, T.; Ishikawa, R.; Wang, Q.; Takata, T.; Chen, S.; Shibata, N.; Ikuhara, Y.; Domen, K. Overall water splitting by Ta_3N_5 nanorod single crystals grown on the edges of $KTaO_3$ particles. *Nat. Catal.* **2018**, *1*, 756. [CrossRef]
41. Li, Y.; Peng, Y.K.; Hu, L.; Zheng, J.; Prabhakaran, D.; Wu, S.; Puchtler, T.J.; Li, M.; Wong, K.-Y.; Taylor, R.A.; et al. Photocatalytic water splitting by N-TiO_2 on MgO(111) with exceptional quantum efficiencies at elevated temperatures. *Nat. Commun.* **2019**, *10*, 4421. [CrossRef]
42. Oshima, T.; Lu, D.; Ishitani, O.; Maeda, K. Intercalation of highly dispersed metal nanoclusters into a layered metal oxide for photocatalytic overall water splitting. *Angew. Chem. Int. Ed.* **2015**, *54*, 2698–2702. [CrossRef]
43. Noda, Y.; Lee, B.; Domen, K.; Kondo, J.N. Synthesis of crystallized mesoporous tantalum oxide and its photocatalytic activity for overall water splitting under ultraviolet light irradiation. *Chem. Mater.* **2008**, *20*, 5361–5367. [CrossRef]
44. Liu, J.; Liu, Y.; Liu, N.; Han, Y.; Zhang, X.; Huang, H.; Lifshitz, Y.; Lee, S.-T.; Zhong, J.; Kang, Z. Metal-free efficient photocatalyst for stable visible water splitting via a two-electron pathway. *Science* **2015**, *347*, 970–974. [CrossRef] [PubMed]
45. Kato, H.; Asakura, K.; Kudo, A. Highly efficient water splitting into H_2 and O_2 over lanthanum-doped $NaTaO_3$ photocatalysts with high crystallinity and surface nanostructure. *J. Am. Chem. Soc.* **2003**, *125*, 3082–3089. [CrossRef] [PubMed]
46. Bai, S.; Yin, W.; Wang, L.; Li, Z.; Xiong, Y. Surface and interface design in cocatalysts for photocatalytic water splitting and CO_2 reduction. *RSC Adv.* **2016**, *6*, 57446–57463. [CrossRef]
47. Yang, J.; Wang, D.; Han, H.; Li, C. Roles of cocatalysts in photocatalysis and photoelectrocatalysis. *Acc. Chem. Res.* **2013**, *46*, 1900–1909. [CrossRef]
48. McKone, J.R.; Lewis, N.S.; Gray, H.B. Will solar-driven water-splitting devices see the light of day? *Chem. Mater.* **2013**, *26*, 407–414. [CrossRef]

49. Chu, S.; Li, W.; Yan, Y.; Hamann, T.; Shih, I.; Wang, D.; Mi, Z. Roadmap on solar water splitting: Current status and future prospects. *Nano Futures* **2017**, *1*, 022001. [CrossRef]
50. Duonghong, D.; Borgarello, E.; Graetzel, M. Dynamics of light-induced water cleavage in colloidal systems. *J. Am. Chem. Soc.* **1981**, *103*, 4685–4690. [CrossRef]
51. Maeda, K.; Lu, D.; Domen, K. Direct water splitting into hydrogen and oxygen under visible light by using modified TaON photocatalysts with d^0 electronic configuration. *Chem. Eur. J.* **2013**, *19*, 4986–4991. [CrossRef]
52. Hara, M.; Kondo, T.; Komoda, M.; Ikeda, S.; Kondo, J.N.; Domen, K.; Shinohara, K.; Tanaka, A. Cu_2O as a photocatalyst for overall water splitting under visible light irradiation. *Chem. Commun.* **1998**, 357–358. [CrossRef]
53. Liao, L.; Zhang, Q.; Su, Z.; Zhao, Z.; Wang, Y.; Li, Y.; Lu, X.; Wei, D.; Feng, G.; Yu, Q.; et al. Efficient solar water-splitting using a nanocrystalline CoO photocatalyst. *Nat. Nanotechnol.* **2014**, *9*, 69. [CrossRef] [PubMed]
54. Zhang, G.; Lan, Z.-A.; Lin, L.; Lin, S.; Wang, X. Overall water splitting by Pt/g-C_3N_4 photocatalysts without using sacrificial agents. *Chem. Sci.* **2016**, *7*, 3062–3066. [CrossRef] [PubMed]
55. Suresh, C.; Biju, V.; Mukundan, P.; Warrier, K. Anatase to rutile transformation in sol-gel titania by modification of precursor. *Polyhedron* **1998**, *17*, 3131–3135. [CrossRef]
56. Beltran, A.; Gracia, L.; Andres, J. Density functional theory study of the brookite surfaces and phase transitions between natural titania polymorphs. *J. Phys. Chem. B* **2006**, *110*, 23417–23423. [CrossRef] [PubMed]
57. Zhang, H.; Banfield, J.F. Understanding polymorphic phase transformation behavior during growth of nanocrystalline aggregates: Insights from TiO_2. *J. Phys. Chem. B* **2000**, *104*, 3481–3487. [CrossRef]
58. Gouma, P.I.; Dutta, P.K.; Mills, M.J. Structural stability of titania thin films. *Nanostruct. Mater.* **1999**, *11*, 1231–1237. [CrossRef]
59. Gouma, P. Nanostructured polymorphic oxides for advanced chemosensors. *Rev. Adv. Mater. Sci.* **2003**, *5*, 122.
60. Tay, Q.; Liu, X.; Tang, Y.; Jiang, Z.; Sum, T.C.; Chen, Z. Enhanced photocatalytic hydrogen production with synergistic two- phase anatase/brookite TiO_2 nanostructures. *J. Phys. Chem. C* **2013**, *117*, 14973–14982. [CrossRef]
61. Di Paola, A.; Bellardita, M.; Palmisano, L. Brookite, the least known TiO_2 photocatalyst. *Catalysts* **2013**, *3*, 36–73. [CrossRef]
62. Cromer, D.T.; Herrington, K. The structures of anatase and rutile. *J. Am. Chem. Soc.* **1955**, *77*, 4708–4709. [CrossRef]
63. Grant, F. Properties of rutile (titanium dioxide). *Rev. Mod. Phys.* **1959**, *31*, 646. [CrossRef]
64. Linsebigler, A.L.; Lu, G.; Yates, J.T., Jr. Photocatalysis on TiO_2 surfaces: Principles, mechanisms, and selected results. *Chem. Rev.* **1995**, *95*, 735–758. [CrossRef]
65. Bünzli, J.-C.G.; Piguet, C. Taking advantage of luminescent lanthanide ions. *Chem. Soc. Rev.* **2005**, *34*, 1048–1077. [CrossRef] [PubMed]
66. Ivanova, A. Physicochemical and catalytic properties of systems based on CeO_2. *Kinet. Catal.* **2009**, *50*, 797–815. [CrossRef]
67. Wang, Y.; Cheng, H.; Hao, Y.; Ma, J.; Li, W.; Cai, S. Photoelectrochemical properties of metal-ion-doped TiO_2 nanocrystalline electrodes. *Thin Solid Film.* **1999**, *349*, 120–125. [CrossRef]
68. Wang, Y.; Cheng, H.; Hao, Y.; Ma, J.; Xu, B.; Li, W.; Cai, S. The photoelectrochemistry of Nd_3+-doped TiO_2 nanocrystalline electrodes. *J. Mater. Sci. Lett.* **1999**, *18*, 127–129. [CrossRef]

catalysts

MDPI

Review

Intensification of Photo-Assisted Advanced Oxidation Processes for Water Treatment: A Critical Review

Jorge Rodríguez-Chueca [1], Jaime Carbajo [2,*] and Patricia García-Muñoz [1]

1 Department of Industrial Chemical & Environmental Engineering, Escuela Técnica Superior de Ingenieros Industriales, Universidad Politécnica de Madrid, Calle José Gutiérrez Abascal 2, 28006 Madrid, Spain
2 Chemical and Materials Engineering Department, Universidad Complutense de Madrid, Pl. de las Ciencias, 2, 28040 Madrid, Spain
* Correspondence: jaime.carbajo@ucm.es

Abstract: In recent decades, ultraviolet-assisted advanced oxidation processes (UV-AOPs) have been successfully applied to remove a wide range of contaminants from polluted water. Despite this, their extended use on an industrial scale is still far from being a reality, largely limited by the operational costs that these processes still entail. In recent years, many researchers have been working to increase UV-AOP efficiency and reduce capital and operating costs. This work aims to review different strategies devoted to the intensification of UV-AOPs. Firstly, the optimization of operational parameters, such as catalyst loading, pH, temperature, or oxidant concentration, has been reviewed as a strategy to augment the efficiency of the photocatalytic processes and reduce reagent consumption and/or treatment time. The review also discusses the development of photocatalytic materials to intensify the UV-AOPs process, and finally, the combination or integration of different UV-AOPs for the treatment of pollutants is also examined.

Keywords: intensification processes; advanced oxidation processes; UV assisted; water treatment; contaminants of emerging concern

1. Introduction

In recent decades, sustained and ubiquitous emission of persistent organic pollutants (i.e., pharmaceuticals, pesticides, personal hygiene products, among others) into the environment has triggered the demand for more efficient wastewater treatment technologies. In this regard, advanced oxidation processes (AOPs) have emerged as promising alternatives to conventional treatments [1]. The advantages of these processes include rapid and non-selective reactions of $HO^\bullet$ with pollutants, the capability of hydroxyl radicals to mineralise organic pollutants into stable inorganic compounds such as water, carbon dioxide and salts, and mild operating conditions. Furthermore, when AOPs are assisted or mediated by UV light (UV-AOPs), the ability to treat persistent pollutants or provide efficient disinfection is enhanced by increasing the efficacy of oxidant agents (H_2O_2, $S_2O_8^{2-}$, O_3, etc.) conversion into active free radicals ($HO^\bullet$, $SO_4^{\bullet-}$).

There is a wide spectrum of UV-AOPs treatments which may be classified as homogeneous or heterogeneous, as shown in Figure 1. Photo-assisted homogeneous or heterogeneous Fenton ($H_2O_2/Fe^{2+}/UV$) uses UV–Vis irradiation for the photolysis of Fe^{3+} complexes by regenerating Fe^{2+}, which subsequently decompose H_2O_2 into active $HO^\bullet$. In heterogeneous photocatalysis (TiO_2/UV), the absorption of a photon with energy equal to or higher than the band-gap of the semiconductor leads to the formation of electron–hole pairs (e^-/h^+) and subsequently to hydroxyl radicals. For their part, O_3- or H_2O_2-based AOPs can also be combined with UV to form active hydroxyl radicals (O_3/UV; $O_3/H_2O_2/UV$ or H_2O_2/UV). In addition, UV-AOPs can use sulfate radicals ($SO_4^{\bullet-}$) as powerful oxidants with a high redox potential of 2.6 V (approximate to $HO^\bullet$ (2.8 V)) [2] by activating potassium/sodium persulfate (PS) or potassium peroxymonosulfate (PMS) salts.

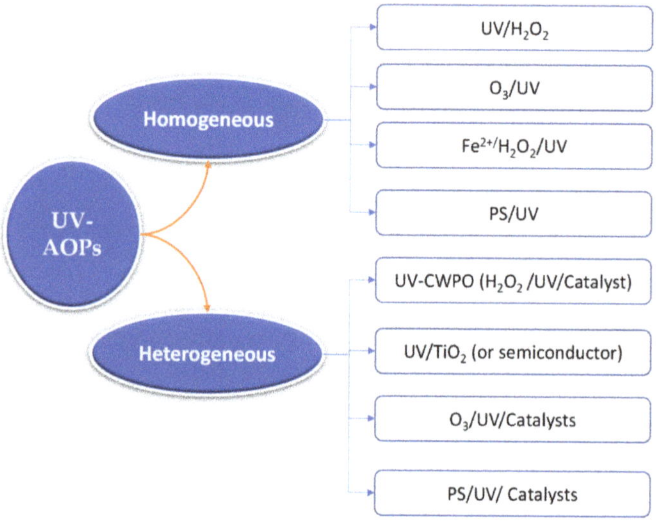

Figure 1. UV-AOPs classification.

Despite the undeniable advantages of UV-AOPs already described, there are significant limitations to the implementation of these processes on an industrial scale. Perhaps, the latter may be explained by the relatively high operating costs (light energy, chemical reagent consumption, etc.) and capital costs (photo-reactors design complexity, even when solar light is employed). Other specific limitations may also be cited, such as optimisation requirements of oxidant dosage to avoid excessive $HO^{\bullet}$ formation, the acidic conditions and iron removal requirements in photo-Fenton processes, or design photoreactor limitations, as UV light penetration is strongly attenuated by turbidity [3]. Therefore, full-scale UV-AOPs application is very limited, and their implementation is still far from being a reality [4,5].

In this context, intensification of UV-AOPs (i.e., setting innovative improvements in the process to increase efficiency or lower capital and operating costs [6]) is undoubtedly a promising avenue to improve the feasibility of these processes on an industrial scale. In recent years, many researchers have been working to increase the efficiency of UV-AOPs and reduce capital and operating costs. Nevertheless, currently, very few review works address effective strategies for improving or intensifying AOPs processes, and to the best of our knowledge, none of these reviews focus specifically on UV-assisted AOPs. Therefore, this review attempts to compile some of the main strategies that can be implemented to optimise UV-AOPs, including the optimisation of operation conditions, the employment of novel and bi-functional catalysts, and the synergistic combination of UV-AOPs.

One strategy to intensify the efficiency of UV-AOPs processes is to optimise operating conditions to reduce reagent consumption and minimise the time needed for pollutant removal and mineralisation. The efficiency of these processes is affected by different operational parameters such as catalyst loading, pH, temperature, and oxidant concentration. Many recent studies have focused on optimising these parameters to intensify the photocatalytic process. In the case of catalyst loading, increasing it can improve the pollutant removal rates, but exceeding the maximum optimal value can slow the rate of mineralisation. Different studies have found varying optimal concentrations for different catalysts and conditions, and some have developed models to better understand the dependence of reaction rates on catalyst loadings.

Another section is focused on the use of bifunctional catalysts used for the photocatalytic treatment of polluted water as a strategy for the process intensification. In recent years, much effort has been spent for developing photocatalytic materials to improve the process efficiency. This intensification has mainly been conducted in two directions: to reduce physical limitations of the reaction and to increase the overall yield of the treatment, which is also related to chemical limitations. The main approaches to augment the overall yield via photocatalysts can be divided into two main categories: design orientated to reduce mass and photon transfer limitations, and design of bifunctional catalysts that increase the generation of oxidising species. These latest, which are bifunctional catalysts, play a dual role in the process. A bifunctional catalyst can be considered for every non-consumed material with a double role in the reaction, i.e., bulk semiconductor nature and with surface catalytic sites, or the combination of a high adsorption and a photocatalytic role. These strategies are further discussed, including the main approaches found in the literature for each of them.

Finally, this review discusses the combination of different photo-assisted UV-AOPs, for the treatment of pollutants as intensification strategy. The combination of processes is often carried out with the aim of obtaining a positive synergistic effect [7], but this is not always the case, and antagonistic effects could be observed [8]. The text describes hybrid (or single-step combined process) and sequential (or two-step combined process) systems that combine AOPs with other technologies, for instance with other AOPs or with other kind of technologies (i.e., physical, chemical or biological). The combination of processes is often associated with a further step in the maturity of the technology and its proximity to application on a larger scale. It is for this reason that, throughout this section, a greater number of close-to-reality works using pilot scales are observed. However, more research is needed to understand the mechanisms of these combinations and to improve their applicability and efficiency at larger scales.

2. Intensification of Operational Conditions in UV-AOPs

Among all intensification strategies, probably one of the most obvious and effective is the optimization of operating conditions that aim to either reduce reagent consumption or minimize the time required to achieve the appropriate pollutant removal and/or mineralization. This section specifically covers recent approaches devoted to the optimisation of the most relevant operational conditions. In this regard, the efficiency of the UV-AOPs processes is severally affected by different operational parameters such as catalyst loading, pH, temperature and or oxidant concentration, among others [9]. In this section, we focus on those variables that can be intensified in operation, leaving aside those that are given by the nature of the effluent to be treated (pollutant load, presence of salts) or by the reaction system (reactor configuration, nature, and intensity of the irradiation). Many works have addressed new strategies for optimising these parameters in recent years to intensify the photocatalytic process, some of which are listed in Tables 1–4. Furthermore, Figure 2 shows the key points to be considered for the optimisation of operating conditions in UV-AOPs.

2.1. Catalyst Loading

In catalytic processes, the increase in catalyst loading usually boosts the formation of active radicals, which is translated into higher pollutant depletion rates. Nevertheless, the catalyst load can only increase up to a certain maximum value in ultraviolet (UV) catalytic processes. Exceeding this optimum results in a shielding phenomenon in which a large part of incident radiation is reflected by scattering, preventing most photons from reaching the catalyst particles, causing the rate of mineralisation to slow down [10].

In UV-TiO_2 processes, the optimum catalyst mass has usually been empirically fixed. In this regard, many works have optimised catalyst concentration for different reactor geometries and applications such as dyes (methylene blue) [11,12], emerging pollutants (caffeine) [13], pharmaceuticals (diclofenac, cefotaxime, ibuprofen) [14–16] or pesticides [17], among others. Similarly, optimal catalyst loadings have also been observed in other UV-activated heterogeneous catalytic systems, such as UV light-assisted persulfate activation by Cu^0-Cu_2O for the degradation of sulfamerazine [18] or persulfate activation by polymeric photocatalysts based on g-C_3N_4 [19]. As can be seen in Table 1, the optimum concentrations vary greatly (0.3 g/L to 2.5 g/L) depending on incident irradiation, path length, and type of catalyst. Optimal loading changes completely with different types of catalysts, such as TiO_2 and ZnO, under the same conditions [15].

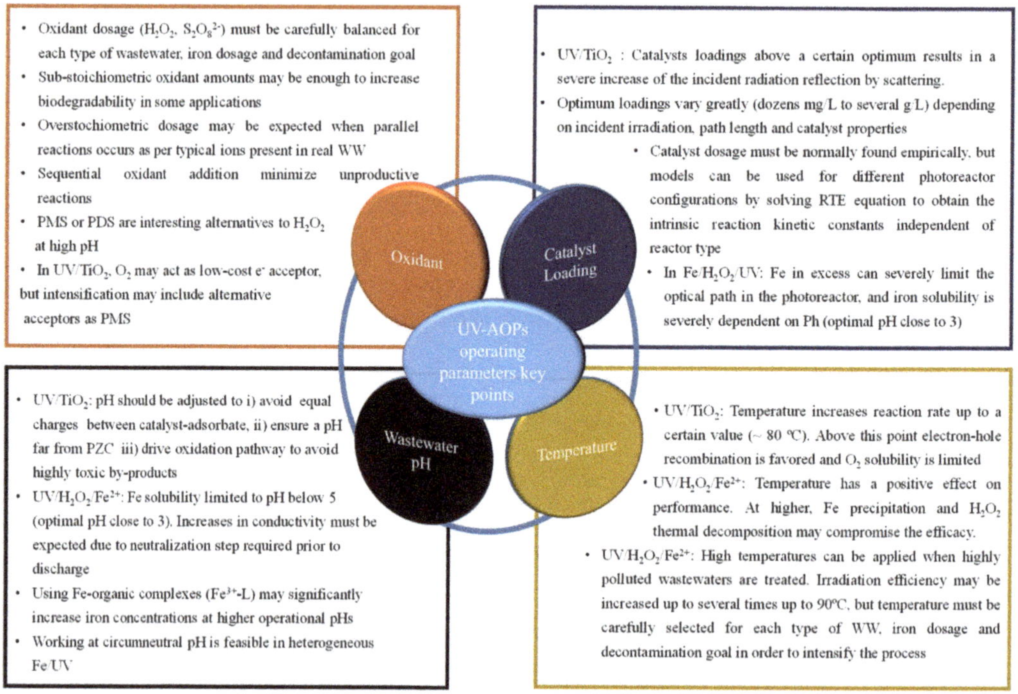

Figure 2. Key aspects for the optimisation of operating conditions in UV-AOPs.

Finding the optimal concentration for each photocatalytic system is not a practical strategy for engineering purposes in the design and optimisation of large-scale photocatalytic reactors [20]. For this reason, some works have focused their efforts on the description of a model including, in an explicit way, the dependence of the reaction rate on the volumetric rate of photon absorption in catalyst suspensions. Therefore, Carbajo and Tolosana et al. [21,22] used reflectance measurements to obtain the extinction coefficients in the UV-A range and the radiation profiles inside the photoreactor to solve the radiative transfer equation (RTE). The corresponding kinetic parameters were well suited to optimise catalyst loadings for different morphological and hydrodynamic characteristics such as titanium dioxides P25, P25/20, and P90, which represents a significant step forward in the design and optimisation of large-scale photocatalytic reactors. For their part, Casado et al. [20] radiated TiO_2 suspensions with concentrations between 0.005 and 5 $g{\cdot}L^{-1}$ to develop a model considering both computational fluid dynamics (CFD) and the resolution of the RTE. This approach includes not only the effect of fluid mechanics, mass transport and chemical reaction, but also radiation transfer, showing an optimum catalyst loading around 0.1–0.2 $g{\cdot}L^{-1}$. Li puma and co-workers [23,24] studied the radiative transfer phenomena in different reactor configurations (solar annular and rectangular channel flat reactors), by coupling a six-flux radiation absorption-scattering model to the Henyey–Greenstein scattering phase function (SFM-HG). The authors provide a new approach to measure the impact of catalyst loading and optical properties suspensions of two different photocatalysts as TiO_2 (or modified Ag-TiO_2) and goethite.

In UV/Fe/H_2O_2 processes, the amount of Fe also plays a crucial role in the production of active radicals. For example, a few mg/L of iron may be enough for the treatment of low concentrated wastewater containing emerging pollutants such as carbamazepine or

5-Fluorouracil [25,26], but catalyst loading could require up to hundreds of mg/L Fe when treating sour water from a refinery plant [27]. Although, generally speaking, increases in iron concentration always led to increases in disappearance kinetic rates, at a certain level, this improvement is residual and the efficiency decreases significantly due to the ineffective decomposition of H_2O_2 caused by excess iron, which can act as a scavenger of hydroxyl radicals, reducing process efficiency (see Equation (1)):

$$Fe^{2+} + HO^{\bullet} \rightarrow OH^{-} + Fe^{3+} \quad (1)$$

Furthermore, excess Fe can severely limit the optical path in the photoreactor [9], reducing the efficiency of photon absorption inside the photoreactor. The influence of iron and irradiance level on pyrimethanil degradation by the photo-Fenton process was analysed by Cabrera Reina et al. [28]. The authors realised that the efficacy of the treatment clearly improved by increasing the dose of Fe from 8 to 20 mg·L^{-1}; thus, iron was no longer the rate limiting factor as the process improves with irradiance, but another increase to 32 mg·L^{-1} caused severely inefficient H_2O_2 consumption, and similar degradation rates were observed due to the undesirable excess of radical production. Under the conditions of this study, the lowest costs values were found with 20 mg·L^{-1} of Fe, proving the ineffectiveness of increasing Fe doses beyond a certain point. Therefore, in UV/Fe/H_2O_2, the Fe dosage must be specifically balanced with the amount of H_2O_2 and the type of wastewater.

Table 1. UV-AOPs intensification by catalyst loading optimisation.

Photocatalytical System	Target Pollutant	Catalysts Loading Conditions	Remarks	Experimental Remarks	Ref.
TiO_2/UV-A	Methylene Blue (Dye)	[catalyst] = 0.5, 1, 1.5 g/L	Optimum achieved at 1 g/L (k = 0.0801 min^{-1})	MB_0 = 60 mg/L; pH neutral; T = 20 °C, UV-A = 40 W/m^2 Immersion Reactor, Ø = 76 mm	[11]
TiO_2/Solar	Methylene Blue, Dichloroacetic acid, 4-Chlorophenol	[catalyst] = 0.2, 0.25, 0.3, 0.35 g/L.	Optimal photocatalyst loading 0.25 g/L. 99% MB degradation at 8000 J/m^2.	$[MB]_0$ = 10 ppm, flow rate = 24 L/min, pH = natural (7.45). Solar Pilot-scale Offset Multi Tubular Photoreactor (OMTP), Ø = 32 mm	[12]
1% Mg-ZnO-Al_2O_3/UV-A/B	Caffeine	[catalyst] = 0.1–1.5 g/L	Optimal photocatalyst loading 0.3 g/L (1% Mg-ZnO-Al_2O_3) Caffeine degradation was 98.9%	[Caffeine] = 20 mg/L. Cylindrical Pyrex immersion photoreactor (2 L). UV Hg lamp (400 W). T = 25 °C	[13]
TiO_2/UV-A	Diclofenac (DIC)	[catalyst] = 0.1–2 g/L	k_{app}, (0.35 min^{-1}) was optimal at a mass of catalyst of 1.0 g/L	$[DIC]_0$ = 8 mg/L. T = 293 K, pH = 6. 800 mL glass immersion photoreactor (Ø = 7.5 cm) TQ 150 mp Hg lamp (λexc < 366 nm)	[16]
TiO_2/UV ZnO/UV	Cefotaxime	[catalyst] = 1.0–2.5 g/L	≈84.2% of Cefotaxime removal optimized at 2.3 g/L of TiO_2 and around 1.45 g L^{-1} of ZnO.	Sunlight sim. Xe lamp (300–800 nm) 250 W/m^2 [Cefotaxime] = 20 mg/L, T = 35 °C	[15]
TiO_2/UV-LED	Ibuprofen (municipal and pharmaceutical spiked MWW)	[catalyst] = 0.5–4 g/L	100% removal of IBU and 55% removal of DOC. k_{app}, (0.024 min^{-1}) was optimal at 2.5 g/L	Lab-scale photoreactor: 10 W UVhigh LEDs (λ_{max} = 382 nm); V = 250 mL, IBUP (up to 213 mg/L)	[14]
$Na_2S_2O_8$/TiO_2/Solar	Mixture of 8 pesticides in agro-phitosanitary wastewater	[catalyst] = 100 to 500 mg/L	The degradation rate increases up to 300 mg/L of catalyst loading	[Pesticides]: 0.21 mg/L (hexythiazox) to 5.97 mg/L (thiamethoxam) CPC photoreactor, Ø 14.6 cm, V = 180 L (75 L illuminate V)	[17]

Table 1. *Cont.*

Photocatalytical System	Target Pollutant	Catalysts Loading Conditions	Remarks	Experimental Remarks	Ref.
PS/Cu^0-Cu_2O/UVA	sulfamerazine	[catalyst] = 0.06 to 1 g/L	SMZ removal increases to 0.2 g/L	$[SMZ]_0$ = 50 mg/L; $[PS]_0$ = 0.8 g/L, λ_{max} = 382 nm, initial; pH = 7, 25 °C; photo-reactor with Hg lamp (800 W)	[18]
g-C_3N_4/PMS/Vis system	Acid Orange 7 (AO7)	[catalyst] = 0.05 to 1.2 g/L	The degradation rate increased with catalyst load to 0.8 g/L	500 W xenon lamp, [PMS] = 0.4 g/L, [AO7] = 20 mg/L, T = 25 °C	[19]
UV/TiO_2	MeOH	Radiation field was simulated. [TiO_2] = 0.005–5 g/L	Maximum at 0.2 g/L Radiative model was confirmed with MeOH oxidation experiments	Differential photoreactor (ODPR). V = 3.2 mL, quartz cell (Hellma QS-130), optical path = 1 cm; 36 UV-A LEDs (365 nm, Rad flux = 410 mW at 700 mA)	[20]
P25, P25/20 and P90 (TiO_2)/UV	Phenol (Model)	Radiation field was simulated. [TiO_2] = 0.1–1.5 g/L	P25 and P90 optimum catalyst load (0.25–0.50 g/L) was 2 times P25/20 (~0.7–1.0 g/L) The kinetic model was successfully validated by experimental data (phenol oxidation)	1L Pyrex slurry photoreactor (Ø = 76 mm), 15 W × 6 Black Light Blue lamps(15W) $[phenol]_0$ = 50 mg/L, pH_0 = 6.0, irradiance = 38.4 W/m^2	[21,22]
TiO_2, Ag-TiO_2, goethite/ UV–Vis	2-hydroxybenzoic acid (2-HBA)	Initial rate of photon absorption (IRPA) was correlated to optimum catalyst Concentration [TiO_2] = 0.1–2 g/L	The apparent optical thickness: τapp = 4.1–4.4 provides optimum catalyst and reactor performance. Intrinsic kinetic parameters of 2-HBA photocatalytic oxidation were determined.	Flat plane photo-reactor (thickness L = 1 cm) or annular reactor (Ø int = 0.054 m), cylindrical lamp mounted axially [2-HBA] = 0.2 $mmol/dm^3$; T = 22; pH = 4.0 UVA to natural light (UVA up to 89.4 W/m^2	[23,24]
Fe^{2+}/H_2O_2/ UV–Vis	5-fluorouracil	[Fe^{2+}] = 0–100 mg/L	Performance increased to a ferric ion concentration of 4.5 mg/L At higher concentrations, the degradation rate increases marginally	double-walled Pyrex glass reactor, thermostated in a solar with a xenon lamp (1.5 kW, 500 W/m^2) 100 mL of 5-FU solution (10 mg/L); pH = 3; H_2O_2 = 0–90 mg/L	[25]
Fe^{2+}/H_2O_2/ UV-B	Sourwater from petroleum refineries	[$FeSO_4$] = 0.13–0.4 g/L	DOC removal was optimised with the highest amount of iron.	2 reactors operating in continuous mode. Fenton (400 mL) and Photo-Fenton (1600 mL), with UV-B lamp (15 W) [COD] = 850–1020 mg/L; $[H_2O_2]_0$ = 4 g/L, $[FeSO_4]_0$ = 0.1–0.4 g/L.	[27]
Fe^{2+}/H_2O_2/UVA	pyrimethanil	[Fe^{2+}] = 8, 20 and 32 mg/L	At 20 mg/L, the treatment always improved with irradiance (process was photo-limited). At 32 mg/L, the excess of iron was counter-productive.	Lab-scale raceway photo-reactor in a SolarBox equipped with Xenon lamp (300–800 nm) [H_2O_2] = 35 mM, pH = 2.8	[28]

2.2. pH

In heterogeneous photocatalysis, pollutant–photocatalyst interactions are directly linked with the charge of photocatalyst particles, its state of aggregation, and the speciation of pollutants; thus, photocatalytical performance is governed by pH modifications. For example, the point of zero charge (PZC) in TiO_2 is usually in the range between 5.6 and 6.8 [29]; thus, when the pH is below the point of zero charge (pH < pHpzc), the catalytic surface is positively charged, while when the pH of the solution is higher (pH > pHpzc), the TiO_2 surface becomes negatively charged (Equations (2) and (3)):

$$\text{-Ti-OH} + H^+ \rightleftarrows \text{-TiOH}_2^+ \tag{2}$$

$$\text{-Ti-OH} + \text{OH-} \rightleftarrows \text{-TiO}^- + H_2O \tag{3}$$

Therefore, when possible, pH should be adjusted to: (i) avoid equal charges between catalyst and adsorbate, (ii) ensure a pH far from PZC, limiting particle aggregation, and (iii) drive the oxidation pathway to avoid highly toxic by-product formation. Many examples of these strategies are collected in Table 2. Seid and co-workers [30] studied the effect of pH in the treatment of several nitrosamines (nizatidine, trimebutine, and metoclopramide), finding that working at an initial pH $\geq$ 5.6 reduced by half the formation of carcinogenic nitrosamine by-products. For their part, Elhalil et al. [13] found that the photocatalytic degradation of caffeine using 1% Mg-ZnO-Al_2O_3 as the photocatalyst was significantly enhanced operating at pH above PZC, thus favouring the adsorption of the protonated caffeine molecule (pK_a = 10.4) onto the negatively charged catalyst surface. Similarly, the importance of PZC identification was revealed when different TiO_2 was tested during methylene blue degradation [31]; therefore, negatively charged TiO_2 surfaces with TiO_2 NP prepared at pH 7.0–10 clearly improved process efficacy favouring interaction with the cationic dye. Carbajo et al. [32] also analysed the effect of the chemical nature of different substrates on the degradation mechanism and observed that at pH 2.7, the interaction between titania catalysts (positively charged) and the dichloroacetate anion $CHCl_2COO^-$ (pK_a = 1.26) favoured DCA photodegradation by direct attack of photogenerated holes (h^+), while on the contrary, indirect attack OH appears to govern the removal of phenol (pK_a = 9.9) at natural pH.

UV/Fe/H_2O_2 systems should usually operate in a narrow pH range (2.8–3.0) to optimize $HO^\bullet$ production [33]. At higher pH, catalyst precipitation occurs as iron hydroxide, while the consumption of HO radicals by protons is enhanced at pH < 2.5 according to Equation (4):

$$OH + H^+ + e^- \rightarrow H_2O \tag{4}$$

For this reason, many works devoted to the treatment of different pollutants such as antineoplastic agents and cytostatics [34,35], antibiotics [36] or industrial wastewater [37,38] have chosen an acidic pH close to 3 as optimal conditions. Nonetheless, UV/Fe/H_2O_2 feasibility is severely limited by acid/base addition, which implies: (i) higher operational costs due to reagent's consumption, (ii) more complex operation due to the requirement of a neutralization step prior to discharge, and (iii) the increases in conductivity in the final effluent [39].

Among all the strategies for the intensification of the UV//Fe/H_2O_2 process, one of the most effective is the use of iron complexes to increase the phototonic efficiency and maintain iron in solution, even at circumneutral pH. In this regard, iron–organic complexes (Fe^{3+}-L) can maintain iron in solution at much higher concentrations at higher pH, and light can reduce Fe^{3+} to Fe^{2+} to form organic radicals (see Equation (5)), which can then participate in the formation of reactive radicals that favour pollutant degradation [40]:

$$Fe^{3+}\text{-L} + UV \rightarrow Fe^{2+} + L \tag{5}$$

Oxalic acid and citric are probably the most investigated iron ligands, as both present high quantum efficiencies and low toxicity [41]. However, these chelating agents leave moderately acidic conditions in the effluent; thus, other alternatives such as ethylenediamine-N,N'-disuccinic acid (EDDS) have emerged as a promising alternative. In this regard, Silva et al. [42] assessed the performance of photo-Fenton at neutral pH in the presence of different organic iron complexes using hydrogen peroxide or persulfate as oxidants for naproxen degradation. As a main conclusion, the best results were observed in the presence of Fe/EDDS (1:1), although the Fe-citrate complex with H_2O_2 was the most cost-effective.

Heterogeneous photo-Fenton can also be considered as a solution to avoid acidic conditions, and in this regard, a wide range of materials including iron oxides, carbon, clays, and perovskites have been satisfactorily employed at neutral conditions. Examples include the treatment of tetracycline at pH_0 = 6 with $ZnFe_2O_4$ [43], the employment of Fe-g-C_3N_4/graphitized mesoporous carbon composite as an effective Fenton-like catalyst to treat Acid Red 73 in a wide pH range [44] or the employment of metallurgical slag as a Fenton-type photocatalyst for the degradation of diclofenac at pH 7 [45]. Obviously, working with heterogeneous systems is always a tricky situation, as several factors must be considered, i.e., mass transfer limitations, light penetration, catalyst stability, lower reaction rates, catalyst separation, etc.

Working at circumneutral pH can also be feasible using significantly low iron doses, even with natural iron present in wastewater, as the proposal of Buitrago et al. [46], who demonstrated that low amounts of iron, such as those typically found in natural surface or well water, are sufficient to remove amoxicillin operating the process at the initial pH ~ 7.0. Similarly, De la Cruz et al. [47], achieved removals of over 80% for 22 micropollutants in an effluent from a MWTP using Fe^{3+} concentration (1.6 mg·L^{-1}) working at natural pH (unmodified pH remained between 6–7).

Table 2. Intensification of UV-AOPs by pH optimisation.

Photocatalytical System	Target Pollutant	pH Conditions	Remarks	Experimental Remarks	Ref.
1% Mg-ZnO-Al_2O_3/UV-A/B	Caffeine	pH = 3.5, 4.5, 9.5	Photocatalytic activity was enhanced at pH of 9.5 and dramatically decreased at pH of 3.5 At pH > 8.41 (surface negatively charged), favours cationic adsorption	[Caffeine] = 20 mg/L. Cylindrical Pyrex immersion photoreactor (2 L). UV Hg lamp (400 W). T = 25 °C	[13]
Immobilized TiO_2/UVA	Nitrosamines in MWTP, river or eutrophic matrices	pH = 3.0, 5.5, 7.0, 9.5	Optimum pH highly dependent on the proportion and speciation of intermediates during oxidation of each nitrosamine	UVA photocatalytic reactor (Blacklight λ = 315–400 nm) = 0.67 mW/cm^2; T = 22 °C	[30]
TiO_2 NPs/UVA	Methylene blue	Synthesis pH = 1.6, 7.0 and 10	Synthesis pH determined pH_{PZC} photocatalyst Optimum degradation (97%, k = 0.018 min^{-1}) was achieved with TiO_2 NPs prepared at pH 10	1 L photoreactor with 6W Lamp (365 nm) The cell was filled with 0.6 L of 10 mg/L of MB and 100 mg/L of the photocatalyst; T = 25 °C	[31]
TiO_2/Solar	Phenol, dichloroacetic acid, pyrimethanil	Natural pH	At pH 2.7, titania (positively charged) and $CHCl_2COO^-$ (pKa = 1.26) interaction favoured direct DCA degradation. Conversely, indirect ·OH attack govern phenol removal (pKa = 9.9) at natural pH	CPC (Compound Parabolic Collectors) tubes (3.2 m^2 irradiated area) under turbulent flow conditions and solar light. V_T = 35 L	[32]

Table 2. *Cont.*

Photocatalytical System	Target Pollutant	pH Conditions	Remarks	Experimental Remarks	Ref.
H_2O_2/Fe^{2+}/UV	Antibiotics (amoxicillin, ampicillin and cloxacillin)	pH = 2.0, 2.5, 3.0, 3.5 and 4.	Maximum degradation was achieved at pH 3	600 mL Pyrex reactor equipped with a UV lamp (6 W) emitting at 365 nm. [AMX, AMP, CLX]$_0$ = ~100 mg/L, [COD] = 520 mg/L; [H_2O_2] = 16.25 mM	[36]
H_2O_2 or $S_2O_8^{2-}$/UV/ Fe-complex (NTA, FeEDTA or FeCit, or FeOx)	Naproxen in a wastewater effluent collected in a MWWT	pH natural	Photo-Fenton at neutral pH was efficient for naproxen degradation in the presence of all iron complexes	Vis Xe high-intensity discharge lamp (X-HID) [NAP] = 1 μmol/L; [H_2O_2] = 16.3 mmol/L or [$S_2O_8^{2-}$] = 4.9 mmol;/Ligand [Fe^{3+}] = 21.4 μmol/L, pH = 7.5 (natural pH).	[42]
$ZnFe_2O_4$/UV/ H_2O_2	Orange II	Initial at pH$_0$ = 3, 6, 7 and 9	Decolourization efficiency increased slightly with pH, optimum at pH = 6	Xe high intensity discharge lamp (X-HID) (454 nm and 150 W) 50 mL glass beaker [Orange II] = 100 mg/L, [H_2O_2] = 5 mM, [catalyst] = 0.5 g/L, T = 20 °C, pH$_0$ = 6)	[43]
Metallurgical slag as a Fenton-type photocatalyst	Diclofenac	Natural pH = 7	Complete depletion and a partial mineralization were achieved with the COB/H_2O_2/sunlight system at pH 7	[Diclofenac] = 500 mg/L, Sunlight simulator with Xe arc lamp (500 W/m^2) pH = 7, [H_2O_2]$_0$ = 180 mg/L, 1:18 mass ratio of Fe/H_2O_2.	[45]
Fe^{3+}/H_2O_2/UV	Diuron and amoxicillin	Pollutants exhibited a strong degradation keeping the circumneutral pH	Presence of anions (HCO_3^{2-}, HCO_3-, humic acids ...) leads to photo-chemical reactions (dissolved ferric–humic acid complexes, colloidal iron ...) at circumneutral pH	Solar simulator (300 W/m^2) [amoxicillin] = 10 mg/L humic acids (HA) = 2.0 mg/L; [carbonates] = 100 mg/L; [Fe^{3+}] = 0.3 mg/L, [H_2O_2] = 15.2 mg/L, pH$_0$ = 7.0	[46]
UV/H_2O_2 and neutral photo-Fenton	22 micropollutants (including 15 pharmaceuticals) in MWTP	Natural pH	Fe addition to the reactor did not improve the process; degradation was higher using uniquely the Fe present in water (1.6 mg/L)	Continuous operation Reactor 5 LP Hg lamps (254 nm, 150 W each); V = 37 L	[47]

2.3. Temperature

In general terms, because of the relatively low activation energy of photocatalytic reactions in contrast to other conventional reactions, the effect of temperature on reaction rate is not so significant. However, certain aspects must be considered, and many studies have analysed the effect of temperature for the optimisation of operating conditions in UV-AOPs (see Table 3). When temperatures are too high (above 80 °C), the adsorption capacity of the solid is reduced and the oxygen solubility in the solution decreases, while electron–hole recombination is preferred [10]. On the contrary, at low temperatures, desorption becomes the limiting step.

In any case, the temperature will increase the photodegradation within a certain range; thus, there is a certain margin for the boosting of heterogeneous photocatalytic processes by temperature. In this regard, Tambat and co-workers [48] studied Milling yellow photocatalytic degradation with a CeO_2 catalyst; they observed that complete degradation was possible by increasing the temperature from 20 to 35 °C, while an only slightly increased degradation rate was observed heating up to 60 °C. Lin and co-workers [49] found similar behaviour in the decolourization efficiency of Red MX-5B dye by UV-A with a TiO_2/Ag catalyst, where higher reaction rates were achieved, increasing the operating temperature to 40 °C, although at 50 °C, the process efficiency decreased slightly due to the acceleration of electron–hole recombination. Chen and Hsu [50] investigated the effects of reaction temperature on the photocatalytic activity of TiO_2 with cocatalysts of Pd and Cu, concluding that when the reaction temperature is higher than 70 °C, the recombination of charge carriers will increase. For its part, the temperature on nitrate photoreduction with ilmenite was studied in the range of 20–80 °C by Silveira and co-workers [51]; in this case, the authors observed increases in reaction kinetics throughout the temperature range. Increasing temperature diminishes oxygen solubility, which could make photoreduction more feasible, avoiding the need to use inert gas flow to remove O_2 in solution.

Jonathan Z. Bloh [52] argues that in UV/TiO_2 systems, the temperature dependence becomes apparent only at high light irradiation intensities, but that under these conditions, the kinetic bottleneck at the particle occurs due to mass transfer and catalysis surface/or limitations. Thus, effective intensification towards the industrial implementation of UV/TiO_2 at significantly high temperature could only take place by optimization of the photophysical properties of photo-absorber materials (i.e., avoid mass transfer limitations and optimizing substrate surface). However, to this day, in heterogeneous photocatalysis, temperatures of up to 50–80 °C can be considered the ideal temperature for photocatalytic optimisation [10,50].

For its part, in photo-Fenton and UV-Fenton-like processes, temperature plays an important role on reaction rate (see Table 3); thus, appropriate temperature selection may be used for intensification purposes [53]. Many works have reported the increase in reaction rate by temperature, including several applications such as industrial textile wastewater treatment [54,55], nitrophenol removal [56], alachlor degradation [57] or phenolic effluent treatment by heterogeneous photo-Fenton with a Fe_2O_3/Al_2O_3 catalyst [58]. Additionally, moderately high temperatures (up 65 °C) may contribute to synergistic treatment of industrial beverage effluent by the photo-Fenton process intensified by activation of persulfate [59].

Few authors have also observed adverse effects when the temperature increases above a certain value. In this regard, Zapata and co-workers [60] assessed the influence of temperature on photo-Fenton degradation to treat commercial pesticides, while increasing the temperature from 25 to 42 °C required a substantial extension of mineralisation (up to 70%) at shorter irradiation times with more efficient use of hydrogen peroxide, a dramatic decline in mineralisation of DOC was observed at 50 °C. The authors attributed this behaviour to a significant loss of iron by precipitation in the studied conditions.

In any case, a relatively limited number of papers have explored the effect of temperature beyond 50 °C, and it seems that further exploration of the intensification by temperature in photo-Fenton processes could be worthwhile. Thus, Carbajo and colleagues [61] studied the possibility of intensifying homogeneous photo-Fenton by increasing the temperature (up to 90 °C) to treat a real leachate effluent of landfill using the stoichiometric amount of H_2O_2 (2.12 g/g COD) and 10 mg·L^{-1} of Fe^{2+}. The required irradiation time to achieve maximum TOC and COD removals (around 80%) was reduced from 180 to 45 min by increasing the temperature from 50 to 90 °C, and the irradiation efficiency increased four-fold within this temperature range. Furthermore, according to the authors, operating costs at high temperatures more than offset the costs due to wastewater heating; thus, working at 90 °C results in a three-fold reduction of the costs observed at 50 °C.

2.4. Oxidants

In UV/Fe/H_2O_2 processes, increasing the concentration of H_2O_2 concentration usually led to an increase in reaction rate. Although excess H_2O_2 may marginally increase the reaction rate and worsen the efficiency of H_2O_2 consumption, the last can be attributed to scavenging reactions where H_2O_2 competes for hydroxyl radicals, with organic matter producing less powerful radicals (see Equation (6)):

$$H_2O_2 + HO^{\bullet} \rightarrow HO_2^{-} + Fe^{3+} \quad (6)$$

Working with poor hydrogen peroxide conditions may also be inappropriate, as it limits the degree of mineralisation and slows down the reaction rate. Therefore, as can be seen in Table 4, many authors have tried to optimise the use of the amount of oxidant to intensify the UV-AOPs processes.

Usually, the first approach to increase the efficiency of a particular process consists of testing the theoretical stoichiometric amount of H_2O_2 required to completely remove COD or TOC, which are found to be appropriate conditions in some cases [62], although many authors have found optimal conditions working below the stoichiometric conditions to minimise H_2O_2 consumption. For example, in complete degradation of persistent antidepressant sertraline in aqueous solution by solar photo-Fenton, TOC removal was achieved up to 90% with 40% of the stoichiometric amount of H_2O_2 [63]. In addition, substoichiometric amounts of H_2O_2 can be enough to increase the biodegradability of wastewater containing recalcitrant or inhibitory pollutants when photo-Fenton is used as a pretreatment before a biological process.

More typically, an over-stochiometric H_2O_2 dosage is expected when the characteristics of the wastewaters are required, for example, when parallel reactions and inefficiencies occur in the process due to typical ions present in real wastewaters such as Cl^- or CO_3^{2-} which react with OH radicals to give less powerful radicals (Cl, CO_3^-, . . .) [64–66]. Some examples are found in optimisation of the treatment of landfill leachate by photo-Fenton [67], where the use of stoichiometric excess H_2O_2 (up to 4, 5 times) was required to achieve a higher degree of mineralisation of acetylsalicylic acid, decreasing energy-related costs [68]. Similar assumptions have been reported for sulfonamide antibiotics derived from sulfanilamide [69], the degradation of antibiotic mixtures (amoxicillin, ampicillin, and cloxacillin) in aqueous solution, and the treatment of reverse osmosis retentate from a paper mill [64].

Optimization of UV/Fe/H_2O_2 can also be carried out by H_2O_2 dosing. In this regard, the sequential addition avoids high instantaneous concentrations of H_2O_2, minimising unproductive reactions (see Equation (5)) and increasing its consumption efficiency. In this context, the sequential addition of H_2O_2 along the reaction has been reported to improve mineralisation in several works [70–72]. Furthermore, continuous automatic dosage has also been studied with positive results [17], and recent work focusses on finding dosage models and methodologies to optimise hydrogen use [73] or employing (Yu et al., 2020) [68] indirect measurement of H_2O_2, such as evolution of O_2 concentration, for continuous oxidant monitoring dosage [74]. In conclusion, the dose of H_2O_2 must be carefully balanced for each type of wastewater, the dose of iron, and the goal of decontamination in order to intensify the process, i.e., improve kinetics and maintain the best possible efficient oxidant consumption, which is by far the largest operational cost in solar photo-Fenton [75].

UV/Fe processes may also be intensified by the combination of or the use of alternate oxidants. A good example of this is the work of Devi et al. [76], where zero-valent metallic iron (Fe^0) was employed in photo-Fenton methyl violet degradation comparing ozone, peroxymonosulfate (PMS) and peroxydisulfate (PDS) with classical hydrogen peroxide (HP). PMS was found to be a better oxidant compared to H_2O_2 and PDS under higher pH conditions. For its part, Silva et al. [42] concluded that naproxen removal in distilled water was faster in the presence of H_2O_2 compared to $S_2O_8^{2-}$; they believed that better performance of $S_2O_8^{2-}$ was observed in sewage treatment plant effluent.

O_2 acts as a low-cost acceptor for conduction band photogenerated e and is the most common oxidant in UV/TiO_2 systems, as it can be adsorbed on titania from the liquid where it is dissolved following Henry's law by simple aeration systems. However, the process of optimization of UV/TiO_2 may include the use of alternative electron acceptors in the reaction to increase the number of trapped e^- inhibiting recombinations that form more powerful radicals [77]. For example, the introduction of persulfate into the TiO_2 photocatalytic system can increase the separation of electron–hole pairs simultaneously generating sulfate radicals ($SO_4^{\bullet-}$), characterised by high redox potential and longer half-life times [78]. Many works have explored this strategy, including elimination of sulfaclozine from water with $SO_4^{\bullet-}$ radicals [79], the degradation of ibuprofen by UVA-LED/TiO_2/persulfate processes [80] or the visible-light activation of persulfate by TiO_2/g-C_3N_4 photocatalyst in the degradation of micropollutants [81]. H_2O_2 has also been used satisfactorily as a powerful oxidising agent to inhibit e^-/h^+ recombination. Good examples of this strategy are the dosage of H_2O_2 during the photocatalytic degradation of ethidium bromide with iron-doped TiO_2 catalysts, where the authors observed a significant increase in the mineralisation of ethidium bromide due to the minimisation of self-scavenging reactions [82], or the heterogeneous photocatalytic degradation of pharmaceuticals in synthetic and real matrices using a tube-in-tube membrane reactor with radial addition of H_2O_2 [83].

Table 3. Intensification of UV-AOPs by temperature optimisation.

Photocatalytical System	Target Pollutant	Temperature Conditions	Remarks	Experimental Remarks	Ref.
TiO_2/Ag	Procion red MX-5B	R.T. to 50 °C	Operating temperature increased decolourization efficiency from R.T. to 40 °C but decreased at 50 °C as e^-/h^+ recombination accelerates	Photoreactor V_T = 0.5 L; Lamp: 10-W UVA (0.7 μW/cm²); [MX-5B] = 30 ppm	[49]
TiO_2, Pd/TiO_2 or Cu/TiO_2	Methylene blue	0 to 70 °C	At 0–50 °C, TiO_2 and Pd/TiO_2 activity increased with temperature; at 70 °C, rate dropped slightly or became less effective due to recombination rate increase	UVC lamps (λ = 254 nm), TUV PL-L 18 W. V_T = 20 mL; [MB] = 10 mg/L	[50]
$FeTiO_3$	NO_3^- in saline water	range of 20–80 °C	73% total nitrogen reduction was reached at 420′ An increase in the temperature enhanced reaction kinetics. At high T, N_2 bubbling to maintain inert conditions is avoided (lowering O_2 solubility)	Set-up: Magnetically stirred glass jacketed batch reactor (V_T = 700 mL). Lamp: 150 W M.P. Hg lamp (30 W/cm²) Working at [$C_2O_4^{2-}$] = 180 mg/L, [$FeTiO_3$] = 450 mg/L; [HCl] = 13 mM,)	[51]
UV-Vis/H_2O_2/Fe(II)	Textile effluents	25 to 70 °C	Temperatures above 25 °C and up to 70 °C show a beneficial effect on organic load reduction	Lamps: 6 W Black-light and 250 W Xe and Solar light.	[55]
Solar/H_2O_2/Fe(II)	Alachlor	20 to 50 °C	At best operating conditions (maximal iron concentration 2.6 mM, maximal temperature 70 °C) an increase reaction rate 5-fold by raising temperature from 20 to 50 °C	Pilot-plant CPC sunlight operated in batch mode. Collector (CF = 1): 20 Pyrex tubes (Øin = 46.4 mm). $A_{collector}$ = 4.16 m², Vi = 44.6 L	[58]
Solar/H_2O_2/Fe(II)	commercial pesticide mixture	25 to 50 °C	Photo-Fenton efficiency gradually rose with temperature; nevertheless, at 50 °C, efficiency decreases	Pilot-plant CPC with sunlight operated in batch mode. Collector (CF = 1): (Øin = 50.0 mm). $A_{collector}$ = 1.04 m² 20 Pyrex tubes; Vi = 44.6 L [DOC] = 200 mg/L (40 mg/L of each commercial pesticide); pH = 2.7–2.9; [H_2O_2] = 100 to 300 mg/L	[61]
UV/H_2O_2/Fe(II)	Phenolic and landfill leachate wastewater	25 to 90 °C	Time to achieve maximum TOC and COD removals (80%) was reduced from 180 to 45 min from 50 to 90 °C. Irradiation efficiency increased 4-fold within this range	immersion-wall batch jacketed 1 L photoreactor; Lamp: 150 W MP Hg. pH = 2.7–2.9; [H_2O_2] = 100 to 300 mg/L; Fe^{2+} = 10 ppm	[62]

Table 4. UV-AOPs intensification by oxidant usage optimization.

Photocatalytical System	Target Pollutant	Oxidant Type and Conditions	Remarks	Experimental Remarks	Ref.
UV–Vis/H_2O_2/Fe(II)	mixture of 6 emerging pollutants	Stoichiometric H_2O_2 to mineralize mixture (146 mg/L)	Data show that photo-Fenton in high-salinity wastewater at pH = 2.8 and pH = 5.0 was capable to remove all pollutants in 1 h	50 W xenon lamp on open glass reactor; borosilicate glass [Emerging pollutants] = 5 mg·L^{-1} each; [Fe] = 5 mg·L^{-1}	[63]
UV–Vis/H_2O_2/Fe(II)	sertraline	[H_2O_2] = 10–100% Sub-stoichiometric amount	TOC removal up to 90% was achieved at a hydrogen peroxide dose as low as 40% of the stoichiometric amount for mineralization	Lamp: Xenon 550 W m^{-2} (300 to 800 nm). Vr = 500 mL; [Sertraline] = 50 mgL^{-1}. The [Fe^{2+}] = 1–10 mgL^{-1}; [H_2O_2] = 10–100% stoich amount	[66]
UV–Vis/H_2O_2/Fe(II)	Acetylsalicylic acid	[H_2O_2] = Up to 9-fold the stoichiometric Amount	Mineralization around 90% is reached at 10 min with 4.5-fold excess of H_2O_2	2 Parabolic tubular modules in series, Lamp: Black-light UVA (40 W) [Fe^{2+}] = 1.5 mM; [H_2O_2] = 45 Mm [Acetylsalicylic acid]$_0$ = 100 ppm	[69]
UV–Vis/H_2O_2/Fe(II)	Orange II (OII)	Continuous addition of H_2O_2	H_2O_2 continuous dosage optimize photocatalytic efficiency (scavenger effect is minimized); 100% decolouration (95% TOC removal) with continuous addition of peroxide	Solar reactor (50 L); Ai = 2 m^2 (CF = 1); 16 borosilicate-glass tubes (OD = 32 mm) Fe(II) = 2 ppm; Orange II = 20 ppm,	[72]
H_2O_2 or $S_2O_8^{2-}$/UV/Fe-complex (NTA, FeEDTA or FeCit, or FeOx)	Naproxen in a sewage effluent collected at a MWWT	H_2O_2 compared to $S_2O_8^{2-}$	H_2O_2 best performed in ultrapure water, while $S_2O_8^{2-}$ best performed with real WW	Lamp: Xe high int. discharge [NAP] = 1 µmol/L; [H_2O_2] = 16.3 mmol/L or [$S_2O_8^{2-}$] = 4.9 mmol; Ligand [Fe^{3+}] = 21.4 µmol/L, pH = 7.5 (natural pH).	[42]
TiO_2/g-C_3N_4	Acetaminophen	PS dosage increases from 0.5 mM to 2 mM	The addition of PS greatly improved the degradation efficiency (5 mg/L AAP almost degraded; at 30 min; k = 0.061 min^{-1}, X_{TOC} = 82.5%)	Lamp: Xe (300 W, 400 nm cutoff filter) [TiO_2/g-C_3N_4] = 500 mg/L, [AAP] = 5 mg/L [PS] = up to 2 mM	[82]
TiO_2/Fe-TiO_2	Ethidium bromide	[H_2O_2] = 80–160 mg/L, 1–2-fold stoichiom, and continuous dosification was evaluated	196 mg/L H_2O_2 addition was optimized throughout; several dosages maintaining H_2O_2/TOC ratio; performance was maximized, raising 84% of TOC conversion	Pyrex photoreactor with a Hg MP lamp (500 W); Vr = 1 L, [EtBr] = 20 mg/L, [Fe-TiO_2] = 500 mg/L, pHo = 3, [H_2O_2] Total = 196 mg/L	[83]

3. Catalysts Engineering

As part of photo-assisted process intensification, a solid/heterogeneous photocatalyst is commonly required to perform the oxidative treatment of polluted water matrices.

In an archetypical photocatalytic process, the semiconductor par excellence is the titania (TiO_2) due to its properties (good stability and relevant optical and electronic properties). However, its employment has been reduced due to the necessity of UV light for excitation.

In this sense, in this part of the review, the catalysts devoted to act with a double role, thus named bifunctional catalysts, will be considered.

In recent years, many efforts have been made for the development of photocatalytic materials, a key piece for intensifying the entire process. Overall, optimisation has been carried out in two preferred directions, one mainly affecting the physical limitations of the reaction and the other affecting the overall yield of the treatment, also related to the chemical limitations. Until now, the main approaches performed for the intensification of the photocatalysts are globally divided into two elements (Figure 3):

(a) Photocatalyst design is orientated to reduce the limitations in mass and photon transfer. This strategy can be achieved via three steps as follows:
  - Improving photocatalyst activation and preventing deactivation. This fact is related to the shift of the absorption edge from the ultraviolet to the visible range. This item has been extensively studied; thus, it will not be developed as a separate alternative in this review unless combined with other sorts of intensifications.
  - Promotion of the adsorption of reagent onto the catalyst surface: the initial step of a Langmuir–Hinshelwood model.
  - Promotion of the desorption of reactions products from the catalyst: The last step of a Langmuir–Hinshelwood model.

(b) Design of bifunctional catalysts favouring the increase in oxidant species generation yield (mainly $HO_x\cdot$). In the same way, two steps have been performed for this purpose:
  - The combination of several single processes without oxidant addition.
  - The extra-addition of oxidant agent.

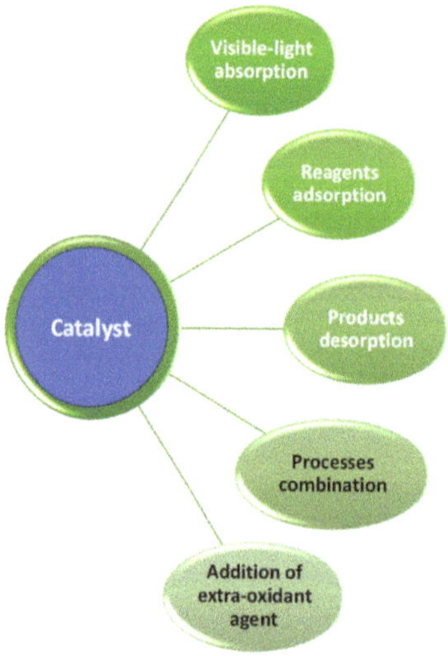

Figure 3. Main strategies developed for process intensification via photocatalyst.

Those strategies are further detailed including the main approaches found in the bibliography for each one of them.

3.1. Design Oriented to Reduce the Physical Limitations

In this part, the mass and photon transfer limitations are meant to be overcome. As mentioned above, the photon absorption strategy will not be developed here due to the large number of works dealing with the visible-assisted photocatalysts existing in the literature. In terms of mass transfer, the issue proceeds with the obtention of photocatalysts characterised by a higher adsorption capacity. This combination favours the initial pre-stage of a Langmuir–Hinshelwood model, typically established in photo-assisted processes. Among the strategies chosen to increase the initial adsorption of the pollutants, the development of mesoporous materials has been elected as the preferred one, not only due to the diffusion of pollutants within the pores in some cases, but also associated with the anchorage of the active phase in the pores of a support, which leads to an inhibition of pollutant diffusion in some cases and therefore to improved photoactivity [84–89]. In addition, we can find works dealing with a direct increase in the surface of the support or works in which the surface has been chemically modified to promote this initial stage [90]. Some work has been reported following these approaches in which a true synergistic effect was demonstrated (Table 5).

Finally, the last and less used strategy consists of chemical modification of the material based on the Bronsted and Lewis acid sites to promote the catalyst–substrate interaction, as was obtained with the $SO_4^{2-}/Fe_{2-x}Zr_xO_3$ catalyst for the photo-Fenton process. The conditions and processes involved are given in detail in Figure 4.

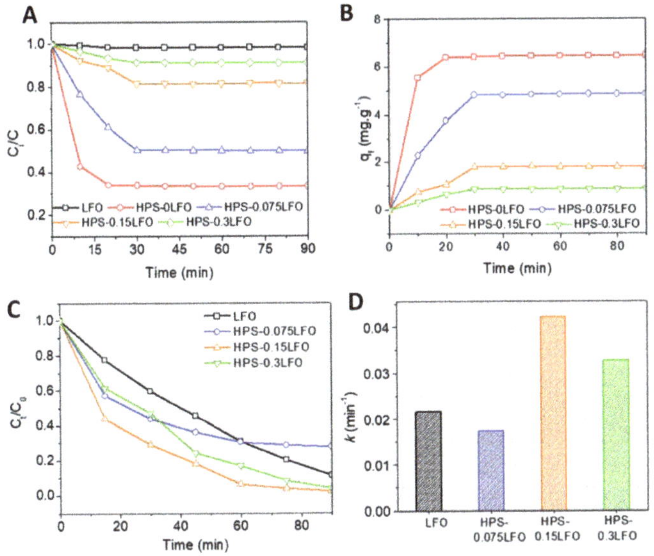

Figure 4. (**A**) Removal of RhB via adsorption by using LFO or HPS_xLFO. (**B**) The corresponding RhB adsorption capacity. (**C**) Removal of RhB in the photo-Fenton reaction using LFO or HPS-xLFO (x = 0.075, 0.15, and 0.3) and (**D**) corresponding degradation rate constants. Operating conditions: T = 25 °C, initial dye concentration = 10 mg/L, catalyst dosage = 1 g/L, initial H_2O_2 concentration = 10 mM and pH_0 = 6). Reproduced with permission from Ref. [87]. Copyright of Royal Society of Chemistry.

In one case, the active phases of the catalysts were immobilised onto surfaces established by a certain porosity. The authors achieved catalytic activities towards pollutant depletion that were higher than those in reactions in which the catalyst was nonsupported. The improvement was basically attributed to two effects: (i) an increase in the amount of adsorption itself and (ii) an enhancement in the transfer of photocharges. Several works have used $LaFeO_3$ as active phase and zeolite [88], silica [87] and resin [91] materials as supported materials (Table 5). In the second case, the mesoporosity was created inside the photocatalyst itself, leading to the configuration of binary materials, e.g., heterojunction. In these cases, in addition to a favoured adsorptive attraction, a substantial increase in oxidation rates was observed. The development of the B/Bi_2WO_6, Au/TiO_2, Fe_2O_3/TiO_2 and Ga_2O_3/TiO_2 mesopores follows with the improvement in charge transfer and the change of the absorption edge towards visible and infrared regions (Figure 3).

In terms of optimisation via the promotion of the product desorption from catalysts, there is evidence that the joint employment of AOP-ultrasound can benefit this step of the process, letting the active phase again be available to the new reactants. However, this combination has not been developed specifically to overcome this limitation. For that, the use of ultrasound will be explained in the following sections of this review.

Table 5. Catalyst used for combining adsorption and photo-assisted processes.

Catalyst	Processes Involved	Light Spectra	Pollutant	Conversions	Advantages	Disadvantages	Ref.
B/Bi_2WO_6	Adsorption/ photocatalysis	Solar	Rhodamine B	B/Bi_2WO_6 = 8.8 k Bi_2WO_6 X_{RhB} = 100% (180 min)	Mesoporous favours adsorption. B presence creates higher pore volume related to mesopores B acts also as electron trap	No TOC	[84]
$LaFeO_3$/Zeolite	Adsorption/ photo-Fenton	Visible	Rhodamine B		Higher adsorption and active sites	No evaluation of TOC parameter	[88]
$LaFeO_3$/Silica	Adsorption/ photo-Fenton		Rhodamine B				[87]
Mesoporous Au/TiO_2	Adsorption/ photocatalysis	Visible	AO7 (dye)				[89]
SO_4^{2-}/ $Fe_{2-x}Zr_xO_3$	Adsorption/ photo-Fenton	Visible					[90]
$LaFeO_3$/Resin	Adsorption/ photo-Fenton/ photocatalysis	Solar	Caffeine (CECs)	80% removal caff in 3 h 60% TOC	Reusability until 6th cycle	Not a complete mineralization	[91]
Mesoporous Fe_2O_3-TiO_2	Adsorption/ photo-Fenton/ photocatalysis	Visible	Norfloxacin antibiotic	100% removal and 97% mineralization in 120 min	Good reusability. New reactor designed with LED light	Low iron leached.	[85]
Mesoporous Ga_2O_3-TiO_2	Adsorption/ photocatalysis	UV	Imazapyr pesticide	98%removal in 180 min	10 and 3 times more activity than Ga_2O_3 and UV100	Loss of activity in 5 cycles	[86]

3.2. Design of (Bifunctional) Catalysts Favouring the Increase in Oxidant Species Generation Yield (Mainly HO_x)

This approach mainly focusses on the development of photoactive materials capable of demonstrating activity, leading to an increase in the generation of oxidising species and thus provoking higher reaction yields in terms of pollutant removal. In the literature, there exists numerous ways of classifying this part; nevertheless, all of them converge in the concept that the synergy is formed by a main process and another contribution that can be either a well-established process or the addition of an extra external oxidising agent.

The increase in the number of reactions devoted to the production of radicals and capable of depollating is the basis of this approach. There are diverse and varied combinations of techniques to promote radical generation: photocatalysis/photo-Fenton, photocatalysis/persulfate activation, photocatalysis/ozonation, or photocatalysis/ultrasound, etc. Among them, this review will focus on those processes in which the catalyst constitutes a key piece for the combination: microwave/photocatalysis, photo-Fenton/photocatalysis, and persulfate addition to photocatalysis.

3.3. Wave-Assisted Photocatalysis

Another approach for intensifying the role of a catalyst is using different kinds of waves. There have been two wave-types mainly reported that are characterised by a different nature and by a varied energetic level of them, and these are ultrasound and microwaves. In the case of ultrasound, the mechanical wave employed has an energy higher than 20 KHz, and in the case of microwaves, the electromagnetic waves are inside the range of 300 MHz to 300 GHz. Despite this important difference, the treatments are generally mentioned as a wave-irradiated process.

One of the first studies reporting the use of a heterogeneous photocatalyst, TiO_2, for this dual process was carried out by the Gogate group [92]. The general basis consists of the creation of localised higher energy and uniform heating that leads to a reduction of the activation energy and bond cleavage of organic molecules. They studied the combination of several AOPs in addition to microwave implementation. The most active synergy resulted from the simultaneous application of microwave, photocatalysis, and H_2O_2 addition. Since microwave irradiation and photocatalysis act via a similar mechanism, their combination suggests a better performance in terms of efficiency. In this work, for the concrete case of microwave and photocatalysis, the reaction rate for the removal of pollutants increased from 6.2×10^{-3} to 3.88×10^{-3} min^{-1}, meaning an improvement of 1.6 in terms of reactivity and passing from 15% to 67% of the oxidation percentage. In another work, a simple modification of a raw titania photocatalyst was developed by N doping to shift absorption spectra into the visible range [93]. Furthermore, titania was coupled to activated carbon and used for the removal of tetracycline antibiotics by a hybrid process [94]. A total synergetic effect was demonstrated with a pollutant removal of 93% and half the mineralisation of the effluent in 180 min of sonophotocatalytic reaction.

As an example of a ternary photocatalyst, CuO-TiO_2/rGO in ultrasonic photocatalysis can be considered [95]. The gain in this case was 3.7 times. The catalyst was used for the removal of methyl orange dye under UV light with a 99% depletion of methyl orange in 90 min, but no TOC evaluation was performed. The last example of this combination deals with the use of a quaternary catalyst, Fe_3O_4/TiO_2–N-GO for ultrasonic photocatalysis. The combination reports several advantages related to different effects. First, ultrasounds favour the catalyst dispersion and thus improve the mass transfer inside the reaction media. Second, ultrasound can generate extra $HO^\bullet$ due to a cavitation effect. Bubbles collapse, leading to high pressure and temperature conditions, provoking the breakdown of H_2O or O_2 molecules on oxidising radicals. These radicals can recombine and produce H_2O_2, which can be taken for a heterogeneous Fenton process. Thus, the Fe_3O_4/TiO_2–N-GO catalyst allows one to take advantage of all the reasons mentioned above with efficiencies higher than 26% compared to the pure photocatalytic system [96]. Recently, another quaternary photocatalyst, N-Cu co-doped TiO_2@CNTs, was developed and employed for the treatment of antibiotic-containing real pharmaceutical wastewater and under UV–visible light. In this case, total depletion of the pollutant was reached with great yields: 93% COD and 89% TOC [97]. The catalyst showed certain stability with a lifetime of more than five cycles of reaction. A full list of works can be found in Table 6.

3.4. Synergistic Effect: Photocatalysis/Photo-Fenton

Photocatalysis/photo-Fenton represents one of the first and most widely used combinations in AOP treatments. The reasons are that the combination provides the advantages of one process with respect to the drawbacks of the other. While photocatalysis allows for total mineralisation of the effluent as a final achievement, the Fenton process provides the drive-in reaction rate to be competitive with other nonadvanced processes. Furthermore, the combination provides a greater number of redox reactions that deal with the increase in the overall yield, not only due to the generation of a radical, but also associated with the capture of the photogenerated charges by H_2O_2 that leads to the inhibition of recombination of the charges, a highly requested issue for the enhancement of the photocatalyst activity.

Table 6. Catalysts devoted to the combination of radiation and photocatalysis.

Catalyst	Processes	Conditions	Pollutant	Yield	Advantages	Disadvantages	Ref.
CuO-TiO_2/rGO	Ultrasound/ photocatalysis	UV light	Methyl orange	99% oxidation in 90 min	Synergy of 3.7 times	No TOC evaluated. Reduction of activity with the reuse	[95]
Fe_3O_4/TiO_2–N-GO	Ultrasound/ photocatalysis	Visible light	Humic acids	93% removal	Surface cleaning, improved mass transfer. 26% higher removal compared to single photocatalysis		[96]
N-Cu co-doped TiO_2@CNTs	Ultrasound/ photocatalysis	Xenon lamp	Sulfamethoxazole in real pharmaceutical wastewater	100% antibiotic 93% COD 89% TOC	Real pharmaceutical water	Reuse until 6th cycle	[97]
TiO_2 decorated on magnetic activated carbon	Ultrasound/ photocatalysis	UV light	tetracycline	93% removal 50% TOC in 180 min		Fe leached measured	[94]
N-doped titania	Ultrasound/ photocatalysis/ filtration	Visible light	Dye	Synergetic effect higher than 20%	Ultrasonic cleans the membrane		[93]
TiO_2	Microwave/ photocatalysis	UV	4-chloro-2-aminophenol	Improvement in removal of more than 50%	Reaction rates of more than one order of magnitude	High energy consumption	[92]

A huge amount of new catalytic materials that increase the movement of charges have been developed in recent years, such as binary, doped, heterojunctions, or recently, network-substituted materials. The most representative photocatalysts and their reaction yields are shown in Table 7. As can be observed, by using this combination, higher depollution grades are obtained compared to the yields achieved by other hybrid processes. In all cases, removal percentages over 90% are reached for the main pollutant. In Table 5 are indicated the main catalysts that have been reported by their employment in the photocatalysis/photo-Fenton process. All of them are characterised by the presence of iron as the main active phase associated with the Fenton part of the process and by its implication as a visible light absorber for the photocatalysis part.

The preferred types of catalyst are heterojunction materials, probably because of the improved charge separation that enables higher reactivity. In this case, several binary and ternary junctions have been reported: $FeVO_4$@BiOCl [98], $LaFeO_3/BiOBr$ [99], $FeOOH/Bi_2MoO_6$ [100], CQDs/FeOOH [101], TiO_2/Graphene oxide/Fe_3O_4 [102] and Fe-Cu oxide/diamond [103]. Ilmenite is formed by two phases: $FeTiO_3$ and rutile provoking a natural heterojunction ready to be used in this hybrid treatment [104]. As a second approach, the doped catalysts exist where graphene oxide remains as the main photocatalyst, and Fe_2O_3 and Fe_3O_4 are inserted into its network to activate H_2O_2 [105,106]. Recently, the last option used for the photocatalytic/Photo-Fenton process is the substitution of the network in the crystal structure, as in $La_{1-x}Ti_xFeO_3$ [107,108] (Figure 5) and $Zn_{1-1.5x}Fe_xS/g\text{-}C_3N_4$ [109]. The complexity of the synthesis to control the percentage substitutions is the reason for the low number of works inside this approach.

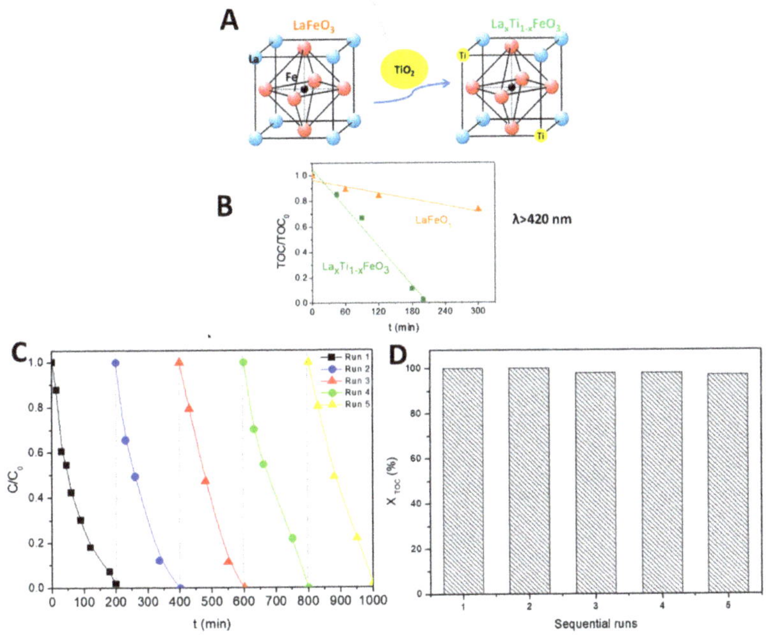

Figure 5. (**A**) Partial substitution of TI by La $LaFeO_3$ crystal structure and (**B**) the resulting TOC removal rate improvement under pure visible light (λ > 420 nm). Reproduced with permission from Ref. [108]. Copyright Elsevier. Reusability tests performed on the $La_{1-X}Ti_XFeO_3$ dual catalyst in the system H_2O_2/visible light: (**C**) Evolution of the relative 4-Chlorophenol concentration and (**D**) TOC mineralization degree obtained after 210 min of test. Operating conditions: $[4\text{-Cl-Phenol}]_0$ = 25 mg/L; $[H_2O_2]_0$ = 125 mg/L; T = 25 °C; [cat] = 0.5 g/L; (λ > 420 nm). Reproduced with permission from Ref. [108]. Copyright Elsevier.

3.5. Persulfate Addition to the Photo-Assisted Process

For the photo-assisted persulfate activation, single, binary, and ternary catalysts have been employed. Despite the substantial increase in mineralisation yield coming from the decomposition of persulfate into a hydroxyl radical and a sulfate anion, the strategy associated with the catalyst design also plays a key role in terms of process efficiency. In Table 8, the most relevant results taking place via catalyst are depicted [110].

For single-phase catalysts, two works have demonstrated an enhancement upon the activity. Zinc oxide and ilmenite ($FeTiO_3$) mineral were used for the activation of the light of persulfate. In the first work, 92% TOC removal was achieved in 1000 min of irradiation under sunlight, while in the second, 96% in 180 min by using ultraviolet light. The main advantages were the improvement in the reaction rates and the final mineralisation degree, while the drawbacks were concerned with all persulfate-treated water and the presence of higher residual sulphate in the solution. In the case of Ilmenite $FeTiO_3$, no leached iron was detected during the reaction time. When binary catalysts are used, the synergy is greatly increased in comparison to single materials. For example, TiO_2/g-C_3N_4 achieved 13 times greater activity than in the absence of persulfate and Co-$Bi_2Fe_4O_9$ more than 3 times compared to the doped catalyst. The last work involving a ternary material, Ag/AgBr/$ZnFe_2O_4$, obtained double reactivity towards the pollutant depletion with a value of 53% under visible light.

Table 7. Catalysts devoted to the combined photocatalytic-photo-Fenton processes.

Catalyst	Processes	Conditions	Pollutant	Yield	Advantages	Disadvantages	Ref.
Fe_2O_3 doped-C_3N_4	Photocatalysis-photo-Fenton	Visible light	Dicamba pesticide		LEDs light	No TOC measured	[106]
$FeVO_4$@BiOCl	Photocatalysis-photo-Fenton-sonophotocatalysis	UV light	p-nitrophenol	89% mineralization in 40 min	Additional OH· by the V^{5+}/V^{4+} cycle	Stability decreased from the 6th cycle	[98]
$LaFeO_3$/BiOBr	Photocatalysis-photo-Fenton	Sunlight	Rhodamine B	99.6% removal RhB in 30 min	Kinetic constant rate is 21 times higher than LFO	Stable just until 4th cycle	[99]
FeOOH/ Bi_2MoO_6	Photocatalysis-photo-Fenton	Visible light	Phenol	100% removal phenol (3 h)	Synergy: removal 1.54 higher than photocatalysis 1.33 higher than Fenton O vacancies improved photocatalysis	After 5th cycle a step decrease in activity Iron leached detected. 48% TOC removal	[100]
Carbon quantum dots/α-FeOOH	Photocatalysis-photo-Fenton	Sunlight	tetracycline	94% removal in 60 min	Oxidation at different pH values; Removal drastically higher than in separate processes	Less H_2O_2 consumption, but necessary for oxidation; stability decreases after 5th cycle	[101]
TiO_2-graphene oxide-Fe_3O_4	Photocatalysis-photo-Fenton	Visible light	Amoxicillin antibiotic	90% TOC pH = 3 50% TOC pH = 5 in 120 min	Magnetic recovery properties	TOC degradation is kept constant until 5th run; leached iron detected in all runs (<1 mg/L)	[102]
Fe-Cu oxide/diamond	Photocatalysis-photo-Fenton	Visible light	Phenol	100% phenol removal in 120 min pH = 4	Iron and copper cycle involved in Fenton reactions	Not good reusability; Expensive support	[103]
$Zn_{1-1.5x}Fe_xS$/ g-C_3N_4	Photocatalysis-photo-Fenton	Visible light	p-nitrophenol	96% removal and 55% TOC in 60 min	Incorporation of Fe in crystal lattice improves degradation rates	Not photocatalytic contribution; Ow reusability of the catalysts	[109]
$FeTiO_3$	Photocatalysis-photo-Fenton	Solar light, visible light	Phenol and sulfonamides	100% removal and 98% mineralization	Synergy between processes; low cost of catalyst.	Small iron concentration leached	[104,111,112]
$FeTiO_3$	Photocatalysis-photo-Fenton	UV–Vis; high temperature	Real hospital wastewater	80% TOC in 300 min	Synergy between processes; low cost of catalyst	Small iron concentration leached	[113]
$La_{1-x}Ti_xFeO_3$	Photocatalysis-photo-Fenton	UV, solar and visible light	4-Cl-phenol	100% and 100% mineralization in 120 min	One single-phase catalyst; the substitution lattice improved redox properties	No iron detected	[107,108,114]
Graphene/ Fe_3O_4	Photocatalysis-photo-Fenton	UV light	Methyl orange	99% removal of dye	Synergy with respect to Fe_3O_4	Slight decrease in activity	[105]

Table 8. Photocatalysts employed for persulfate activation.

Catalyst	Processes	Conditions	Pollutant	Yield	Advantages	Disadvantages	Ref.
TiO_2/g-C_3N_4	PS/photocatalysis	Vis light	Acetaminophen	100% removal in 30 min 86% TOC in 60 min	Persulfate got 13 times activity than single photocatalyst	Reduction of 5 percent of activity after 5 cycles	[81]
ZnO	PS/photocatalysis	Sunlight	Pesticides	92% TOC in 1000 min	Improvement reaction rate	Generation of big loads of sulphate in solution	[115]
Ag/AgBr/ $ZnFe_2O_4$	PS/photocatalysis	Visible light	Carbamazepine	53% removal; double of activity with PS.	LED light; magnetic separation	When using H_2O_2 the Z-scheme was inhibited and a reduction of the degradation	[116]
Co-doped $Bi_2Fe_4O_9$	PS/photocatalysis	Visible light	Levofloxacin		LED light; doped materials have 3.52 times higher than that non-doped;	Iron and cobalt leaching	[117]
Ilmenite	PS/photocatalysis	UV light	Azo dye	>95% of mineralization	improvement reaction rate; no iron leached	Generation of big loads of sulphate in solution	[118]

4. Catalyst Immobilisation: Intensification of Reactor Design

Some of the main previous strategies have also been gathered into a further step for implementation of the photocatalytic process in a continuous mode, which inherently involves the immobilisation of the photoactive material onto solid media. In addition, immobilisation provokes a reduction of energy investment by a decrease in the number of process units. Globally, the support media must integrate some characteristics: (i) to allow high light transmission, (ii) to possess chemical and mechanical stability, and (iii) to have the ability to keep the solid photocatalyst attached to the surface during photoreactions [119].

Until now, there have been two kinds of supports that solve the problem of handling powdery catalysts and make catalyst replacement easier once the reaction is performed. Those are the multichannel monoliths and the open-cell alveolar ceramic foams. In the case of monoliths, they are the substitute of the packed-bed photocatalyst, but they suffer from poor light transmission in their inner parts and do not allow the photoreaction to proceed. While in the case of open-cell alveolar foams, they benefit from acting as a static mixer, which can work under low pressure drop, and they have higher light transmission on the entire surface. For them, works dealing with (β-)SiC, ZrO_2 and metallic foams for attaching the photocatalyst have been reported. In this context, the main studies that report the use of media for photo-assisted wastewater photo-assisted treatment can be seen in Table 9.

Most works are devoted to a pure photocatalytic process in which titania is the supported photocatalyst par excellence [120–123], while only two works have been found in the literature dealing with the photo-Fenton process involving iron-based materials anchoring in a medium [124,125]. It must be highlighted that the works of Rico-Santacruz et al. [122,123] are not characterised by the attachment of the catalyst to the support unless by the one-step synthesis of the whole active phase/support, avoiding in this manner, the leaching of the active phase to the liquid solution. Some details are found in Table 9.

Table 9. Photocatalysts immobilised onto supports for water treatment of AOPs.

Catalyst/Support	Conditions	Advantages	Disadvantages	Ref.
TiO_2/β-SiC	Pesticide pyrimethanil /photocatalysis	Titania coating optimized; no active phase release		[121]
TiO_2/β-SiC	Nanoplastic pollutants /photocatalysis	Titania coating optimized; no active phase release		[120]
TiO_2/ZrO_2	CEC and disinfection /photocatalysis		No evaluation of active phase release	[126]
TiO_2@β-SiC	Diuron pesticide /photocatalysis	Single unique phase; no separation between the active phase and the support.	No active phase release	[122,123]
Perovskite/ monolith	Methylparaben /photo-Fenton	Photo-Fenton at $pH_0 = 7$		[125]
TiO_2-$FeSO_4$/ metallic foam	Malachite green dye/photo-Fenton	LED employment	Small amount of Fe leached	[124]

5. Process Combination

This section offers investigations that reported the combination of processes, either because of a combination of different photo-assisted advanced oxidation processes (AOPs) or because of a combination of at least one AOP with another type of treatment. In the same way, this section will review and classify the combined treatments as those that are hybrid or single-step processes and sequential or two-step processes, as seen in Figure 6.

During the latest years, the scientific community has focused on the study of technologies combination. However, most of these studies have been performed at the lab scale, with very low implementation on a pilot and real scale. Therefore, this section especially summarises the recent advances at the lab scale by combining technologies.

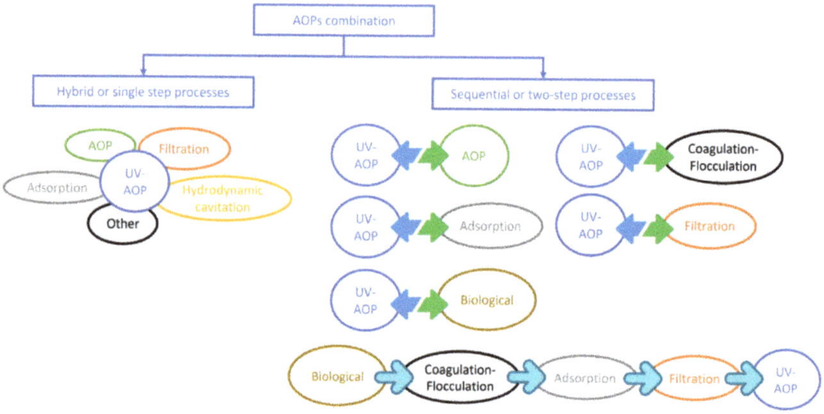

Figure 6. Possibilities of combining UV-AOPs in a single or sequenced stage.

In this way, the combination of processes only has one objective: increasing the treatment capacity of the process, either by an accumulative effect as a consequence of the increase in the production of free radical species, or by a synergistic effect by the positive interaction of the individual processes [127]. The calculation of the synergistic effect has been reported by different authors in different ways. While Dewil et al. [127] reported the calculation following Equation (7), Spuhler et al. [128] followed Equation (8).

$$S = \frac{k_{combined} - \sum_1^n k_i}{k_{combined}} \quad (7)$$

$$S = \frac{k_{combined}}{\sum_i^n k_i} \quad (8)$$

It seems logical to think that a combination of treatments is carried out with the aim of obtaining a positive synergetic effect. However, this does not always happen, and sometimes, this combination does not necessarily have to be beneficial from the point of view of final performance. This effect may be more common when the combination of processes involves different AOPs and is a consequence of the excess of radical species generated, which can produce scavenger effects.

5.1. Hybrid/Integrated Systems including at Least One Photo-Assisted AOP

Therefore, in a first approach, we will describe the recent literature found about the combination processes as a hybrid or integrated system. Most of the research has been focused separately on photo-assisted advanced oxidation processes. However, hybrid or integrated systems are single-step processes that use the advantages of available treatment to overcome the challenges and limitations during the removal of pollutants. These hybrid systems can be divided by a combination of a photo-assisted AOP with other AOPs or with other kinds of technologies (physical, chemical, or biological).

Table 10 summarises a collection of recent hybrid treatments based on the combination of a photo-assisted AOP with other AOPs in a single step. Furthermore, in these combinations, the authors have proved the synergistic effect as a result of combination of two AOPs in a single step with respect to the effect of the different AOPs individually.

Table 10. Hybrid or single-step combination of photo-assisted advanced oxidation processes.

Combination	Target Pollutant	Water Matrix	Maximum Efficiency	Remarks	Ref.
$O_3/H_2O_2/$UV-C $ZnO/Na_2S_2O_8/$UV-C $TiO_2/Na_2S_2O_8/$UV-C	Fluroxypyr and Triclopyr (herbicides)	Drinking water (DW) Leaching water (LW)	92% Fluroxypyr in DW; 62% Triclopyr in DW, reached with $O_3/H_2O_2/$UV-C system (120 min)	H_2O_2 = 75 mg/L; O_3 = 500 mg/h; pH = 7.3 (LW); 8.2 (DW) ZnO = 300 mg/L; $Na_2S_2O_2$ = 300 mg/L; pH = 7.3 (LW); 8.2 (DW) TiO_2 = 300 mg/L; $Na_2S_2O_2$ = 300 mg/L; pH = 7.3 (LW); 8.2 (DW)	[129]
$O_3/H_2O_2/$UV-A	Sulfamethoxazole (antibiotic)	Ultrapure water (UW)	100% (15 min)	H_2O_2 = 45.3 μL; O_3 = 50g/m^3; pH = 5.1	[130]
$O_3/H_2O_2/$UV-A	Methylparaben, ethylparaben, propylparaben, butylparaben and benzylparaben (parabens)	Ultrapure water (UW)	100% (90 min)	H_2O_2 = 13.5 mg/L; O_3 = 0.2 mg/L; pH = 5.7	[131]
Solar Photo-Fenton/O_3	Acetaminophen (drug), antipyrine (analgesic), bisphenol A (additive), caffeine (drug abuse), metoprolol (β-blocker) and testosterone (hormone)	Milli-Q water (UW)	≈100% in average (10 min)	Fe(III) = 2.79 mg/L; H_2O_2/Fe(III) = 6.09 (mass); average incident UV-A solar radiation 41.2 W/m^2; O_3 = 13 mg/L; pH = 3	[132]
TiO_2/Solar radiation/O_3			≈100% in average (15–20 min)	TiO_2 = 200 mg/L; average incident UV-A solar radiation 41.2 W/m^2; Ozone = 13 mg/L; pH = 7	
Solar photo-Fenton/O_3 Solar H_2O_2/O_3 Solar Fe/O_3	Pentachlorophenol (plaguicide), Terbutryn (herbicide), Chlorofenvinphos (insecticide), Diclofenac (drug)	Ultrapure water (UW) Natural water (NW) Simulated wastewater (SW)	80% in average (4.4 kJ/L) in SW	H_2O_2 = 1.5 mM; Fe(II) and Fe(III) = 0.1 mM; O_3 = 0.2–0.6 mg/L; pH = 8	[133]
Solar TiO_2/O_3	DEET (insecticide)	Ultrapure water (UW) Real Wastewater (WW)	≈90% (10 min)	TiO_2 = 150 g/L (supported); pH = 8; O_3 = 15 mg/L; I_{solar} =550 W/m^2	[134]
$TiO_2/UV/O_3$	Dimethyl silanediol, acetic acid; diisobutyl phthalate (odour compounds)	Industrial wastewater (IW)	63.8% Dimethyl silanediol; 41.5% acetic acid; 74.2% diisobutyl phthalate (25 min)	O_3 = 16 mg/L; pH = 7.5; TiO_2 = 3 g/L	[135]
$TiO_2/O_3/$UV-A	Methyl-dopa and famotidine (drugs), and nalidixic acid (antibiotic)	Ultrapure water (UW)	84.93–99.15% (30 min)	O_3 = 6 L/h; TiO_2 (supported) = 12.5 g/L	[136]
$TiO_2/UV/O_3/H_2O_2$	Volatile Organic Compounds mix	Simulated wastewater	98% (30 min)	TiO_2 = 100 mg/L; O_3 = 5.92 g; H_2O_2 = 1.78 g	[137]
PMS/TiO_2/UV-A	Methylene Blue (dye)	Ultrapure water (UW)	>90% (60 min)	PMS = 0.32 mM; TiO_2 = 5 mM; pH = 7	[138]
Solar photo-Fenton/TiO_2	Metronidazole (antibiotic)	Real Wastewater (WW)	≈60% (15 min)	H_2O_2 = 450 mg/L; pH = 3.0–3.5; Fe and TiO_2 supported	[139]
Solar photoelectron-Fenton (Fe_3O_4/ZnO/graphene)	Pyrimethanil and Methomyl mix (pesticides)	Ultrapure water (UW)	>50% (5 min)	Na_2SO_4 = 50 mM; pH = 3.0; j = 74 mA/cm^2; Q = 5.6 L/min; Q_{air} = 10 L/min; Fe(II) = 0.5 mM; 64.9 mg H_2O_2/min	[140]
Sono photo-Fenton	Methylene blue and Congo red (dyes)	Ultrapure water (UW)	100% (<60 min)	40 W UV-C lamps; 40 kHz of US; Fe_3O_4/ZnO/graphene nanocomposites; Ph = 3 and 13; H_2O_2 = 4 mL	[141]
Sono photo-Fenton	Di-n-butyl phthalate (plasticizer)	Ultrapure water (UW)	80% (30 min)	Fe(II) = 0.1 mM; 400 kHz; UV-C; pH = 2–9; H_2O_2 = 0.025–0.2 mM	[142]
CuO_x–$BiVO_4$/SPS/Solar System	Sulfamethoxazole	Ultrapure water (UW), bottled water (BW) secondary wastewater (WW)	UW: 100% (30 min) BW: 100% (60 min) WW: 60% (120 min)	Solar radiation; [SPS] = 500 mg/L; [Catalyst] = 500 mg/L; 0.75, 3.0 and 10.0 Cu.BVO	[143]
CoAl-LDH/BiOBr/PMS/Visible	Ciprofloxacin	Ultrapure water (UW)	96% (30 min)	300 W Xenon lamp; 8wt% CoAl-LDH-BiOBr; [PMS] = 60–100 mg/L; [Catalyst] = 30–50 mg/L	[144]
WO_3/BiOBr/PMS/visible	Tetracycline and enrofloxacin	Ultrapure water (UW)	Tetracycline: 98% (60 min) Enrofloxacin: 87%	300 W Xenon lamp; 2 WO_3 + BiOBr; [PMS] = 30 mg/L; [Catalyst] = 20–60 mg	[145]
PDS/H_2O_2/Fe/Solar radiation	*E. coli* (bacteria); MS2 (bacteriophage); 13 micropollutants mix	Ultrapure water (UW)	>6-Log for *E. coli* (30 min) and MS2 (10 min); >90% for micropollutants (30 min)	H_2O_2 = 10 ppm; 40°CM Fe(II) = 1 ppm; PDS = 9 × 10^{-5} M; 900 W/m^2	[146]
PMS/H_2O_2/Fe/Solar radiation	*E. coli* (bacteria) 13 micropollutants mix	Natural water (NW)	>6-Log for *E. coli* (20 min); >90% for micropollutants (15 min)	H_2O_2 = 10 ppm; 40°CM Fe(II) = 1 ppm; PMS = 3.6 × 10^{-5} M; 900 W/m^2	[147]

As can be observed in Table 10, there are many references exploring this combination of photo-assisted AOPs. In general, the combination comes from the mix of oxidants (O_3, H_2O_2, peroxymonosulfate (PMS) or peroxydisulfate (PDS), etc.) with catalysts (TiO_2, ZnO, Fe_3O_4 or other iron species, etc.) in the presence of different kinds of UV radiation (UV-C, UV-A, UV–Vis, solar radiation), or combined with electro or sono- AOPs. Although there is no clear trend, since for this combination of processes, researchers apply it both to the elimination of pathogens [146,147] and to the removal of micropollutants [129–136,138–142], most of the work focusses on the latter. When micropollutants are used as a target pollutant to test the synergy of the treatments, these are normally tested in ultrapure water matrices [130–134,136,138,140–142,146] with a very low influence of other substances present in water, ensuring very high performance (between 90 and 100% of removal), while when more complex aqueous matrices are used, such as natural water or real or simulated urban or industrial wastewater [129,133–135,137,139,147], the performance is lower or a higher reaction time is required to reach total removal. Most of these works are performed on a laboratory scale. The main reason is that AOPs are still emerging technologies without the maturity level necessary to be implemented on a full scale, and this coupling of AOPs in hybrid systems is one more step in the maturation process of this technology.

Therefore, it is worth highlighting the work carried out on complex water samples. For example, Pérez-Lucas et al. [129] studied the combination of different photo-assisted AOPs in a single step (O_3/H_2O_2/UV-C; $ZnO/Na_2S_2O_8$/UV-C and $TiO_2/Na_2S_2O_8$/UV-C) on the removal of two different herbicides (Fluroxypyr and Triclopyr) present in natural and leaching water. Under these operating conditions, 92 and 62% of removal of fluroxypyr and Triclopyr, respectively, was obtained after 120 min, being reduced to 66 and 13% in leaching water samples due to the complexity of these samples. Roccamante et al. [133] studied the removal of a mixture of four different micropollutants (Pentachlorophenol, Terbutryn, Chlorofenvinphos and Diclofenac) in different water matrices (simulated, natural and simulated wastewater) using solar-driven AOP (solar photo-Fenton, solar H_2O_2 and solar Fe(II)) in combination with ozonation. Under these operating conditions, Roccamante et al. reported an average removal of the mixture of micropollutants of 80% with an accumulated dose of radiation of 4 kJ/L when the treatments were applied over the simulated wastewater sample, which is the most complex sample of the three studied. Talwar et al. [139] described a 60% metronidazole after 15 min of treatment using a combination of solar photo-Fenton applied in a cylindrical parabolic collector (CPC) together with TiO_2 as a photocatalyst. Moving to another field of application, Rodríguez-Chueca et al. [147] reported the simultaneous removal of microorganisms and 13 micropollutants with the PMS/solar photo-Fenton system in natural water samples. In this work, the total removal of *E. coli* was observed in 20 min of reaction, while 15 min was required to reduce the average concentration of more than 90% of the 13 micropollutants.

On the other hand, Table 11 summarises the combination of photo-assisted AOPs with other kind of technologies.

In this case, as observed in Table 10, different physical or chemical technologies are the most adequate to pair in the same step with these AOPs, especially the filtration steps, such as microfiltration, ultrafiltration or nanofiltration [148–150], adsorption with activated carbon [151] or hydrodynamic cavitation [152]. The combination of photo-assisted AOPs with a filtration system is particularly noteworthy. The maturity level of this system can be considered higher than the hybrid combination of AOPs shown in Table 10, especially in well-known photocatalytic membrane reactors or, in other words, in photocatalytic membrane reactors. This higher level of maturity is demonstrated in the greater application of these technologies over complex water matrices, such as simulated or real wastewater samples.

Table 11. Hybrid or single-step combination photo-assisted advanced oxidation processes with other technologies.

Combination	Target Pollutant	Water Matrix	Maximum Efficiency	Remarks	Ref.
TiO_2/UV-C/Microfiltration	*E. coli* (bacteria), *Enterococcus* sp. (bacteria), *Candida albicans* (fungi)	Simulated wastewater (SW)	4-Log	316-L porous stainless-steel membranes; 0.2 gTiO_2/membrane; UV-C (254 nm)	[148]
Photo-Fenton/ Ultrafiltration	Chemical Oxygen Demand (COD); Total Organic Carbon (TOC)	Industrial wastewater (IW)	85.9% COD; 74.5% TOC	Under UV-A: pH = 4.00; H_2O_2/Fe^{2+} (g/g) = 10.75 Under UV-C: pH = 4.44; H_2O_2/Fe^{2+} (g/g) = 7.27	[149]
TiO_2/UV/ Ultrafiltration	Chemical Oxygen Demand (COD); Total Organic Carbon (TOC)	Industrial wastewater (IW)	87.4% COD; 70% TOC (360 min)	UV-A (340 nm); 10 g TiO_2 immobilized to Halloysite nanotubes; Hollow fibre ultrafiltration membrane	[150]
Activated Carbon/US/UV/ H_2O_2	Chemical Oxygen Demand (COD)	Petrochemical wastewater (IW)	87% (80 min)	pH = 4.0 ± 0.2; Activated Carbon = 0.4 g/L; H_2O_2 8.0 mM; 320 W US	[151]
Hydrodynamic cavitation/photo-Fenton	Imidacloprid (insecticide)	Ultrapure water (DW)	99.23% (15 min)	H_2O_2 = 3.91 mM; Fe(II):H_2O_2 = 1:40; UV-A (364 nm); inlet pressure from 5 to 20 bar	[152]
Hydrodynamic cavitation/UV-A			45.56% (120 min)	UV-A (364 nm); inlet pressure from 5 to 20 bar	
Hydrodynamic cavitation/photocatalysis			55.18% (120 min)	Nb_2O_5 = 200 mg/L; UV-A (364 nm); inlet pressure from 5 to 20 bar	

Rodríguez-Chueca et al. [148] reduced the microbial population in simulated wastewater samples using a photocatalytic membrane reactor, combining TiO_2 with UV-C radiation and microfiltration. Under the reported operating conditions, 4-log units of *E. coli*, *Enterococcus* sp., and *Candida albicans* were reduced in simulated wastewater samples using a very short contact time. In a photocatalytic membrane reactor, Moslehyani et al. [150] reported the successful reduction in organic matter content in terms of COD and TOC (87.4 and 70%, respectively) in industrial wastewater, using ultrafiltration and TiO_2 exposed to UV-A radiation. Similar to photocatalytic membrane reactors but working in a homogeneous phase, the combined system proposed by Aydiner et al. [149], in which photo-Fenton (with UV-A or UV-C radiation) was combined with ultrafiltration to reduce the content of organic matter (COD and TOC) in industrial wastewater samples, reached similar results to the reported by Moslehyani et al. with TiO_2 [150]. In the same level of reduction of the organic matter content (87% of COD) in industrial wastewater are the results reported by Kakavandi and Ahmadi [151]; however, in this case, photo-assisted AOP was not combined with a filtration step, but also, a combination of activated carbon with sonication and H_2O_2/UV was used. Therefore, in general terms, it is clearly observed that hybrid systems combining AOPs with other technologies have proven to be mature enough to reduce organic content and organic micropollutants in complex water matrices, demonstrating the synergistic effect caused by the different treatments performed by separation.

5.2. Sequential or Two-Step Combination Processes

Not always is the combination of processes in a single step technically possible; even this combination should be the best one in terms of efficiency. For this reason, the combination of processes in sequential or two-step systems is also studied. This is also called a treatment chain or treatment train.

In the first approach, this section summarises those research reports where treatment trains combine different AOPs (with at least one photo-assisted). This information is shown in Table 12.

Table 12. Sequential or two-step combination of advanced oxidation processes.

Combination	Target Pollutant	Water Matrix	Efficiency	Remarks	Ref.
Ozonation/UV/ H_2O_2	N,N-Diethyl-p-phenylenediamine sulfate, nitrobenzene, benzoic acid	Ultrapure water (UW) with Dissolved Organic Matter	Benzoic acid 80% (30 min)	O_3 = 5 mg/L; UV-C (254 nm); H_2O_2 = 1 mM; pH = 7	[153]
Ozonation/UV/ $S_2O_8^{2-}$			Benzoic acid > 90% (30 min)	O_3 = 5 mg/L; UV-C (254 nm); $S_2O_8^{2-}$ = 1 mM; pH = 7	
Ozonation/UV/ HClO			Benzoic acid > 20% (20 min)	O_3 = 5 mg/L UV-C (254 nm); HClO = 5 mg/L; pH = 7	
Electrocoagulation/ Electrooxidation/ PMS/UV/$CuFe_2O_4$	Chemical Oxygen Demand (COD); Total Organic Carbon (TOC); Biochemical Oxygen Demand (BOD), Ammonia (NH_4^+)	Landfill leachate (LL)	COD 95.6%; TOC 90.5%; BOD 91.6%; NH_4^+ 99.8%	Current density = 50 mA/cm^2; PbO_2 anode; pH = 5; PMS = 15 mM; $CuFe_2O_4$ = 0.15 g/L	[154]

As can be seen in Table 10, the sequential combination of AOPs is not common in the literature. It is not clear, the reason to explain this, except to think that AOPs are such efficient treatments from a contaminant removal point of view, it does not seem to make much sense to combine more than one treatment in sequence. Normally, the combination of treatments in sequence aims to remove contaminants of a different nature (organic, inorganic, metals, microorganisms, etc.), and therefore, treatments of a different nature are used. However, the use of AOPs is exclusively focused on the chemical oxidation of organic and biological contaminants; thus, if these are reduced in great quantities in a first stage, it does not seem to make much sense to include a new subsequent stage with the same nature. However, Sun et al. [153] reported the removal of N,N-Diethyl-p- phenylenediamine sulphate, nitrobenzene and benzoic acid through a sequential combination of ozonation and different photo-assisted AOPs (such as H_2O_2/UV, $S_2O_8^{2-}$/UV and HClO/UV). Under the operating conditions reported by Sun et al. [153], acid benzoic was reduced by more than 90% in the $O_3/S_2O_8^{2-}$/UV system, followed by O_3/H_2O_2/UV with a yield of 80%. Ghanbari et al. [154] reported a reduction in organic matter in more than 90% (COD and TOC) in landfill leachate by sequential combination of electrocoagulation/electrooxidation and the $CuFe_2O_4$/PMS/UV system.

In addition, as mentioned above, it makes more sense to combine, in sequence, AOPs with other different technologies, in order to reach synergies and to maximise the removal of pollutants in water. In addition, AOPs are always considered the last step in a chain treatment because of their ability to remove organic micropollutants or pathogenic germs, and because the cleaner the water matrix, the more effective it is. Therefore, these AOPs may be coupled with biological treatments in order to reduce previously the content of organic matter, such as with coagulation–flocculation to reduce the colloidal matter in water, with adsorption, or with filtration systems to maximise the removal of a certain organic substance or a group of them. Table 13 summarises recent references in the literature studying this treatment train that include at least one photo-assisted AOP. These treatment trains are more focused on the treatment of real water samples, especially focused on the treatment of industrial or urban wastewater, and at a larger scale than those treatments shown in the previous tables.

Table 13. Sequential or two-step combination processes (one photo-assisted AOP + other technology).

Combination	Target Pollutant	Water Matrix	Efficiency	Remarks	Ref.
Biological/Coagulation–flocculation/Photo-Fenton/Biological	Chemical Oxygen Demand (COD); Biochemical Oxygen Demand (BOD); Ammonium; Alkalinity; chromium; total suspended solids (TSS); recalcitrant organic compounds	Leather tannery landfill leachate (LL)	The efficiency for each target pollutant is not reported in the global system	Continuous-flow SBR: 2.0 g/L < MLVSS < 4.0 g/L; sludge volume index (SVI) of 53 mL/g; 200–400 mg/L $FeCl_3$ as coagulant at pH 3; H_2O_2 = 400 mg/L; Dissolved iron = 150 mg/L	[155]
Biological/Coagulation–flocculation/Photo-electroFenton/Biological				Continuous-flow SBR: 2.0 g/L < MLVSS < 4.0 g/L; sludge volume index (SVI) of 53 mL/g; 200–400 mg/L $FeCl_3$ as coagulant at pH 3; H_2O_2 = 400 mg/L; Dissolved iron = 150 mg/L; Current density = 300 mA/cm^2	
Solar photo-Fenton/Activated sludge	Chemical Oxygen Demand (COD); Total Organic Carbon (TOC)	Low strength Industrial wastewater (IW)	84% (COD)	pH = 3; H_2O_2 = 0.25 M; Fe(II) = 0.05 M; 20% (*v*/*v*) sludge concentration	[156]
		High strength Industrial wastewater (IW)	82% (COD)	pH = 3; H_2O_2 = 1 M; Fe(II) = 0.1 M; 25% (*v*/*v*) sludge concentration	
Electrocoagulation/Adsorption/photo-Fenton-like	Chemical Oxygen Demand (COD); Total Organic Carbon (TOC); Turbidity; Colour; Suspended Solids (SS)	Textile Wastewater (IW)	87% (TOC); 49% (COD); 96% (Turbidity); 90% (Colour); 95–97% (SS)	Electrocoagulation: Al Electrode; current density = 25 mA/cm^2; pH = 5. Adsorption: 1 g/L corncob. Photo-Fenton-like: $BiNiO_3$ = 0.75 g/L; H_2O_2 = 2–8 mM; pH = 7–7.5	[157]
Activated sludge/H_2O_2/UV-C	Carbamazepine; Clarithromycin; Diclofenac; Metoprolol; Benzotriazole; Mecoprop.	Urban Wastewater (WW)	Average removal micropollutants: 25% (activated sludge) + 100% (10 min)	Activated sludge: hydraulic retention time = 4 h; sludge retention time = 2 d; H_2O_2 = 25 mg/L	[158]
Activated sludge/solar photo-Fenton			Average removal micropollutants: 25% (activated sludge) + 28% (60 min)	Activated sludge: hydraulic retention time = 4 h; sludge retention time = 2 d Fe(II) = 5 mg/L; H_2O_2 = 25 mg/L	
Moving bed bioreactor/H_2O_2/UV-C			Average removal micropollutants: 40% (moving bed bioreactor) + 100% (10 min)	Moving bed bioreactor: no remarks; H_2O_2 = 25 mg/L	
Moving bed bioreactor/solar photo-Fenton			Average removal micropollutants: 40% (moving bed bioreactor) + 31% (60 min)	Moving bed bioreactor: no remarks; Fe(II) = 5 mg/L; H_2O_2 = 25 mg/L	
Coagulation–Flocculation/H_2O_2/UV-C			Average removal micropollutants: 20% (coagulation–flocculation) + 100% (30 min)	Coagulation–flocculation = $FeCl_3$ (40%) as coagulant. H_2O_2 = 25 mg/L	
Coagulation–Flocculation/solar photo-Fenton			Average removal micropollutants: 20% (coagulation–flocculation) + 11% (60 min)	Coagulation–flocculation = $FeCl_3$ (40%) as coagulant. Fe(II) = 5 mg/L; H_2O_2 = 25 mg/L	

Table 13. *Cont.*

Combination	Target Pollutant	Water Matrix	Efficiency	Remarks	Ref.
TiO_2/UV/ Ultrafiltration	Dissolved Organic Carbon (DOC)	Simulated fresh water	>80% (120 min)	10 kDa-flat sheet polyethersulfone (PES) membrane 41.8 cm^2; TiO_2 = 0.4–0–6 g/L; UV-A (354 nm)	[159]
O_3/H_2O_2/Carbon based biofilter/ UV-C/H_2O_2	13 detected micropollutants	Treated real wastewater (WW)	O_3/H_2O_2 = 78%; Carbon based biofilter = 87%; UV/H_2O_2 = 43%	O_3 = 13 ± 0.5 mg/L, H_2O_2 = 11 ± 0.4 mg/L for the O_3/H_2O_2 process, and UV = 410 ± 63.5 mJ/cm^2, H_2O_2 = 5 mg/L for the UV-C/H_2O_2 process	[160]
O_3/H_2O_2/ Limestone Based Biofilter/UV-C/H_2O_2			O_3/H_2O_2 = 78%; Limestone based biofilter = 67%; UV/H_2O_2 = 43%		
O_3/H_2O_2/ Ultrafiltration/ UV-C/H_2O_2/ Reverse Osmosis			O_3/H_2O_2 = 78%; Ultrafiltration = 0%; UV/H_2O_2 = 43%; Reverse osmosis = 99%		
Coagulation–flocculation/ solar photo-Fenton/aerobic bio-treatment	Chemical Oxygen Demand (COD); Dissolved Organic Carbon (DOC); Total polyphenol content (TPC)	Cork boiling wastewater (IW)	Coagulation–flocculation/solar photo-Fenton = 93,4% (COD); 92,8% (DOC); 94,5% (TPC)	Coagulation–flocculation: $FeCl_3$ as coagulant (3 min 100 rpm + 30 min 30 rpm + 30 min). Solar photo-Fenton: Fe(III) from coagulation = 46–80 mg/L; H_2O_2 = 1–2.5 g/L; Sequencing Batch Bioreactor (Activated sludge)	[161]
Moving bed bioreactor/ photo-Fenton	Chemical Oxygen Demand (COD); Total Organic Carbon (TOC); Oil and grease	Industrial Wastewater (IW)	>95% (COD)	Fe(II):H_2O_2 = 250:800 (mg/L)	[162]
Coagulation–flocculation/ UV-A-LED/ Photo-Fenton	Chemical Oxygen Demand (COD)	Industrial wastewater (IW)	74% (360 min)	H_2O_2 = 5459 mg/L; Fe(III) = 286 mg/L; UV-A LED 85 W/m^2	[163]
Photo-Fenton/ Chemical Addition Dissolved Air Flotation (CA-DAF)	Chemical Oxygen Demand (COD); Total Petroleum Hydrocarbon (TPH)	Industrial wastewater (IW)	99.85% for COD and 98.9% for TPH	pH = 3; $FeSO_4$ = 100 mg/L; H_2O_2 = 17.8 g/L. For DAF unit volume and loading rate were 7 m^3 and 35–40 L/min. Aeration rate 15–20 L/min, and pressure was set at 3 bar and the saturation time of 30 min	[164]
Coagulation–flocculation/ Photo-Fenton	*E. coli*; *Enterococcus* sp.; *Pseudomonas aeruginosa*	Simulated wastewater (SW)	>4 log in all bacteria (210 min)	pH = 5; Fe(III) = 5 mg/L; H_2O_2 = 25 mg/L	[165]
Coagulation–Flocculation/ H_2O_2/UV-C	Detected micropollutants	Treated urban wastewater (WW)	Average removal: Coagulation–flocculation < 10%; 55% (H_2O_2/UV-C)	Coagulation–flocculation = 1.1 kg/m^3 of polyelectrolyte; H_2O_2 = 0.5 mM; pH = natural	[166]
Coagulation–Flocculation/ PDS/UV-C			Average removal: Coagulation–flocculation < 10%; <20% (PDS/UV-C)	Coagulation–flocculation = 1.1 kg/m^3 of polyelectrolyte; PDS = 0.5 mM; pH = natural	
Coagulation–Flocculation/ PMS/UV-C			Average removal: Coagulation–flocculation < 10%; 48% (H_2O_2/UV-C)	Coagulation–flocculation = 1.1 kg/m^3 of polyelectrolyte; PMS = 0.5 mM; pH = natural	

As it has been previously mentioned and as seen in Table 13, these treatment trains are normally tested over real water samples, especially over samples with a higher complexity such as urban and industrial wastewater, due to the combination of mature technology (biological, coagulation–flocculation, adsorption or filtration systems) with an emerging technique such as AOPs. In addition, in this case, research is focused not only on the removal of emerging pollutants but also on the reduction of organic matter content or other parameters usually controlled for in regular wastewater treatment facilities, such as nitrogen, phosphorus, turbidity, total suspended solids, or even microorganisms. The works presented in Table 13 are not comparable with each other when analysing different water matrices, with different operating conditions, even with different types of unitary processes within the treatment chain. However, it can be concluded that the combination of processes is what increases the results of the elimination of the pollutants in question.

6. Concluding Remarks

In conclusion, UV-AOPs have been successfully applied to the removal of a wide range of contaminants from polluted water. However, their application at an industrial scale remains a challenge due to the high operational costs associated with these processes. To address this issue, intensifying UV-AOPs through optimisation of operating conditions and development of bifunctional catalysts is a promising approach for improving their feasibility and effectiveness on an industrial scale.

Recent research has focused on optimising different operational parameters, such as catalyst loading, pH, temperature, and oxidant concentration, to augment the photocatalytic process efficiency and reduce reagent consumption and treatment time. The efficiency of UV-AOPs is sensitive to these parameters, and while increasing catalyst loading can improve pollutant removal rates, exceeding the maximum optimal value can slow the rate of mineralisation. Different studies have found varying optimal concentrations for different catalysts and conditions, and some have developed models to better understand the dependence of reaction rates on catalyst loadings.

The review also discusses the development of photocatalytic materials for intensifying the UV-AOPs process. This optimisation has primarily been in two directions: to reduce the physical limitations of the reaction and to increase the overall yield of the treatment, which is also related to the chemical limitations. The main approaches to intensifying photocatalysts can be divided into two main categories: design orientated to reduce mass and photon transfer limitations and design of bifunctional catalysts that increase the generation of oxidising species. These strategies are further discussed, including the main approaches found in the literature for each of them.

Additionally, the review examines the combination of different photo-assisted AOPs for the treatment of pollutants. The combination of processes is often carried out with the aim of achieving a positive synergistic effect, but this is not always the case. The text also describes hybrid and integrated systems that combine AOPs with other technologies and reports on recent research in this area. However, while much research has been on AOPs on the laboratory scale, there has been little implementation on pilot or real scales. Therefore, more research is needed to better understand the mechanisms of these combinations and to improve their applicability at larger scales.

In conclusion, intensifying UV-AOPs through optimisation of operating conditions and development of bifunctional catalysts are promising approaches for improving their feasibility and effectiveness on an industrial scale. Further research is also necessary to better understand the mechanisms of combining different photo-assisted AOPs and to improve their applicability at larger scales.

Author Contributions: Conceptualization, J.R.-C., J.C. and P.G.-M.; methodology, J.R.-C., J.C. and P.G.-M.; formal analysis, J.R.-C., J.C. and P.G.-M.; investigation, J.R.-C., J.C. and P.G.-M.; writing—original draft preparation, J.R.-C., J.C. and P.G.-M.; writing—review and editing, J.R.-C., J.C. and P.G.-M.; visualization, J.R.-C., J.C. and P.G.-M. All authors have equally contributed to this manuscript. All authors have read and agreed to the published version of the manuscript.

Funding: J.R.-C. acknowledges Comunidad de Madrid for funding by the pluriannual agreement with the Polytechnic University of Madrid in the line of action Programme of Excellence for University Teaching Staff (M190020074BJJRC). P.G.-M. acknowledges Comunidad de Madrid through the call Research Grants for Young Investigators from Universidad Politécnica de Madrid for funding the research project SUNCAT4PLAST (APOYO-JOVENES-21-JV4DEB-3-Q2WGKV). J.R.-C. and P.G.-M. acknowledge the Spanish Ministry of Science and Innovation for funding the research project PHOTORAS (PID2021-128165OA-I00).

Informed Consent Statement: Not Applicable.

Data Availability Statement: Data sharing is not applicable to this article.

Conflicts of Interest: The authors declare no conflict of interest.

References

1. Pandis, P.K.; Kalogirou, C.; Kanellou, E.; Vaitsis, C.; Savvidou, M.G.; Sourkouni, G.; Zorpas, A.A.; Argirusis, C. Key Points of Advanced Oxidation Processes (AOPs) for Wastewater, Organic Pollutants and Pharmaceutical Waste Treatment: A Mini Review. *Chem. Eng.* **2022**, *6*, 8. [CrossRef]
2. Ushani, U.; Lu, X.; Wang, J.; Zhang, Z.; Dai, J.; Tan, Y.; Wang, S.; Li, W.; Niu, C.; Cai, T.; et al. Sulfate radicals-based advanced oxidation technology in various environmental remediation: A state-of-the–art review. *Chem. Eng. J.* **2020**, *402*, 126232. [CrossRef]
3. Brienza, M.; Katsoyiannis, I.A. Sulfate Radical Technologies as Tertiary Treatment for the Removal of Emerging Contaminants from Wastewater. *Sustainability* **2017**, *9*, 1604. [CrossRef]
4. Salazar, L.M.; Grisales, C.M.; Garcia, D.P. How does intensification influence the operational and environmental performance of photo-Fenton processes at acidic and circumneutral pH. *Environ. Sci. Pollut. Res.* **2018**, *26*, 4367–4380. [CrossRef] [PubMed]
5. Spasiano, D.; Marotta, R.; Malato, S.; Fernandez-Ibañez, P.; Di Somma, I. Solar photocatalysis: Materials, reactors, some commercial, and pre-industrialized applications. A comprehensive approach. *Appl. Catal. B Environ.* **2015**, *170–171*, 90–123. [CrossRef]
6. Dimian, A.C.; Bildea, C.S.; Kiss, A.A. Process Intensification. *Comput. Aided Chem. Eng.* **2014**, 397–448. [CrossRef]
7. Huang, K.; Wang, S.-J.; Shan, L.; Zhu, Q.; Qian, J. Seeking synergistic effect—A key principle in process intensification. *Sep. Purif. Technol.* **2007**, *57*, 111–120. [CrossRef]
8. Mesones, S.; Mena, E.; López-Muñoz, M.J.; Adán, C.; Marugán, J. Synergistic and antagonistic effects in the photoelectrocatalytic disinfection of water with TiO_2 supported on activated carbon as a bipolar electrode in a novel 3D photoelectrochemical reactor. *Sep. Purif. Technol.* **2020**, *247*, 117002. [CrossRef]
9. Malato, S.; Fernández-Ibáñez, P.; Maldonado, M.; Blanco, J.; Gernjak, W. Decontamination and disinfection of water by solar photocatalysis: Recent overview and trends. *Catal. Today* **2009**, *147*, 1–59. [CrossRef]
10. Herrmann, J.-M. Photocatalysis fundamentals revisited to avoid several misconceptions. *Appl. Catal. B Environ.* **2010**, *99*, 461–468. [CrossRef]
11. Azzaz, A.; Jellali, S.; Hamed, N.; El Jery, A.; Khezami, L.; Assadi, A.; Amrane, A. Photocatalytic Treatment of Wastewater Containing Simultaneous Organic and Inorganic Pollution: Competition and Operating Parameters Effects. *Catalysts* **2021**, *11*, 855. [CrossRef]
12. Ochoa-Gutiérrez, K.S.; Tabares-Aguilar, E.; Mueses, M.; Machuca-Martínez, F.; Puma, G.L. A Novel Prototype Offset Multi Tubular Photoreactor (OMTP) for solar photocatalytic degradation of water contaminants. *Chem. Eng. J.* **2018**, *341*, 628–638. [CrossRef]
13. Elhalil, A.; Elmoubarki, R.; Farnane, M.; Machrouhi, A.; Sadiq, M.; Mahjoubi, F.; Qourzal, S.; Barka, N. Photocatalytic degradation of caffeine as a model pharmaceutical pollutant on Mg doped ZnO-Al_2O_3 heterostructure. *Environ. Nanotechnol. Monit. Manag.* **2018**, *10*, 63–72. [CrossRef]
14. Jallouli, N.; Pastrana-Martínez, L.M.; Ribeiro, A.R.; Moreira, N.F.; Faria, J.L.; Hentati, O.; Silva, A.M.; Ksibi, M. Heterogeneous photocatalytic degradation of ibuprofen in ultrapure water, municipal and pharmaceutical industry wastewaters using a TiO_2/UV-LED system. *Chem. Eng. J.* **2018**, *334*, 976–984. [CrossRef]
15. León, D.E.; Zúñiga-Benítez, H.; Peñuela, G.A.; Mansilla, H.D. Photocatalytic Removal of the Antibiotic Cefotaxime on TiO_2 and ZnO Suspensions Under Simulated Sunlight Radiation. *Water Air Soil Pollut.* **2017**, *228*, 361. [CrossRef]
16. Martínez, C.; Fernández, M.I.; Santaballa, J.A.; Faria, J. Aqueous degradation of diclofenac by heterogeneous photocatalysis using nanostructured materials. *Appl. Catal. B Environ.* **2011**, *107*, 110–118. [CrossRef]
17. Garrido, I.; Flores, P.; Hellín, P.; Vela, N.; Navarro, S.; Fenoll, J. Solar reclamation of agro-wastewater polluted with eight pesticides by heterogeneous photocatalysis using a modular facility. A case study. *Chemosphere* **2020**, *249*, 126156. [CrossRef]
18. Wang, B.; Fu, T.; An, B.; Liu, Y. UV light-assisted persulfate activation by Cu^0-Cu_2O for the degradation of sulfamerazine. *Sep. Purif. Technol.* **2020**, *251*, 117321. [CrossRef]
19. Tao, Y.; Ni, Q.; Wei, M.; Xia, D.; Li, X.; Xu, A. Metal-free activation of peroxymonosulfate by g-C_3N_4 under visible light irradiation for the degradation of organic dyes. *RSC Adv.* **2015**, *5*, 44128–44136. [CrossRef]

20. Casado, C.; Marugán, J.; Timmers, R.; Muñoz, M.; van Grieken, R. Comprehensive multiphysics modeling of photocatalytic processes by computational fluid dynamics based on intrinsic kinetic parameters determined in a differential photoreactor. *Chem. Eng. J.* **2017**, *310*, 368–380. [CrossRef]
21. Carbajo, J.; Tolosana-Moranchel, A.; Casas, J.A.; Faraldos, M.; Bahamonde, A. Analysis of photoefficiency in TiO_2 aqueous suspensions: Effect of titania hydrodynamic particle size and catalyst loading on their optical properties. *Appl. Catal. B Environ.* **2018**, *221*, 1–8. [CrossRef]
22. Tolosana-Moranchel, A.; Casas, J.; Carbajo, J.; Faraldos, M.; Bahamonde, A. Influence of TiO_2 optical parameters in a slurry photocatalytic reactor: Kinetic modelling. *Appl. Catal. B Environ.* **2017**, *200*, 164–173. [CrossRef]
23. Acosta-Herazo, R.; Mueses, M.; Puma, G.L.; Machuca-Martínez, F. Impact of photocatalyst optical properties on the efficiency of solar photocatalytic reactors rationalized by the concepts of initial rate of photon absorption (IRPA) dimensionless boundary layer of photon absorption and apparent optical thickness. *Chem. Eng. J.* **2018**, *356*, 839–849. [CrossRef]
24. Grčić, I.; Puma, G.L. Six-flux absorption-scattering models for photocatalysis under wide-spectrum irradiation sources in annular and flat reactors using catalysts with different optical properties. *Appl. Catal. B Environ.* **2017**, *211*, 222–234. [CrossRef]
25. Koltsakidou, A.; Antonopoulou, M.; Sykiotou, M.; Evgenidou, E.; Konstantinou, I.; Lambropoulou, D. Photo-Fenton and Fenton-like processes for the treatment of the antineoplastic drug 5-fluorouracil under simulated solar radiation. *Environ. Sci. Pollut. Res.* **2016**, *24*, 4791–4800. [CrossRef] [PubMed]
26. Villota, N.; Ferreiro, C.; Qulatein, H.; Lomas, J.; Lombraña, J. Turbidity Changes during Carbamazepine Oxidation by Photo-Fenton. *Catalysts* **2021**, *11*, 894. [CrossRef]
27. Coelho, A.; Castro, A.V.; Dezotti, M.; Sant'Anna, G. Treatment of petroleum refinery sourwater by advanced oxidation processes. *J. Hazard. Mater.* **2006**, *137*, 178–184. [CrossRef] [PubMed]
28. Cabrera-Reina, A.; Miralles-Cuevas, S.; López, J.C.; Pérez, J.S. Pyrimethanil degradation by photo-Fenton process: Influence of iron and irradiance level on treatment cost. *Sci. Total Environ.* **2017**, *605-606*, 230–237. [CrossRef]
29. Ahmed, S.; Rasul, M.G.; Martens, W.N.; Brown, R.; Hashib, M.A. Heterogeneous photocatalytic degradation of phenols in wastewater: A review on current status and developments. *Desalination* **2010**, *261*, 3–18. [CrossRef]
30. Seid, M.G.; Byun, J.; Kim, W.; Cho, K.; Hong, S.W. Changes in levels of N-nitrosamine formed from amine-containing compounds during chloramination via photocatalytic pretreatment with immobilized TiO_2: Effect of source water and pH. *J. Hazard. Mater.* **2021**, *424*, 127398. [CrossRef]
31. Azeez, F.; Al-Hetlani, E.; Arafa, M.; Abdelmonem, Y.; Nazeer, A.A.; Amin, M.O.; Madkour, M. The effect of surface charge on photocatalytic degradation of methylene blue dye using chargeable titania nanoparticles. *Sci. Rep.* **2018**, *8*, 7104. [CrossRef] [PubMed]
32. Carbajo, J.; Jiménez, M.; Miralles, S.; Malato, S.; Faraldos, M.; Bahamonde, A. Study of application of titania catalysts on solar photocatalysis: Influence of type of pollutants and water matrices. *Chem. Eng. J.* **2016**, *291*, 64–73. [CrossRef]
33. Mirzaei, A.; Chen, Z.; Haghighat, F.; Yerushalmi, L. Removal of pharmaceuticals from water by homo/heterogonous Fenton-type processes—A review. *Chemosphere* **2017**, *174*, 665–688. [CrossRef]
34. Cavalcante, R.P.; Sandim, L.D.R.; Bogo, D.; Barbosa, A.M.J.; Osugi, M.E.; Blanco, M.; De Oliveira, S.C.; Matos, M.D.F.C.; Machulek, A.; Ferreira, V.S. Application of Fenton, photo-Fenton, solar photo-Fenton, and UV/H_2O_2 to degradation of the antineoplastic agent mitoxantrone and toxicological evaluation. *Environ. Sci. Pollut. Res.* **2012**, *20*, 2352–2361. [CrossRef] [PubMed]
35. Governo, M.; Santos, M.S.F.; Alves, A.; Madeira, L.M. Degradation of the cytostatic 5-Fluorouracil in water by Fenton and photo-assisted oxidation processes. *Environ. Sci. Pollut. Res.* **2016**, *24*, 844–854. [CrossRef]
36. Elmolla, E.S.; Chaudhuri, M. Degradation of the antibiotics amoxicillin, ampicillin and cloxacillin in aqueous solution by the photo-Fenton process. *J. Hazard. Mater.* **2009**, *172*, 1476–1481. [CrossRef]
37. El Shahawy, A.; Mohamadien, R.H.; El-Fawal, E.M.; Moustafa, Y.M.; Dawood, M.M.K. Hybrid Photo-Fenton oxidation and biosorption for petroleum wastewater treatment and optimization using Box–Behnken Design. *Environ. Technol. Innov.* **2021**, *24*, 101834. [CrossRef]
38. Tawfik, A.; Alalm, M.G.; Awad, H.M.; Islam, M.; Qyyum, M.A.; Al-Muhtaseb, A.H.; Osman, A.I.; Lee, M. Solar photo-oxidation of recalcitrant industrial wastewater: A review. *Environ. Chem. Lett.* **2022**, *20*, 1839–1862. [CrossRef]
39. Arzate, S.; Campos-Mañas, M.; Miralles-Cuevas, S.; Agüera, A.; Sánchez, J.G.; Pérez, J.S. Removal of contaminants of emerging concern by continuous flow solar photo-Fenton process at neutral pH in open reactors. *J. Environ. Manag.* **2020**, *261*, 110265. [CrossRef] [PubMed]
40. Prada-Vásquez, M.A.; Estrada-Flórez, S.E.; Serna-Galvis, E.A.; Torres-Palma, R.A. Developments in the intensification of photo-Fenton and ozonation-based processes for the removal of contaminants of emerging concern in Ibero-American countries. *Sci. Total Environ.* **2020**, *765*, 142699. [CrossRef]
41. Clarizia, L.; Russo, D.; Di Somma, I.; Marotta, R.; Andreozzi, R. Homogeneous photo-Fenton processes at near neutral pH: A review. *Appl. Catal. B Environ.* **2017**, *209*, 358–371. [CrossRef]
42. Silva, G.D.; Marson, E.O.; Batista, L.L.; Ueira-Vieira, C.; Starling, M.C.V.; Trovó, A.G. Contrasting the performance of photo-Fenton at neutral pH in the presence of different organic iron-complexes using hydrogen peroxide or persulfate as oxidants for naproxen degradation and removal of antimicrobial activity. *Process. Saf. Environ. Prot.* **2021**, *147*, 798–807. [CrossRef]
43. Cai, C.; Zhang, Z.; Liu, J.; Shan, N.; Zhang, H.; Dionysiou, D.D. Visible light-assisted heterogeneous Fenton with $ZnFe_2O_4$ for the degradation of Orange II in water. *Appl. Catal. B Environ.* **2016**, *182*, 456–468. [CrossRef]

44. Ma, J.; Yang, Q.; Wen, Y.; Liu, W. Fe-g-C_3N_4/graphitized mesoporous carbon composite as an effective Fenton-like catalyst in a wide pH range. *Appl. Catal. B Environ.* **2017**, *201*, 232–240. [CrossRef]
45. Arzate-Salgado, S.-Y.; Morales-Pérez, A.-A.; Solís-López, M.; Ramírez-Zamora, R.-M. Evaluation of metallurgical slag as a Fenton-type photocatalyst for the degradation of an emerging pollutant: Diclofenac. *Catal. Today* **2016**, *266*, 126–135. [CrossRef]
46. Buitrago, J.L.; Sanabria, J.; Gútierrez-Zapata, H.M.; Urbano-Ceron, F.J.; García-Barco, A.; Osorio-Vargas, P.; Rengifo-Herrera, J.A. Photo-Fenton process at natural conditions of pH, iron, ions, and humic acids for degradation of diuron and amoxicillin. *Environ. Sci. Pollut. Res.* **2019**, *27*, 1608–1624. [CrossRef]
47. De la Cruz, N.; Esquius, L.; Grandjean, D.; Magnet, A.; Tungler, A.; de Alencastro, L.; Pulgarín, C. Degradation of emergent contaminants by UV, UV/H_2O_2 and neutral photo-Fenton at pilot scale in a domestic wastewater treatment plant. *Water Res.* **2013**, *47*, 5836–5845. [CrossRef] [PubMed]
48. Tambat, S.; Umale, S.; Sontakke, S. Photocatalytic degradation of Milling Yellow dye using sol–gel synthesized CeO. *Mater. Res. Bull.* **2016**, *76*, 466–472. [CrossRef]
49. Lin, Y.-C.; Lee, H.-S. Effects of TiO_2 coating dosage and operational parameters on a TiO_2/Ag photocatalysis system for decolorizing Procion red MX-5B. *J. Hazard. Mater.* **2010**, *179*, 462–470. [CrossRef]
50. Chen, Y.-W.; Hsu, Y.-H. Effects of Reaction Temperature on the Photocatalytic Activity of TiO_2 with Pd and Cu Cocatalysts. *Catalysts* **2021**, *11*, 966. [CrossRef]
51. Silveira, J.E.; Garcia-Costa, A.L.; Carbajo, J.; Ribeiro, A.R.; Pliego, G.; Paz, W.S.; Zazo, J.A.; Casas, J.A. Nitrate removal in saline water by photo-reduction using natural $FeTiO_3$ as catalyst. *Chem. Eng. J. Adv.* **2022**, *12*, 100387. [CrossRef]
52. Bloh, J.Z. Intensification of Heterogeneous Photocatalytic Reactions Without Efficiency Losses: The Importance of Surface Catalysis. *Catal. Lett.* **2021**, *151*, 3105–3113. [CrossRef]
53. Prieto-Rodríguez, L.; Spasiano, D.; Oller, I.; Fernández-Calderero, I.; Agüera, A.; Malato, S. Solar photo-Fenton optimization for the treatment of MWTP effluents containing emerging contaminants. *Catal. Today* **2013**, *209*, 188–194. [CrossRef]
54. Pérez, M.; Torrades, F.; Domènech, X.; Peral, J. Fenton and photo-Fenton oxidation of textile effluents. *Water Res.* **2001**, *36*, 2703–2710. [CrossRef]
55. Torrades, F.; Pérez, M.; Mansilla, H.D.; Peral, J. Experimental design of Fenton and photo-Fenton reactions for the treatment of cellulose bleaching effluents. *Chemosphere* **2003**, *53*, 1211–1220. [CrossRef] [PubMed]
56. Sagawe, G.; Lehnard, A.; Lübber, M.; Bahnemann, D. The insulated solar fenton hybrid process: Fundamental investi-gations. *Helv. Chim. Acta* **2001**, *84*, 3742–3759. [CrossRef]
57. Gernjak, W.; Fuerhacker, M.; Fernández-Ibañez, P.; Blanco, J.; Malato, S. Solar photo-Fenton treatment—Process parameters and process control. *Appl. Catal. B Environ.* **2006**, *64*, 121–130. [CrossRef]
58. Zhang, J.; Zhang, X.; Wang, Y. Degradation of phenol by a heterogeneous photo-Fenton process using Fe/Cu/Al catalysts. *RSC Adv.* **2016**, *6*, 13168–13176. [CrossRef]
59. Expósito, A.J.; Monteagudo, J.M.; Díaz, I.; Durán, A. Photo-fenton degradation of a beverage industrial effluent: Intensification with persulfate and the study of radicals. *Chem. Eng. J.* **2016**, *306*, 1203–1211. [CrossRef]
60. Zapata, A.; Oller, I.; Bizani, E.; Sánchez-Pérez, J.; Maldonado, M.; Malato, S. Evaluation of operational parameters involved in solar photo-Fenton degradation of a commercial pesticide mixture. *Catal. Today* **2009**, *144*, 94–99. [CrossRef]
61. Carbajo, J.; Silveira, J.; Pliego, G.; Zazo, J.; Casas, J. Increasing Photo-Fenton process Efficiency: The effect of high temperatures. *Sep. Purif. Technol.* **2021**, *271*, 118876. [CrossRef]
62. Vallés, I.; Santos-Juanes, L.; Amat, A.; Moreno-Andrés, J.; Arques, A. Effect of Salinity on UVA-Vis Light Driven Photo-Fenton Process at Acidic and Circumneutral pH. *Water* **2021**, *13*, 1315. [CrossRef]
63. Pliego, G.; Xekoukoulotakis, N.; Venieri, D.; Zazo, J.A.; Casas, J.A.; Rodriguez, J.J.; Mantzavinos, D. Complete degradation of the persistent anti-depressant sertraline in aqueous solution by solar photo-Fenton oxidation. *J. Chem. Technol. Biotechnol.* **2014**, *89*, 814–818. [CrossRef]
64. Hermosilla, D.; Merayo, N.; Ordóñez, R.; Blanco, Á. Optimization of conventional Fenton and ultraviolet-assisted oxidation processes for the treatment of reverse osmosis retentate from a paper mill. *Waste Manag.* **2012**, *32*, 1236–1243. [CrossRef]
65. Pouran, S.R.; Aziz, A.A.; Daud, W.M.A.W. Review on the main advances in photo-Fenton oxidation system for recalcitrant wastewaters. *J. Ind. Eng. Chem.* **2015**, *21*, 53–69. [CrossRef]
66. Sirtori, C.; Zapata, A.; Gernjak, W.; Malato, S.; Lopez, A.; Agüera, A. Solar photo-Fenton degradation of nalidixic acid in waters and wastewaters of different composition. Analytical assessment by LC–TOF-MS. *Water Res.* **2011**, *45*, 1736–1744. [CrossRef]
67. Hermosilla, D.; Cortijo, M.; Huang, C.P. Optimizing the treatment of landfill leachate by conventional Fenton and photo-Fenton processes. *Sci. Total Environ.* **2009**, *407*, 3473–3481. [CrossRef]
68. Cunha-Filho, F.J.; Mota-Lima, A.; Ratkievicius, L.A.; Silva, D.J.; Silva, D.N.; Chiavone-Filho, O.; Nascimento, C.A.O.D. Rapid mineralization rate of acetylsalicylic acid in a tubular photochemical reactor: The role of the optimized excess of H_2O_2. *J. Water Process. Eng.* **2019**, *31*, 100856. [CrossRef]
69. Batista, A.P.S.; Nogueira, R.F.P. Parameters affecting sulfonamide photo-Fenton degradation–Iron complexation and substituent group. *J. Photochem. Photobiol. A Chem.* **2012**, *232*, 8–13. [CrossRef]
70. Chu, W.; Chan, K.; Kwan, C.; Choi, K. Degradation of atrazine by modified stepwise-Fenton's processes. *Chemosphere* **2007**, *67*, 755–761. [CrossRef]

71. Monteagudo, J.; Durán, A.; Martín, I.S.; Aguirre, M. Effect of continuous addition of H_2O_2 and air injection on ferrioxalate-assisted solar photo-Fenton degradation of Orange II. *Appl. Catal. B Environ.* **2009**, *89*, 510–518. [CrossRef]
72. Prieto-Rodríguez, L.; Oller, I.; Zapata, A.; Agüera, A.; Malato, S. Hydrogen peroxide automatic dosing based on dissolved oxygen concentration during solar photo-Fenton. *Catal. Today* **2011**, *161*, 247–254. [CrossRef]
73. Yu, X.; Somoza-Tornos, A.; Graells, M.; Pérez-Moya, M. An experimental approach to the optimization of the dosage of hydrogen peroxide for Fenton and photo-Fenton processes. *Sci. Total Environ.* **2020**, *743*, 140402. [CrossRef] [PubMed]
74. Saldaña-Flores, K.E.; Flores-Estrella, R.A.; Alcaraz-Gonzalez, V.; Carissimi, E.; de Souza, B.G.; Ruotolo, L.A.M.; Urquieta-Gonzalez, E. Regulation of Hydrogen Peroxide Dosage in a Heterogeneous Photo-Fenton Process. *Processes* **2021**, *9*, 2167. [CrossRef]
75. Silva, T.F.; Fonseca, A.; Saraiva, I.; Boaventura, R.A.; Vilar, V.J. Scale-up and cost analysis of a photo-Fenton system for sanitary landfill leachate treatment. *Chem. Eng. J.* **2015**, *283*, 76–88. [CrossRef]
76. Devi, L.G.; Srinivas, M.; ArunaKumari, M. Heterogeneous advanced photo- Fenton process using peroxymonosulfate and peroxydisulfate in presence of zero valent metallic iron: A comparative study with hydrogen peroxide photo-Fenton process. *J. Water Process. Eng.* **2016**, *13*, 117–126. [CrossRef]
77. Giannakis, S.; Lin, K.-Y.A.; Ghanbari, F. A review of the recent advances on the treatment of industrial wastewaters by Sulfate Radical-based Advanced Oxidation Processes (SR-AOPs). *Chem. Eng. J.* **2020**, *406*, 127083. [CrossRef]
78. Ge, M.; Hu, Z.; Wei, J.; He, Q.; He, Z. Recent advances in persulfate-assisted TiO_2-based photocatalysis for wastewater treatment: Performances, mechanism and perspectives. *J. Alloys Compd.* **2021**, *888*, 161625. [CrossRef]
79. Ismail, L.; Ferronato, C.; Fine, L.; Jaber, F.; Chovelon, J.-M. Elimination of sulfaclozine from water with SO_4−radicals: Evaluation of different persulfate activation methods. *Appl. Catal. B Environ.* **2017**, *201*, 573–581. [CrossRef]
80. Ding, H.; Hu, J. Degradation of ibuprofen by UVA-LED/TiO_2/persulfate process: Kinetics, mechanism, water matrix effects, intermediates and energy consumption. *Chem. Eng. J.* **2020**, *397*, 125462. [CrossRef]
81. Du, X.; Bai, X.; Xu, L.; Yang, L.; Jin, P. Visible-light activation of persulfate by TiO_2/g-C_3N_4 photocatalyst toward efficient degradation of micropollutants. *Chem. Eng. J.* **2019**, *384*, 123245. [CrossRef]
82. Carbajo, J.; Adán, C.; Rey, A.; Martínez-Arias, A.; Bahamonde, A. Optimization of H_2O_2 use during the photocatalytic degradation of ethidium bromide with TiO_2 and iron-doped TiO_2 catalysts. *Appl. Catal. B Environ.* **2010**, *102*, 85–93. [CrossRef]
83. Cuervo Lumbaque, E.; Sirtori, C.; Vilar, V.J. Heterogeneous photocatalytic degradation of pharmaceuticals in synthetic and real matrices using a tube-in-tube membrane reactor with radial addition of H_2O_2. *Sci. Total Environ.* **2020**, *743*, 140629. [CrossRef]
84. Fu, Y.; Chang, C.; Chen, P.; Chu, X.; Zhu, L. Enhanced photocatalytic performance of boron doped Bi2WO6 nanosheets under simulated solar light irradiation. *J. Hazard. Mater.* **2013**, *254*, 185–192. [CrossRef]
85. García-Muñoz, P.; Zussblatt, N.P.; Pliego, G.; Zazo, J.A.; Fresno, F.; Chmelka, B.F.; Casas, J.A. Evaluation of photoassisted treatments for norfloxacin removal in water using mesoporous Fe_2O_3-TiO_2 materials. *J. Environ. Manag.* **2019**, *238*, 243–250. [CrossRef]
86. Ismail, A.A.; Abdelfattah, I.; Faisal, M.; Helal, A. Efficient photodecomposition of herbicide imazapyr over mesoporous Ga_2O_3-TiO_2 nanocomposites. *J. Hazard. Mater.* **2018**, *342*, 519–526. [CrossRef]
87. Phan, T.T.N.; Nikoloski, A.N.; Bahri, P.A.; Li, D. Adsorption and photo-Fenton catalytic degradation of organic dyes over crystalline $LaFeO_3$-doped porous silica. *RSC Adv.* **2018**, *8*, 36181–36190. [CrossRef]
88. Phan, T.T.N.; Nikoloski, A.N.; Bahri, P.A.; Li, D. Enhanced removal of organic using $LaFeO_3$-integrated modified natural zeolites via heterogeneous visible light photo-Fenton degradation. *J. Environ. Manag.* **2018**, *233*, 471–480. [CrossRef]
89. Wang, G.; Wang, X.; Liu, J.; Sun, X. Mesoporous Au/TiO_2 Nanocomposite Microspheres for Visible-Light Photocatalysis. *Chem. Eur. J.* **2012**, *18*, 5361–5366. [CrossRef]
90. Wu, Y.; Qin, L.; Zhang, G.; Chen, L.; Guo, X.; Liu, M. Porous Solid Superacid SO_4^{2-}/$Fe_{2-x}Zr_xO_3$ Fenton Catalyst for Highly Effective Oxidation of X-3B under Visible Light. *Ind. Eng. Chem. Res.* **2013**, *52*, 16698–16708. [CrossRef]
91. Simsek, E.B.; Tuna, Ö.; Balta, Z. Construction of stable perovskite-type $LaFeO_3$ particles on polymeric resin with boosted photocatalytic Fenton-like decaffeination under solar irradiation. *Sep. Purif. Technol.* **2019**, *237*, 116384. [CrossRef]
92. Barik, A.J.; Gogate, P.R. Degradation of 4-chloro 2-aminophenol using combined strategies based on ultrasound, photolysis and ozone. *Ultrason. Sonochem.* **2016**, *28*, 90–99. [CrossRef]
93. Sheydaei, M.; Zangouei, M.; Vatanpour, V. Coupling visible light sono-photocatalysis and sono-enhanced ultrafiltration processes for continuous flow degradation of dyestuff using N-doped titania nanoparticles. *Chem. Eng. Process. Process. Intensif.* **2019**, *143*. [CrossRef]
94. Kakavandi, B.; Bahari, N.; Kalantary, R.R.; Fard, E.D. Enhanced sono-photocatalysis of tetracycline antibiotic using TiO_2 decorated on magnetic activated carbon (MAC@T) coupled with US and UV: A new hybrid system. *Ultrason. Sonochem.* **2019**, *55*, 75–85. [CrossRef]
95. Babu, S.G.; Karthik, P.; John, M.C.; Lakhera, S.K.; Ashokkumar, M.; Khim, J.; Neppolian, B. Synergistic effect of sono-photocatalytic process for the degradation of organic pollutants using CuO-TiO_2/rGO. *Ultrason. Sonochem.* **2018**, *50*, 218–223. [CrossRef]
96. Geng, N.; Chen, W.; Xu, H.; Lin, T.; Ding, M.; Wang, Y.; Tao, H.; Hu, K. Preparation of Fe_3O_4/TiO_2-N-GO sonocatalyst and using for humic acid removal with the assist of ultrasound. *Mater. Sci. Semicond. Process.* **2019**, *102*, 104593. [CrossRef]
97. Isari, A.A.; Hayati, F.; Kakavandi, B.; Rostami, M.; Motevassel, M.; Dehghanifard, E. N, Cu co-doped TiO_2@functionalized SWCNT photocatalyst coupled with ultrasound and visible-light: An effective sono-photocatalysis process for pharmaceutical wastewaters treatment. *Chem. Eng. J.* **2019**, *392*, 123685. [CrossRef]

98. Eshaq, G.; Wang, S.; Sun, H.; Sillanpää, M. Core/shell $FeVO_4$@BiOCl heterojunction as a durable heterogeneous Fenton catalyst for the efficient sonophotocatalytic degradation of p-nitrophenol. *Sep. Purif. Technol.* **2019**, *231*, 115915. [CrossRef]
99. Guan, S.; Yang, H.; Sun, X.; Xian, T. Preparation and promising application of novel $LaFeO_3$/BiOBr heterojunction photocatalysts for photocatalytic and photo-Fenton removal of dyes. *Opt. Mater.* **2019**, *100*, 109644. [CrossRef]
100. Hu, J.; Li, J.; Cui, J.; An, W.; Liu, L.; Liang, Y.; Cui, W. Surface oxygen vacancies enriched FeOOH/Bi_2MoO_6 photocatalysis- fenton synergy degradation of organic pollutants. *J. Hazard. Mater.* **2019**, *384*, 121399. [CrossRef]
101. Huang, S.; Zhang, Q.; Liu, P.; Ma, S.; Xie, B.; Yang, K.; Zhao, Y. Novel up-conversion carbon quantum dots/α-FeOOH nanohybrids eliminate tetracycline and its related drug resistance in visible-light responsive Fenton system. *Appl. Catal. B Environ.* **2019**, *263*, 118336. [CrossRef]
102. Li, Q.; Kong, H.; Li, P.; Shao, J.; He, Y. Photo-Fenton degradation of amoxicillin via magnetic TiO_2-graphene oxide-Fe_3O_4 composite with a submerged magnetic separation membrane photocatalytic reactor (SMSMPR). *J. Hazard. Mater.* **2019**, *373*, 437–446. [CrossRef]
103. Manickam-Periyaraman, P.; Espinosa, J.C.; Ferrer, B.; Subramanian, S.; Álvaro, M.; García, H.; Navalón, S. Bimetallic iron-copper oxide nanoparticles supported on nanometric diamond as efficient and stable sunlight-assisted Fenton photocatalyst. *Chem. Eng. J.* **2020**, *393*, 124770. [CrossRef]
104. García-Muñoz, P.; Pliego, G.; Zazo, J.; Bahamonde, A.; Casas, J. Ilmenite ($FeTiO_3$) as low cost catalyst for advanced oxidation processes. *J. Environ. Chem. Eng.* **2016**, *4*, 542–548. [CrossRef]
105. Arshad, A.; Iqbal, J.; Ahmad, I.; Israr, M. Graphene/Fe_3O_4 nanocomposite: Interplay between photo-Fenton type reaction, and carbon purity for the removal of methyl orange. *Ceram. Int.* **2018**, *44*, 2643–2648. [CrossRef]
106. Desipio, M.M.; Van Bramer, S.E.; Thorpe, R.; Saha, D. Photocatalytic and photo-fenton activity of iron oxide-doped carbon nitride in 3D printed and LED driven photon concentrator. *J. Hazard. Mater.* **2019**, *376*, 178–187. [CrossRef]
107. Garcia-Muñoz, P.; Lefevre, C.; Robert, D.; Keller, N. Ti-substituted $LaFeO_3$ perovskite as photoassisted CWPO catalyst for water treatment. *Appl. Catal. B Environ.* **2019**, *248*, 120–128. [CrossRef]
108. Garcia-Muñoz, P.; Fresno, F.; Lefevre, C.; Robert, D.; Keller, N. Highly robust $La_{1-x}Ti_xFeO_3$ dual catalyst with combined photocatalytic and photo-CWPO activity under visible light for 4-chlorophenol removal in water. *Appl. Catal. B Environ.* **2019**, *262*, 118310. [CrossRef]
109. Wang, Q.; Wang, P.; Xu, P.; Li, Y.; Duan, J.; Zhang, G.; Hu, L.; Wang, X.; Zhang, W. Visible-light-driven photo-Fenton reactions using $Zn_{1-1.5}Fe$ S/g-C_3N_4 photocatalyst: Degradation kinetics and mechanisms analysis. *Appl. Catal. B Environ.* **2020**, *266*, 118653. [CrossRef]
110. Yang, L.; Jiao, Y.; Xu, X.; Pan, Y.; Su, C.; Duan, X.; Sun, H.; Liu, S.; Wang, S.; Shao, Z. Superstructures with Atomic-Level Arranged Perovskite and Oxide Layers for Advanced Oxidation with an Enhanced Non-Free Radical Pathway. *ACS Sustain. Chem. Eng.* **2022**, *10*, 1899–1909. [CrossRef]
111. García-Muñoz, P.; Pliego, G.; Zazo, J.A.; Casas, J.A. Photocatalytic wet peroxide oxidation process at circumneutral pH using ilmenite as catalyst. *J. Environ. Chem. Eng.* **2018**, *6*, 7312–7317. [CrossRef]
112. García-Muñoz, P.; Pliego, G.; Zazo, J.; Bahamonde, A.; Casas, J. Sulfonamides photoassisted oxidation treatments catalyzed by ilmenite. *Chemosphere* **2017**, *180*, 523–530. [CrossRef] [PubMed]
113. García-Muñoz, P.; Pliego, G.; Zazo, J.; Barbero, B.; Bahamonde, A.; Casas, J. Modified ilmenite as catalyst for CWPO-Photoassisted process under LED light. *Chem. Eng. J.* **2017**, *318*, 89–94. [CrossRef]
114. Garcia-Muñoz, P.; Fresno, F.; Lefevre, C.; Robert, D.; Keller, N. Synergy effect between photocatalysis and heterogeneous photo-Fenton catalysis on Ti-doped $LaFeO_3$ perovskite for high efficiency light-assisted water treatment. *Catal. Sci. Technol.* **2020**, *10*, 1299–1310. [CrossRef]
115. Grilla, E.; Matthaiou, V.; Frontistis, Z.; Oller, I.; Polo, I.; Malato, S.; Mantzavinos, D. Degradation of antibiotic trimethoprim by the combined action of sunlight, TiO_2 and persulfate: A pilot plant study. *Catal. Today* **2018**, *328*, 216–222. [CrossRef]
116. Yentür, G.; Dükkancı, M. Fabrication of magnetically separable plasmonic composite photocatalyst of Ag/AgBr/$ZnFe_2O_4$ for visible light photocatalytic oxidation of carbamazepine. *Appl. Surf. Sci.* **2020**, *510*, 145374. [CrossRef]
117. Zhong, X.; Zou, Z.-S.; Wang, H.-L.; Huang, W.; Zhou, B.-X. Enhanced Activation of Persulfate by Co-Doped Bismuth Ferrite Nanocomposites for Degradation of Levofloxacin Under Visible Light Irradiation. *Materials* **2019**, *12*, 3952. [CrossRef]
118. Silveira, J.E.; Paz, W.S.; Garcia-Muñoz, P.; Zazo, J.A.; Casas, J.A. UV-LED/ilmenite/persulfate for azo dye mineralization: The role of sulfate in the catalyst deactivation. *Appl. Catal. B Environ.* **2017**, *219*, 314–321. [CrossRef]
119. Robert, D.; Keller, V.; Keller, N. Immobilization of a Semiconductor Photocatalyst on Solid Supports: Methods, Materials, and Applications. In *Photocatalysis and Water Purification: From Fundamentals to Recent Applications*; Wiley-VCH Verlag GmbH & Co. KGaA: Weinheim, Germany, 2013; pp. 145–178. [CrossRef]
120. Allé, P.H.; García-Muñoz, P.; Adouby, K.; Keller, N.; Robert, D. Efficient photocatalytic mineralization of polymethylmethacrylate and polystyrene nanoplastics by TiO_2/β-SiC alveolar foams. *Environ. Chem. Lett.* **2020**, *19*, 1803–1808. [CrossRef]
121. M'Bra, I.C.; García-Muñoz, P.; Drogui, P.; Keller, N.; Trokourey, A.; Robert, D. Heterogeneous photodegradation of Pyrimethanil and its commercial formulation with TiO_2 immobilized on SiC foams. *J. Photochem. Photobiol. A Chem.* **2019**, *368*, 1–6. [CrossRef]
122. Rico-Santacruz, M.; García-Muñoz, P.; Marchal, C.; Batail, N.; Pham, C.; Robert, D.; Keller, N. Coating-free TiO_2@β-SiC alveolar foams as a ready-to-use composite photocatalyst with tunable adsorption properties for water treatment. *RSC Adv.* **2020**, *10*, 3817–3825. [CrossRef] [PubMed]

123. Rico-Santacruz, M.; García-Muñoz, P.; Keller, V.; Batail, N.; Pham, C.; Robert, D.; Keller, N. Alveolar TiO_2-β-SiC photocatalytic composite foams with tunable properties for water treatment. *Catal. Today* **2018**, *328*, 235–242. [CrossRef]
124. Abdi, P.; Farzi, A.; Karimi, A. Application of a hybrid enzymatic and photo-fenton process for investigation of azo dye decolorization on TiO_2/metal-foam catalyst. *J. Taiwan Inst. Chem. Eng.* **2017**, *71*, 137–144. [CrossRef]
125. Orak, C.; Atalay, S.; Ersöz, G. Photocatalytic and photo-Fenton-like degradation of methylparaben on monolith-supported perovskite-type catalysts. *Sep. Sci. Technol.* **2017**, *52*, 1310–1320. [CrossRef]
126. Martín-Sómer, M.; Pablos, C.; de Diego, A.; van Grieken, R.; Encinas, A.; Monsalvo, V.M.; Marugán, J. Novel macroporous 3D photocatalytic foams for simultaneous wastewater disinfection and removal of contaminants of emerging concern. *Chem. Eng. J.* **2019**, *366*, 449–459. [CrossRef]
127. Dewil, R.; Mantzavinos, D.; Poulios, I.; Rodrigo, M.A. New perspectives for Advanced Oxidation Processes. *J. Environ. Manag.* **2017**, *195*, 93–99. [CrossRef]
128. Spuhler, D.; Rengifo-Herrera, J.A.; Pulgarin, C. The effect of Fe^{2+}, Fe^{3+}, H_2O_2 and the photo-Fenton reagent at near neutral pH on the solar disinfection (SODIS) at low temperatures of water containing *Escherichia coli* K12. *Appl. Catal. B Environ.* **2010**, *96*, 126–141. [CrossRef]
129. Pérez-Lucas, G.; Aliste, M.; Vela, N.; Garrido, I.; Fenoll, J.; Navarro, S. Decline of fluroxypyr and triclopyr residues from pure, drinking and leaching water by photo-assisted peroxonation. *Process. Saf. Environ. Prot.* **2020**, *137*, 358–365. [CrossRef]
130. Martini, J.; Orge, C.A.; Faria, J.L.; Pereira, M.F.R.; Soares, O.S.G. Sulfamethoxazole degradation by combination of advanced oxidation processes. *J. Environ. Chem. Eng.* **2018**, *6*, 4054–4060. [CrossRef]
131. Gmurek, M.; Gomes, J.F.; Martins, R.C.; Quinta-Ferreira, R.M. Comparison of radical-driven technologies applied for paraben mixture degradation: Mechanism, biodegradability, toxicity and cost assessment. *Environ. Sci. Pollut. Res.* **2019**, *26*, 37174–37192. [CrossRef]
132. Quiñones, D.H.; Álvarez, P.M.; Rey, A.; Contreras, S.; Beltrán, F.J. Application of solar photocatalytic ozonation for the degradation of emerging contaminants in water in a pilot plant. *Chem. Eng. J.* **2015**, *260*, 399–410. [CrossRef]
133. Roccamante, M.; Salmerón, I.; Ruiz, A.; Oller, I.; Malato, S. New approaches to solar Advanced Oxidation Processes for elimination of priority substances based on electrooxidation and ozonation at pilot plant scale. *Catal. Today* **2019**, *355*, 844–850. [CrossRef]
134. Rodríguez, E.M.; Rey, A.; Mena, E.; Beltrán, F.J. Application of solar photocatalytic ozonation in water treatment using supported TiO_2. *Appl. Catal. B Environ.* **2019**, *254*, 237–245. [CrossRef]
135. Buyukada, M. Removal, potential reaction pathways, and overall cost analysis of various pollution parameters and toxic odor compounds from the effluents of turkey processing plant using TiO_2–assisted UV/O_3 process. *J. Environ. Manag.* **2019**, *248*, 109298. [CrossRef] [PubMed]
136. Fathinia, M.; Khataee, A.; Naseri, A.; Aber, S. Monitoring simultaneous photocatalytic-ozonation of mixture of pharmaceuticals in the presence of immobilized TiO_2 nanoparticles using MCR-ALS: Identification of intermediates and multi-response optimization approach. *Spectrochim. Acta Part A Mol. Biomol. Spectrosc.* **2015**, *136*, 1275–1290. [CrossRef] [PubMed]
137. Fernandes, A.; Gągol, M.; Makoś, P.; Khan, J.A.; Boczkaj, G. Integrated photocatalytic advanced oxidation system (TiO_2/UV/O_3/H_2O_2) for degradation of volatile organic compounds. *Sep. Purif. Technol.* **2019**, *224*, 1–14. [CrossRef]
138. Rodríguez-Chueca, J.; Alonso, E.; Singh, D.N. Photocatalytic Mechanisms for Peroxymonosulfate Activation through the Removal of Methylene Blue: A Case Study. *Int. J. Environ. Res. Public Health* **2019**, *16*, 198. [CrossRef]
139. Talwar, S.; Verma, A.K.; Sangal, V.K.; Štangar, U.L. Once through continuous flow removal of metronidazole by dual effect of photo-Fenton and photocatalysis in a compound parabolic concentrator at pilot plant scale. *Chem. Eng. J.* **2020**, *388*, 124184. [CrossRef]
140. Salmerón, I.; Plakas, K.V.; Sirés, I.; Oller, I.; Maldonado, M.I.; Karabelas, A.J.; Malato, S. Optimization of electrocatalytic H_2O_2 production at pilot plant scale for solar-assisted water treatment. *Appl. Catal. B Environ.* **2018**, *242*, 327–336. [CrossRef]
141. Saleh, R.; Taufik, A. Degradation of methylene blue and congo-red dyes using Fenton, photo-Fenton, sono-Fenton, and sonophoto-Fenton methods in the presence of iron(II,III) oxide/zinc oxide/graphene (Fe_3O_4/ZnO/graphene) composites. *Sep. Purif. Technol.* **2018**, *210*, 563–573. [CrossRef]
142. Xu, L.; Chu, W.; Graham, N. Degradation of di-n-butyl phthalate by a homogeneous sono–photo–Fenton process with in situ generated hydrogen peroxide. *Chem. Eng. J.* **2014**, *240*, 541–547. [CrossRef]
143. Kouvelis, K.; Kampioti, A.A.; Petala, A.; Frontistis, Z. Degradation of Sulfamethoxazole Using a Hybrid CuO_x–$BiVO_4$/SPS/Solar System. *Catalysts* **2022**, *12*, 882. [CrossRef]
144. Liu, C.; Mao, S.; Shi, M.; Wang, F.; Xia, M.; Chen, Q.; Ju, X. Peroxymonosulfate activation through 2D/2D Z-scheme CoAl-LDH/ /BiOBr photocatalyst under visible light for ciprofloxacin degradation. *J. Hazard. Mater.* **2021**, *420*, 126613. [CrossRef]
145. Liu, C.; Mao, S.; Wang, H.; Wu, Y.; Wang, F.; Xia, M.; Chen, Q. Peroxymonosulfate-assisted for facilitating photocatalytic degradation performance of 2D/2D WO3/BiOBr S-scheme heterojunction. *Chem. Eng. J.* **2021**, *430*, 132806. [CrossRef]
146. Marjanovic, M.; Giannakis, S.; Grandjean, D.; de Alencastro, L.F.; Pulgarin, C. Effect of μM Fe addition, mild heat and solar UV on sulfate radical-mediated inactivation of bacteria, viruses, and micropollutant degradation in water. *Water Res.* **2018**, *140*, 220–231. [CrossRef]
147. Rodríguez-Chueca, J.; Giannakis, S.; Marjanovic, M.; Kohantorabi, M.; Gholami, M.R.; Grandjean, D.; de Alencastro, L.F.; Pulgarín, C. Solar-assisted bacterial disinfection and removal of contaminants of emerging concern by Fe^{2+}-activated HSO5-vs. $S_2O_8{}^{2-}$ in drinking water. *Appl. Catal. B Environ.* **2019**, *248*, 62–72. [CrossRef]

148. Rodríguez-Chueca, J.; Mesones, S.; Marugán, J. Hybrid UV-C/microfiltration process in membrane photoreactor for wastewater disinfection. *Environ. Sci. Pollut. Res.* **2018**, *26*, 36080–36087. [CrossRef] [PubMed]
149. Aydiner, C.; Mert, B.K.; Dogan, E.C.; Yatmaz, H.C.; Dagli, S.; Aksu, S.; Tilki, Y.M.; Goren, A.Y.; Balci, E. Novel hybrid treatments of textile wastewater by membrane oxidation reactor: Performance investigations, optimizations and efficiency comparisons. *Sci. Total Environ.* **2019**, *683*, 411–426. [CrossRef] [PubMed]
150. Moslehyani, A.; Mobaraki, M.; Matsuura, T.; Ismail, A.; Othman, M.; Chowdhury, M. Novel green hybrid processes for oily water photooxidation and purification from merchant ship. *Desalination* **2016**, *391*, 98–104. [CrossRef]
151. Kakavandi, B.; Ahmadi, M. Efficient treatment of saline recalcitrant petrochemical wastewater using heterogeneous UV-assisted sono-Fenton process. *Ultrason. Sonochem.* **2019**, *56*, 25–36. [CrossRef]
152. Raut-Jadhav, S.; Saharan, V.K.; Pinjari, D.V.; Saini, D.R.; Sonawane, S.H.; Pandit, A.B. Intensification of degradation of imidacloprid in aqueous solutions by combination of hydrodynamic cavitation with various advanced oxidation processes (AOPs). *J. Environ. Chem. Eng.* **2013**, *1*, 850–857. [CrossRef]
153. Sun, B.; Wang, Y.; Xiang, Y.; Shang, C. Influence of pre-ozonation of DOM on micropollutant abatement by UV-based advanced oxidation processes. *J. Hazard. Mater.* **2020**, *391*, 122201. [CrossRef] [PubMed]
154. Ghanbari, F.; Wu, J.; Khatebasreh, M.; Ding, D.; Lin, K.-Y.A. Efficient treatment for landfill leachate through sequential electrocoagulation, electrooxidation and PMS/UV/$CuFe_2O_4$ process. *Sep. Purif. Technol.* **2020**, *242*, 116828. [CrossRef]
155. Webler, A.D.; Moreira, F.C.; Dezotti, M.W.; Mahler, C.F.; Segundo, I.D.B.; Boaventura, R.A.; Vilar, V.J. Development of an integrated treatment strategy for a leather tannery landfill leachate. *Waste Manag.* **2019**, *89*, 114–128. [CrossRef]
156. Changotra, R.; Rajput, H.; Dhir, A. Treatment of real pharmaceutical wastewater using combined approach of Fenton applications and aerobic biological treatment. *J. Photochem. Photobiol. A Chem.* **2019**, *376*, 175–184. [CrossRef]
157. Bener, S.; Atalay, S.; Ersöz, G. The hybrid process with eco-friendly materials for the treatment of the real textile industry wastewater. *Ecol. Eng.* **2020**, *148*. [CrossRef]
158. Giannakis, S.; Gamarra Vives, F.A.; Grandjean, D.; Magnet, A.; de Alencastro, L.F.; Pulgarin, C. Effect of advanced oxidation processes on the micropollutants and the effluent organic matter contained in municipal wastewater previously treated by three different secondary methods. *Water Res.* **2015**, *84*, 295–306. [CrossRef]
159. Ly, Q.V.; Kim, H.C.; Hur, J. Tracking fluorescent dissolved organic matter in hybrid ultrafiltration systems with TiO_2/UV oxidation via EEM-PARAFAC. *J. Membr. Sci.* **2018**, *549*, 275–282. [CrossRef]
160. Piras, F.; Santoro, O.; Pastore, T.; Pio, I.; de Dominicis, E.; Gritti, E.; Caricato, R.; Lionetto, M.G.; Mele, G.; Santoro, D. Controlling micropollutants in tertiary municipal wastewater by O_3/H_2O_2, granular biofiltration and UV254/H_2O_2 for potable reuse applications. *Chemosphere* **2020**, *239*, 124635. [CrossRef]
161. Ponce-Robles, L.; Oller, I.; Polo-López, M.I.; Rivas-Ibáñez, G.; Malato, S. Microbiological evaluation of combined advanced chemical-biological oxidation technologies for the treatment of cork boiling wastewater. *Sci. Total Environ.* **2019**, *687*, 567–576. [CrossRef]
162. De Oliveira Gonçalves, L.; Starling, M.C.V.M.; Leal, C.D.; Oliveira, D.V.M.; Araújo, J.C.; Leão, M.M.D.; Amorim, C.C. Enhanced biodiesel industry wastewater treatment via a hybrid MBBR combined with advanced oxidation processes: Analysis of active microbiota and toxicity removal. *Environ. Sci. Pollut. Res.* **2019**, *26*, 4521–4536. [CrossRef] [PubMed]
163. Rodríguez-Chueca, J.; Amor, C.; Fernandes, J.R.; Tavares, P.B.; Lucas, M.S.; Peres, J.A. Treatment of crystallized-fruit wastewater by UV-A LED photo-Fenton and coagulation-flocculation. *Chemosphere* **2016**, *145*, 351–359. [CrossRef] [PubMed]
164. Amin, M.M.; Golbini Mofrad, M.M.; Pourzamani, H.; Sebaradar, S.M.; Ebrahim, K. Treatment of industrial wastewater contaminated with recalcitrant metal working fluids by the photo-Fenton process as post-treatment for DAF. *J. Ind. Eng. Chem.* **2017**, *45*, 412–420. [CrossRef]
165. Rodríguez-Chueca, J.; Morales, M.; Mosteo, R.; Ormad, M.P.; Ovelleiro, J.L. Inactivation of Enterococcus faecalis, Pseudomonas aeruginosa and Escherichia coli present in treated urban wastewater by coagulation-flocculation and photo-Fenton processes. *Photochem. Photobiol. Sci.* **2013**, *12*, 864–871. [CrossRef] [PubMed]
166. Rodríguez-Chueca, J.; Laski, E.; García-Cañibano, C.; Martín de Vidales, M.J.; Encinas, B.; Kuch, J. Marugán, Micropollutants removal by full-scale UV-C/sulfate radical based Advanced Oxidation Processes. *Sci. Total Environ.* **2018**, *630*, 1216–1225. [CrossRef] [PubMed]

MDPI AG
Grosspeteranlage 5
4052 Basel
Switzerland
Tel.: +41 61 683 77 34

Catalysts Editorial Office
E-mail: catalysts@mdpi.com
www.mdpi.com/journal/catalysts

www.ingramcontent.com/pod-product-compliance
Lightning Source LLC
LaVergne TN
LVHW070456061126
851858LV00011B/245
* 9 7 8 3 7 2 5 8 8 4 8 1 0 *